Cell Chemistry and Physiology: Part III

PRINCIPLES OF MEDICAL BIOLOGY
A Multi-Volume Work, Volume 4

Editors: **E. EDWARD BITTAR,** *Department of Physiology,*
University of Wisconsin, Madison
NEVILLE BITTAR, *Department of Medicine,*
University of Wisconsin, Madison

Principles of Medical Biology
A Multi-Volume Work

Edited by **E. Edward Bittar,** *Department of Physiology, University of Wisconsin, Madison and* **Neville Bittar,** *Department of Medicine University of Wisconsin, Madison*

This work provides:

* A holistic treatment of the main medical disciplines. The basic sciences including most of the achievements in cell and molecular biology have been blended with pathology and clinical medicine. Thus, a special feature is that departmental barriers have been overcome.

* The subject matter covered in preclinical and clinical courses has been reduced by almost one-third without sacrificing any of the essentials of a sound medical education. This information base thus represents an integrated core curriculum.

* The movement towards reform in medical teaching calls for the adoption of an integrated core curriculum involving small-group teaching and the recognition of the student as an active learner.

* There are increasing indications that the traditional education system in which the teacher plays the role of expert and the student that of a passive learner is undergoing reform in many medical schools. The trend can only grow.

* Medical biology as the new profession has the power to simplify the problem of reductionism.

* Over 700 internationally acclaimed medical scientists, pathologists, clinical investigators, clinicians and bioethicists are participants in this undertaking.

Cell Chemistry and Physiology: Part III

Edited by **E. EDWARD BITTAR**
Department of Physiology
University of Wisconsin
Madison, Wisconsin

NEVILLE BITTAR
Department of Medicine
University of Wisconsin
Madison, Wisconsin

 **JAI PRESS INC.**

Greenwich, Connecticut   *London, England*

Library of Congress Cataloging-in-Publication Data

Cell chemistry and physiology / edited by E. Edward Bittar, Neville
   Bittar.
      p. cm.—(Principles of medical biology ; v. 4)
      Includes index.
      ISBN 1-55938-807-6
      1. Cytochemistry.   2. Cell physiology.   I. Bittar, E. Edward.
   II. Bittar, Neville.   III. Series.
      [DNLM: 1. Cells—chemistry.   2. Cells—physiology.   QH 581.2 C392
   1996]
   QH611.C4214   1996
   574.87—cd20
   for Library of Congress   94-37215
                                                            CIP

Copyright © 1996 by JAI PRESS INC.
55 Old Post Road, No. 2
Greenwich, Connecticut 06836

JAI PRESS LTD.
The Courtyard
29 High Street
Hampton Hill, Middlesex TW12 1PD
England

ISBN: 1-55938-807-2
Library of Congress Catalog No.: 95-33561

Transferred to digital printing 2006
Printed and bound by CPI Antony Rowe, Eastbourne

CONTENTS

LIST OF CONTRIBUTORS

Jens P. Andersen

Institute of Physiology
University of Aarhus
Aarhus, Denmark

Kevin M. Brindle

Department of Biochemistry
University of Cambridge
Cambridge, England

Charles F. Burant

Department of Medicine
University of Chicago
Chicago, Illinois

Rainer Callies

Department of Biochemistry
University of Cambridge
Cambridge, England

Frederick L. Crane

Department of Biological Sciences
Purdue University
West Lafayette, Indiana

Ruth A. Crowe

Department of Biological Sciences
Purdue University
West Lafayette, Indiana

Christian Frelin

Institut de Pharmacologie
Moléculaire et Cellulaire
Université de Nice Sophia
Antipolis
Valbonne, France

Dieter Häussinger

Abteilung Innere Medizin
Düsseldorf Universitat
Düsseldorf, Germany

ix

H.K. Huang   Department of Radiology
University of California
Medical Center
San Francisco, California

Rose M. Johnstone   Department of Biochemistry
McGill University
Montreal, Quebec, Canada

Richard B. Kemp   Department of Biological Sciences
The University of Wales
Wales, United Kingdom

Hans Löw   Department of Endocrinology
Karolinska Institute
Stockholm, Sweden

John I. McCormick   Department of Biochemistry
McGill University
Montreal, Quebec, Canada

Emanuel E. Strehler   Department of Biochemistry and
Molecular Biology
Mayo Clinic
Rochester, Minnesota

Iris L. Sun   Department of Biological Sciences
Purdue University
West Lafayette, Indiana

Paul Vigne   Institut de Pharmacologie
Moléculaire et Cellulaire
Université de Nice Sophia
Antipolis
Valbonne, France

Bente Vilsen   Institute of Physiology
University of Aarhus
Aarhus, Denmark

Ingemar Wadsö   Chemical Center
University of Lund
Lund, Sweden

PREFACE

The first section of this volume consists of five chapters relating to the nature of membrane transport systems. A chapter on secondary active glucose transport has been omitted because this topic is slated to appear in the Nephrobiology module. Chapter 6 deals with oxidase control of plasma membrane proton transport, while chapter 7 addresses the question of how cell volume is regulated. Although we chose not to have a separate chapter covering additional co-transport systems namely, Na^+-K^+-$2Cl^-$, KCl, Cl^--HCO_3^-, as well as Cl^--HCO_3^- exchange and K^+ and Cl^- movements through channels, the role of each in cell volume regulation is emphasized in chapter 7.

Instead of devoting an entire section to the thermodynamics of metabolism, we thought it desirable to have the subjects of medical imaging and NMR of cell metabolism discussed in some detail in two chapters. These are followed by a chapter on the thermodynamic instrument—the calorimeter. Calorimetry allows the measurement of net changes of heat in cells, tissues, organs and whole body. As will be recognized, heat dissipation does not arise only from chemical reactions but also from interactions between macromolecules and conformational changes in protein complexes and mass Ca^{2+} movement such as that occurring in contracting skeletal muscle. The last chapter provides an account of equilibrium and non-equilibrium thermodynamics and the enthalpy balance method. It reveals that calorimetric measurements are useful in studies of clinical and toxicological problems.

Our thanks are due to the contributing authors for their cooperation and patience. We also wish to thank Ms. Lauren Manjoney and the production staff of JAI Press for their skill and courtesy.

E. EDWARD BITTAR
NEVILLE BITTAR

Chapter 1

Primary Ion Pumps

JENS P. ANDERSEN AND BENTE VILSEN

Principles of Medical Biology, Volume 4
Cell Chemistry and Physiology: Part III, pages 1–66.
Copyright © 1996 by JAI Press Inc.
All rights of reproduction in any form reserved.
ISBN: 1-55938-807-2

INTRODUCTION

The term ion pump, synonymous with active ion-transport system, is used to refer to a protein that translocates ions across a membrane, uphill against an electrochemical potential gradient. The primary pumps do so by utilization of energy derived from various types of chemical reactions such as ATP hydrolysis, electron transfers (redox processes), and decarboxylations, or from the absorption of light (Table 1). Secondary pumps are symport and antiport systems that derive the energy for uphill movement of one species from a coupled downhill movement of another species. The electrochemical gradient driving the latter movement is often created by a primary pump.

In animal cells, the Na^+-K^+-ATPase in the plasma membrane is responsible for the generation of primary Na^+ gradients used to drive the secondary pumps involved in nutrient uptake, in cellular Ca^{2+} and H^+ homeostasis, and in transport of water and electrolytes across epithelia. Moreover, the passive flux of K^+ out of the cell, down the K^+ gradient set up by the Na^+-K^+-ATPase, provides the major motive power for generation of the membrane potential. Inhibition of the Na^+-K^+-ATPase by cardiac glycosides (digitalis) is used clinically to produce a positive inotropic effect on the heart. The inhibition of the Na^+-K^+-pump reduces the Na^+ gradient, thereby slowing the outward transport of Ca^{2+} through the Na^+/Ca^{2+} antiport present in the plasma membrane of cardiac myocytes. The resulting rise in intracellular Ca^{2+} produces the increase in contractile force. The Na^+-K^+-ATPase belongs to the family of ion-motive ATPases named P-type by reference to the covalent phosphoenzyme intermediate formed in the course of the reaction cycle by transfer of the γ-phosphoryl group of ATP to the protein (Table 2). This phosphoenzyme

Table 1. Some Primary Ion Pumps

Pump	Ions	Energy Source	Occurrence
Bacteriorhodopsin	H^+	Light	Halobacteria
Halorhodopsin	Cl^-	Light	Halobacteria
Cytochrome oxidase	H^+	Redox energy	Mitochondria
NADH dehydrogenase	H^+	Redox energy	Mitochondria
Ion-transporting decarboxylases	Na^+	Decarboxylation	Bacteria
H^+-pyrophosphatase	H^+	Hydrolysis of pyrophosphate	Plant vacuoles
Cation-motive ATPases		Hydrolysis of ATP	
P-type	Na^+,K^+ H^+,Ca^{2+} Cd^{2+}		Animal and plant cells, bacteria
F-type	H^+,Na^+		Bacteria, mitochondria
V-type	H^+		Animal and plant cells, bacteria
Oxyanion-motive ATPase	AsO_2^-, AsO_4^{3-}	Hydrolysis of ATP	Bacteria

Table 2. Transport Stoichiometry and Cation Substrates in Some P-Type Pumps

	Cations Transported by the Pump[a]		Usual Stoichiometry a:c:ATP
Pump	a-Type	c-Type	
Na$^+$-K$^+$-ATPase	Na$^+$, H$^+$, Li$^+$	K$^+$, Tl$^+$, Rb$^+$,NH$_4^+$, Cs$^+$, Li$^+$, Na$^+$, H$^+$	3:2:1
H$^+$-K$^+$-ATPase	H$^+$, Na$^+$	K$^+$, Tl$^+$, Rb$^+$,NH$_4^+$, Cs$^+$, Li$^+$, Na$^+$	2:2:1 or 1:1:1
SR(ER) Ca^{2+}-ATPase[b]	Ca^{2+}, Sr^{2+}	H$^+$ (?)	2:2(?):1
PM Ca^{2+}-ATPase[c]	Ca^{2+}, Sr^{2+}	H$^+$ (?)	?
Plant PM H$^+$-ATPase	H$^+$		1:1

Notes: [a]Transport away from the cytoplasm is defined as a-type. Transport towards the cytoplasm is defined as c-type (countertransport).
[b]SR = sarcoplasmic reticulum. ER = endoplasmic reticulum.
[c]PM = plasma membrane.

intermediate is critical to the function of the P-type ATPases, and these pumps are susceptible to inhibition by vanadate, which acts as a transition state analog of the phosphoryl group. In addition to Na$^+$-K$^+$-ATPase, the P-type ATPases include the Ca^{2+}-transport ATPases of plasma membranes (PM Ca^{2+}-ATPases) and the Ca^{2+}-ATPases of intracellular organellar membranes, which produce Ca^{2+} gradients that are essential to cytoplasmic and nuclear Ca^{2+} signaling systems. The initiation of muscle contraction by Ca^{2+} release from sarcoplasmic reticulum (SR) and the initiation of exocytosis—leading to secretion of enzymes, hormones, and neurotransmitters—by Ca^{2+} release from endoplasmic reticulum (ER) in nonmuscle cell types only represent a fraction of the physiological functions that depend critically on the Ca^{2+} gradients set up by the P-type Ca^{2+}-pumps. The H$^+$-K$^+$-ATPases responsible for acid secretion by the gastric mucosa and for H$^+$:K$^+$ exchange processes in the epithelia of the distal colon and nephron also belong to the family of P-type pumps, and so do the ATPases that generate H$^+$-gradients driving secondary active transport in plants and fungi (Serrano, 1988).

In most prokaryotes, the primary ion gradient in the cell is the H$^+$-gradient produced by pumps that use light or redox energy instead of ATP. Due to the presence of Na$^+$/H$^+$-antiport systems, the H$^+$-gradients give rise to secondary Na$^+$-gradients, and the uptake of nutrients in prokaryotes occurs by coupled systems that use either the H$^+$-gradient or the Na$^+$-gradient as an energy source. In addition, the primary H$^+$-gradient in prokaryotes is used to drive the production of ATP by ATP-dependent H$^+$-pumps operating in the reverse mode as ATP-synthases. The eubacterial type of ATP-synthase (F-type ATPase, F for coupling *f*actor) has persisted in the mitochondria and chloroplasts of eukaryotic cells, where it serves to supply the ATP spent in many of the energy consuming reactions of the cell, including those that involve ion-motive ATPases running in the forward direction (Figure 1).

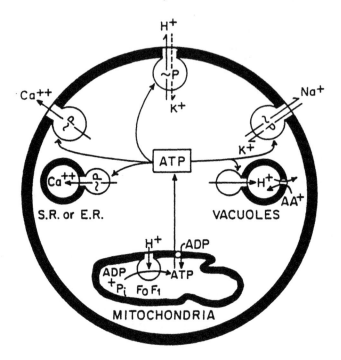

Figure 1. Localization of the major types of ion-motive ATPases in the eukaryotic cell. Na^+-K^+, H^+-K^+, and Ca^{2+}-ATPases of P-type transport the respective cations across the plasma membrane or into sarcoplasmic (SR) or endoplasmic (ER) reticulum. H^+-ATPase of V-type acidifies different types of vacuoles and vesicles allowing their secondary uptake of amino acids and amines (AA^+). H^+-ATPase (working as ATP synthase) of F-type (F_0F_1) generates ATP in the mitochondria. Modified from Pedersen and Carafoli, 1987.

Eukaryotic cells contain a third type of cation-motive ATPase in addition to those of the P- and F-types, the so-called V-type ATPase (V for vacuolar; see Figure 1). This ATPase pumps H^+ into organelles such as tonoplasts of plant cells and chromaffin granules, synaptic vesicles, clathrin-coated vesicles, and lysosomes of animal cells, thereby providing the acidification required for such diverse processes as vesicular accumulation of amines, uncoupling of internalized receptors from their ligands, and activation of lysosomal enzymes (Forgac, 1989). Moreover, ATPases of V-type occur in the plasma membranes of kidney distal tubule interca-lated cells, where they participate in the final urine acidification, and in the plasma membrane of osteoclasts, where they are responsible for bone resorption. The V-type ATPases are more closely related to the F-type ATPases than to the P-type ATPases (Pedersen and Carafoli, 1987). Both V- and F-type ATPases undergo a reaction cycle in which ATP is hydrolyzed without any detectable formation of a covalent phosphoenzyme intermediate, and they are not inhibited by vanadate. In addition, both V- and F-type ATPases consist of multiple subunits unlike the single catalytic peptide of the P-type ATPases. The eukaryotic V-type ATPases can only operate in the forward direction as H^+-pumps, but archaebacteria contain V-type ATPases that function as ATP-synthases in the same way as the F-type ATPases of eubacteria, and it is believed that V- and F-type ATPases have evolved from a common ancestral ATPase gene (Nelson, 1992).

New types of primary pumps are currently being discovered, and the spectrum of ATP-driven pumps now covers transporters of small anions as well as more complex molecules, some of which have to be extruded from the cells because of their toxicity. In view of the wide scope of the topic, we have found it necessary to narrow the discussion to selected examples that illustrate the principles pertaining to the primary ion pumps and the methodology used for studying them. Hence, the present chapter deals most extensively with the Na^+-K^+- and Ca^{2+}-ATPases of P-type, whose mechanisms and structures have been studied in great detail.

REACTION CYCLE

Basic Principles

Although the number of structural and functional differences between the P-type ATPases and V/F-type ATPases seems too large to warrant speculations about a common ancestor of all ion-motive ATPases, these pump families do have a number of fundamental characteristics in common, which if evolved independently must represent the only suitable way of constructing an ion pump. A large hydrophilic head protruding into the cytoplasm is responsible for the scalar reactions (ATP-hydrolysis/synthesis), whereas the vectorial transport processes are carried out by a more slender membrane-buried part most likely consisting of a bundle of transmembrane α-helices. The coupling between the scalar and vectorial reactions requires long distance communication between the cytoplasmic and transmembrane parts of the pump protein through conformational changes. It is symbolic that in the F-type ATPases the two components are separable as a soluble F_1 particle with ATPase activity and a membrane-bound F_0 part with proton channel activity, and it has been suggested that evolution combined a soluble kinase-type of enzyme with a preexisting channel. It has recently been shown that a similar functional subdivision exists for the single catalytic peptide chain of P-type ATPases by site-directed mutagenesis analysis of the functions of individual amino acid residues (Figure 16) and by use of proteolytic enzymes to selectively remove the cytoplasmic domains (Karlish et al., 1990). Moreover, a common characteristic of the F-type and P-type ATPases seems to be that their catalytic sites exist in at least two different functional states. One of these (E_1) is characterized by the conservation of the high energy of the phosphoryl compound present in the site, be it noncovalently bound ATP (in F-type ATPases), or the acylphosphorylated amino acid residue derived by reaction of the enzyme with ATP (in P-type ATPases). In the other conformation of the catalytic site (E_2), the bound phosphoryl compound, formerly of high energy, is able to exist in reversible equilibrium with low energy inorganic phosphate (P_i). Energy transduction depends on the coupling of the switch between these two states of the catalytic site with the translocation of ions. There are speculations that in P-type ATPases, as well as in V/F-type ATPases, the E_1-E_2 interconversion involves a hydrophilic–hydrophobic transition (a closure) of the catalytic site. The basic consideration is that the free energy of hydrolysis is high, when ATP or an acylphosphorylated amino acid residue is in a hydrophilic (aqueous) environment, whereas it is low in a hydrophobic environment, which is comparable to a gas phase (De Meis, 1989). The utilization of the binding energy derived from noncovalent interaction between the bound phosphoryl compounds and the enzyme to lower the free energy of their hydrolysis in the E_2 state has also been emphasized as a possible explanation of the reversibility of the hydrolytic reactions in both pump types (Jencks, 1989a).

The transport process begins with the binding of the ion at a specific uptake site in the pump protein. This is followed by a series of reactions that finally lead to dissociation of the translocated ion from a discharge site facing the opposite side of the membrane. The discharge sites are generally found to display lower affinities for the transported ions than the uptake sites. In principle, uphill transport should be feasible even in the absence of any difference between the binding affinities of the uptake and discharge sites, provided there is a strongly unidirectional equilibrium in another part of the reaction cycle. In practice, however, the limitations on rate constants of the partial reaction steps make it impossible to maintain an adequate overall pumping rate, if the steady-state levels of the various reaction intermediates become too disparate. This means that the steady-state concentrations of filled and unfilled ion-binding sites must be roughly comparable under physiological conditions where the pump is working against an ion gradient, implying different binding constants at the uptake and discharge sides of the membrane (Tanford, 1982b). Theoretical analysis also predicts that a movement of the translocated species from a high affinity site to a preformed low affinity site within the pump protein would impose an unacceptably slow step on the reaction cycle (Tanford, 1983). Therefore, an essential feature of the pumps seems to be the ability to undergo a conformational rearrangement that reduces the affinity of the binding site, while the transported ion remains attached. Figure 2 shows a hypothetical model of how a conformational rearrangement of the transmembrane helices may reorientate the ion-binding site towards the discharge side of the membrane and simultaneously reduce the binding affinity by removing some of the liganding groups, thereby turning the high-affinity uptake site into a low-affinity discharge site.

Uptake side of membrane

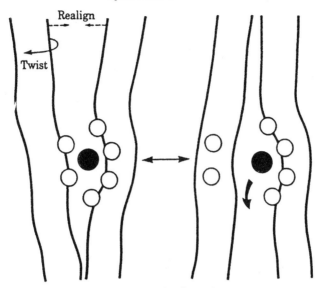

Discharge side of membrane

Figure 2. Hypothetical model showing how a conformational rearrangement of the transmembrane helices may simultaneously change the accessibility of the cation binding site from one side of the membrane to the other and reduce the affinity of the binding site. The open circles represent protein ligands that constitute the binding site. The closed circles represent the ion. Modified from Tanford, 1982a.

The Na^+-K^+, H^+, H^+-K^+ and Ca^{2+}-pumps located in the plasma membrane work under the influence of a membrane potential; some of them are electrogenic since they pump more positive charges out than into the cell (Table 2). Conceptually, transport against an electrical potential gradient is not fundamentally different from transport against a chemical potential gradient. If the binding site involved in ion movement is connected with the adjacent aqueous medium through an access channel, such as that hypothesized for the Na^+-K^+-pump (Figure 7), the local ion concentration "seen" by the binding site depends on the voltage gradient across the part of the membrane in which the access channel is located, and a so-called ion well is created. The effective ion concentration at the binding site is equal to that prevailing at the surface modified by the factor exp(VF/RT), where V is the voltage difference between the membrane surface and the binding site, F is Faraday's constant, R is the gas constant, and T is the absolute temperature (Läuger, 1991). Accordingly, a voltage change has kinetically the same effect as a change of the aqueous ion concentration. The effect of the membrane potential on the kinetics of ion movement is most pronounced if the access channel is of low conductance ("narrow"), so that the corresponding voltage drop constitutes a significant fraction of the membrane potential. Charge movement resulting from conformational changes of the pump protein with the ion attached to its binding site may also contribute to the dependence of transport kinetics on membrane potential. For such effects to be generated, it is not necessary that the bound ion or charges intrinsic to the protein actually move relative to the electrical potential gradient. The charges may remain stationary, while the potential gradient moves. This could occur as a consequence of a protein conformational change that modifies the relative conductances in access channels to uptake and discharge sites as illustrated in Figure 3.

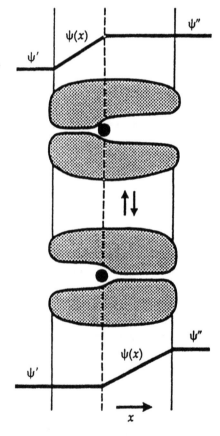

Figure 3. Model showing how a protein conformational change altering the conductances in access channels may allow the electric field to move past the bound ion (closed circle), which remains stationary. In either conformation the electrical potential gradient is located corresponding to a narrow access channel of low conductance. $\psi(x)$ indicates electrical potential.

The E_1-E_2 Model of P-Type Pumps

The P-type pumps seem to be able to work at thermodynamic efficiencies well above 50%, and like the F-type ATP-synthases, the P-type pumps can be run in the reverse direction, synthesizing ATP from ADP and inorganic phosphate (P_i) at the expense of dissipation of a previously created ion gradient. To ensure efficient utilization of the free energy derived from ATP hydrolysis in ion translocation, reactions that would lead to unproductive ATP hydrolysis without net transport must be forbidden (kinetically blocked). The conformational changes that bring about ion movement are, therefore, linked with the catalytic steps in ATP processing according to certain coupling rules that describe the interdependence of catalytic and vectorial specificities (Jencks, 1989a). One of the aims of research in this field has been to clarify these rules, and, ultimately, to describe their structural basis. Results are often presented and used in the form of reaction schemes, in which the arrows specify the partial reaction steps that are permitted (Figs. 4 and 11).

As already noted, the transport function of P-type pumps depends on the formation of a phosphorylated enzyme intermediate, which is part of the mechanism for ATP hydrolysis. The phosphate is covalently attached to an aspartyl β-carboxyl group in the cytoplasmic domain of the pump protein. Due to its stability in acid, the phosphorylated protein can easily be isolated, and quantitation of the phosphorylation is simple, if radioactive ^{32}P-labeled substrate has been included in the reaction. Using this approach, the properties of the phosphoenzyme intermediate have been intensely studied under a variety of conditions. This led quite early to formulation of similar schemes for the reaction cycles of the Na^+-K^+-ATPase and the SR Ca^{2+}-ATPase, and much of the later discussion has revolved around this so-called Post-Albers or E_1-E_2 model (Figure 4) and modified versions of it, such as those shown in Figure 11. Here, we will touch upon a few important points. For more details on the reaction mechanisms of Na^+-K^+-, H^+-K^+- and Ca^{2+}-ATPases the reader is referred to the vast review literature (Na^+-K^+-ATPase: Glynn, 1985, 1993; Nørby and Klodos, 1988; Cornelius, 1991; Skou and Esmann, 1992; H^+-K^+-ATPase: Rabon and Reuben, 1990; SR Ca^{2+}-ATPase: De Meis and Vianna, 1979; De Meis, 1981; Tanford, 1984; Inesi, 1985; Andersen, 1989; Jencks, 1989b; PM Ca^{2+}-ATPase: Garrahan and Rega, 1990; Carafoli, 1991; Na^+-K^+-ATPase and Ca^+-ATPase: Jørgensen and Andersen, 1988; Glynn and Karlish, 1990; Läuger, 1991).

The first essential coupling rule states that the phosphorylation from ATP requires prior binding at the cytoplasmically oriented high-affinity uptake sites of those ions that are to be transported away from the cytoplasm (henceforth designated a-type ions, see Table 2). As long as the extracytoplasmic discharge sites have not yet been vacated by release of the translocated a-type ions, the phosphorylation reaction is fully reversible. After extracytoplasmic dissociation of the a-type ion, reversal of pump phosphorylation with resulting ATP synthesis can still

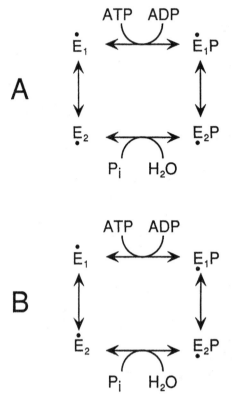

Figure 4. Two versions of the E_1-E_2 model illustrating the uncertainty with respect to the orientation of cation-binding sites in E_1P and E_2. The orientation is indicated by a dot (pointing upwards for cytoplasmic orientation and downwards for extracytoplasmic orientation). In **A** the orientation of the cation-binding site with respect to the membrane changes simultaneously with a change in catalytic specificity of the ATP-binding/phosphorylation site. In **B** it is the phosphorylation that determines orientation of the cation binding site.

be achieved by addition of a high concentration of a-type ion to the extracytoplasmic side of the membrane if ADP is available. Using sided vesicular membrane preparations of Na^+-K^+-ATPase, the accompanying ionic fluxes can be demonstrated with isotopes as: Na^+:Na^+-exchange without net hydrolysis (Figure 5), and similar Ca^{2+}:Ca^{2+}-exchange is observed with closed membrane vesicles containing Ca^{2+}-ATPase.

    In the absence of a-type ions from their cytoplasmic uptake sites, the pump protein displays an alternative catalytic specificity: the aspartyl residue can no longer react with ATP, but it readily phosphorylates from inorganic phosphate

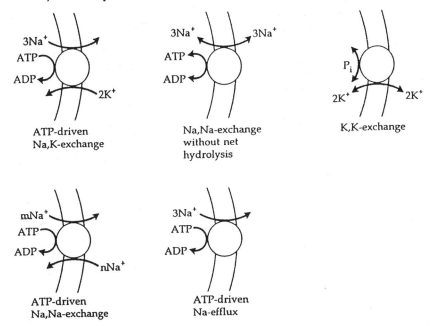

Figure 5. Different modes of operation of the Na$^+$-K$^+$-pump that can be detected by addition of radioactive isotopes to sealed vesicular preparations of defined sidedness. The canonical flux mode is the ATP-driven exchange of three cytoplasmic Na$^+$ for two extracellular K$^+$. In Na$^+$-Na$^+$-exchange without net hydrolysis, only the Na$^+$ limb of the Na$^+$-K$^+$-ATPase cycle is used. In the K$^+$:K$^+$-exchange mode, only the K$^+$-limb of the Na$^+$-K$^+$-ATPase cycle is used. In ATP-driven Na$^+$:Na$^+$-exchange, the whole cycle is involved, but extracellular Na$^+$ substitutes for K$^+$ (m = 3, n = 2). In ATP-driven Na$^+$-efflux, the whole cycle is involved, but the transport sites return empty from the extracellular to the cytoplasmic surface. Reproduced from Läuger, 1991 with permission from Sinauer Associates, Inc.

(backdoor phosphorylation), forming an acyl phosphate bond indistinguishable by chemical methods from that obtained with ATP. The phosphoenzyme formed from P$_i$ reacts reversibly with water, but cannot transfer the phosphoryl group directly to ADP (Figure 4). The enzyme form with specificity for catalysis of phosphoryl transfer between the aspartyl residue and ADP is denoted E$_1$ (the term ADP-sensitive or E$_1$P is used to denote the phosphoenzyme), whereas the enzyme form with specificity for catalysis of phosphoryl transfer between the aspartyl residue and water is denoted E$_2$ (the phosphoenzyme is ADP-insensitive and is denoted E$_2$P). The ADP-sensitive and ADP-insensitive phosphoenzyme intermediates are interconvertible, and it is believed that in the normal transport cycle initiated by phosphorylation of E$_1$ from ATP, dephosphorylation occurs through hydrolysis of E$_2$P after conversion of E$_1$P to E$_2$P. The measured time courses of the appearance and disappearance of E_1P and E_2P seem to be consistent with the hypothesis that these intermediates occur sequentially during the pump cycle, although it should be noted that for the Na^+-K^+-ATPase there is still controversy with respect to this point (Nørby and Klodos, 1988).

The second essential coupling rule states that the hydrolysis of the phosphoenzyme intermediate does not occur, until the a-type ion taken up from the cytoplasm has been released from the extracellular low-affinity discharge site. Measurements on the SR Ca^{2+}-ATPase have shown that the nonhydrolyzable ADP-sensitive E_1P form transforms into the hydrolyzable ADP-insensitive E_2P form almost simultaneously with the dissociation of the two translocated calcium ions at the extracytoplasmic side (see Figure 6), suggesting that either concerted conformational changes in the ion-binding domain and the catalytic domain lead to the discharge of the calcium ions, as well as to the change in the catalytic specificity, or that it is the mere presence of the calcium ions on the enzyme that confers ability of the phosphoenzyme to react with ADP and inability to react with water. The Na^+-K^+-ATPase behaves somewhat differently from the Ca^{2+}-ATPase, since the three translocated sodium ions are released consecutively in at least two steps, and dephosphorylation can occur as soon as the first sodium ion has been released (see below).

In those P-type pumps that countertransport K^+, such as the Na^+-K^+-ATPase and the H^+-K^+-ATPase, K^+ binds with high affinity at extracytoplasmic uptake sites and stimulates the rate of hydrolysis of the E_2P phosphoenzyme intermediate more than 100-fold, with resulting translocation of K^+ to the cytoplasmic side. Also these steps are fully reversible (in the presence of inorganic phosphate to phosphorylate "backdoor"), and potassium ion fluxes in both directions (K^+-K^+-exchange, see Figure 5) can be demonstrated by use of isotopes in experiments with sided vesicular enzyme preparations. The specificity of this so-called K^+-limb of the Na^+-K^+-pump cycle is somewhat relaxed relative to that pertaining to the Na^+-limb, and a number of ions other than K^+ (including Na^+) may serve as counterions to Na^+ (henceforth designated c-type ions, see Table 2).

Extracellular Na^+ release is an early electrogenic event in the Na^+-K^+-pump cycle, which can occur in the complete absence of K^+, and it is commonly assumed that only a single set of cation transport sites exists. These sites alternate between two possible orientations in a consecutive (ping-pong) mechanism, so that in the normal ATP-driven $3Na^+/2K^+$ exchange mode all three translocated sodium ions are released extracellularly prior to the binding and translocation of the two potassium ions by the vacated (and structurally altered) sites (Figure 7). Hydrolysis of the Na^+-K^+-ATPase phosphoenzyme can occur with two sodium ions substituting for potassium at the extracellular sites (ATP-driven Na^+:Na^+-exchange, Figure 5), or without any ion attached (ATP-driven Na^+-efflux, Figure 5), but in the physiological situation the rate of these side-reactions is negligible compared to that of K^+-stimulated hydrolysis.

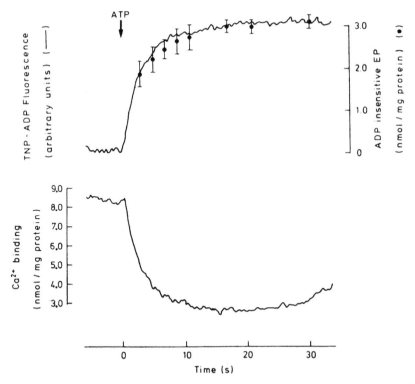

Figure 6. Time course of change in catalytic specificity (upper panel) and Ca^{2+} dissociation from extracytoplasmic low affinity sites (lower panel) following phosphorylation of the SR Ca^{2+}-ATPase with ATP. The amount of ADP-insensitive phosphoenzyme (E_2P) was measured in two ways: (I) $[\gamma\text{-}^{32}P]$ATP was included in the reaction mixture and the radioactivity incorporated into the enzyme was determined after acid quenching at various time intervals. To remove the ADP-sensitive phosphoenzyme so that only ADP-insensitive phosphoenzyme was measured, ADP was added 4 sec before the quench (upper panel, right scale); (2) by the enhancement of fluorescence from a trinitrophenyl-derivative of ADP bound in the catalytic site in exchange with ADP after the phosphorylation (upper panel, left scale). The change in Ca^{2+} binding was measured indirectly by use of murexide as an indicator of free Ca^{2+} in the medium. The data show that Ca^{2+} dissociates simultaneously with formation of E_2P. The data points were taken from Andersen et al., 1985.

For the SR Ca^{2+}-ATPases, it is commonly assumed that the extracytoplasmic sites on E_2P from which the two translocated calcium ions have been released, either remain empty during recirculation to the cytoplasmic side or bind and countertransport 2–3 protons (Levy et al., 1990). In the former case, there would be a clear difference from the sodium pump, since the hydrolysis of the empty E_2P intermediate would occur at a much higher rate in Ca^{2+}-ATPase. It has yet to be established

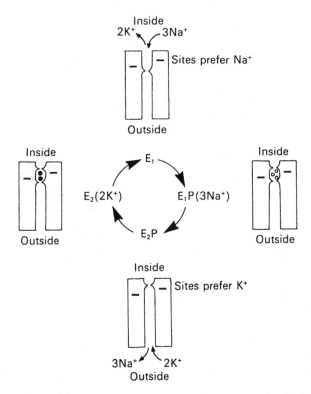

Figure 7. Outline of the current concepts in cation transport by the Na⁺-K⁺-pump. Three Na⁺ and two K⁺ are thought to be transported consecutively by protein sites containing two negative carboxylate groups. The ions become occluded during the transport process by the simultaneous closure of cytoplasmic and extracytoplasmic gates. A low conductance access channel connects the sites to the extracytoplasmic surface, resulting in an ion well with voltage sensitivity of extracytoplasmic ion binding and dissociation. Reproduced with permission from Glynn, 1993.

whether in analogy with the effect of K⁺-binding on Na⁺-K⁺-ATPase, phosphoenzyme hydrolysis in calcium pumps is activated by protons binding at extracytoplasmic sites. The plasma membrane Ca^{2+}-ATPases are thought to follow a scheme with exchange of one or two H⁺ for one Ca^{2+}. A stoichiometry of 2H⁺ for $2Ca^{2+}$ as implicated for the SR Ca^{2+}-ATPases is, however, not excluded.

One version of the E_1-E_2 model for Ca^{2+}-ATPase (De Meis and Vianna, 1979; Tanford, 1984; see Figure 4A) rationalized the two types of catalytic specificity in the phosphoryl transfer reactions in terms of a single global conformational change of the pump protein, involving not only the catalytic site, but also the ion-binding sites. The latter were supposed to change their orientation (vectorial specificity) relative to the membrane simultaneously with the rearrangements in the catalytic site. Hence, this model postulated the existence of two major conformations of the pump protein: E_1 forms with cytoplasmically orientated ion-binding sites and catalytic specificity for ADP-ATP exchange, and E_2 forms with extracytoplasmically orientated ion-binding sites and catalytic specificity for P_i-H_2O exchange. The two-state model was based on studies of the phosphorylation and dephosphorylation reactions taking place in the catalytic site, and it is no wonder that the assumptions regarding the orientation of the ion-binding sites have turned out to be inaccurate, if not erroneous. Kinetic experiments monitoring the rate of Ca^{2+} binding to the SR Ca^{2+}-ATPase in vesicles retaining the native sidedness of the membrane have demonstrated a consecutive binding pattern, in which there is ready access for one calcium ion to a cytoplasmic uptake site, whereas the second Ca^{2+}-binding site is inaccessible until after the occurrence of a relatively slow step triggered by the binding of the first calcium ion. This has been interpreted in terms of the existence of a cytoplasmically oriented uptake site in the unphosphorylated E_2 form (Jencks, 1989b). Another possibility would be to assume a fast E_2-E_1 transition before binding of the first ion to E_1, but a slow conformational change would then have to be interposed between the E_2-E_1 transition and the phosphorylation by ATP (cf., Figure 8). Both interpretations are in disagreement with the concept that the E_2-E_1 conformational change is global, involving a slow reorientation of both ion-binding sites and a simultaneous change in catalytic specificity. Another point is that the calcium ions bound at the cytoplasmic uptake sites in the E_1 form of Ca^{2+}-ATPase in tight vesicular preparations become inaccessible to exchange with the free ions in the cytoplasm immediately upon phosphorylation with ATP. This seems to indicate that a reorientation of the ion-binding sites with respect to the membrane occurs in relation to phosphorylation and not in relation to the ensuing E_1P-E_2P transition. This and other evidence led to the proposal of an alternative reaction scheme for the Ca^{2+}-ATPase, which is based on the sort of permutational matrix of four states shown in Figure 4B (Jencks, 1989b). The catalytic specificities defining E_1 and E_2 are the same as in the original E_1-E_2 model, but each of these forms can assume either type of vectorial specificity of the ion-binding sites. It was proposed that the phosphorylation acts as a switch determining the vectorial specificity, so that all phosphorylated forms possess extracytoplasmically oriented Ca^{2+} sites, whereas non-phosphorylated forms possess cytoplasmically oriented sites.

Cation Occlusion

There are, however, also problems with a symmetrical 4-state model, since studies of Ca^{2+}-ATPase as well as Na^+-K^+-ATPase have made it clear that the phosphorylation of E_1 from ATP is a much faster reaction than the extracytoplasmic appearance of the translocated a-type ions. The latter event constitutes a major rate-limiting step in the pump cycle, and can in fact be further slowed down by

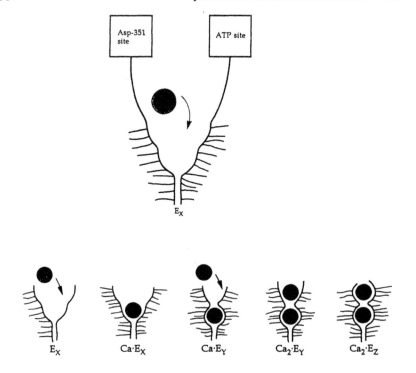

Figure 8. Model for the occlusion of two calcium ions at the cytoplasmic uptake sites in the SR Ca^{2+}-ATPase. The upper part shows the coupling between Ca^{2+} binding and phosphorylation. The lower part shows various subconformations of the Ca^{2+} binding sites related to the consecutive binding mechanism. Ca^{2+} binding draws the phosphorylation and nucleotide—binding subdomains together as we pass from E_x to Ca_2E_z, permitting transfer of the γ-phosphoryl group of ATP to the protein. The phosphorylation, on the other hand, stabilizes the occluded state of the Ca^{2+} binding cavity.

specific modifications of amino acid residues and peptide bonds in the pump protein, without a corresponding reduction of the rate of phosphorylation (see below). In the Na^+-K^+-ATPase, the translocation of K^+ from the extracytoplasmic side to the cytoplasm constitutes another major rate-limiting step, and the cytoplasmic appearance of the translocated K^+ is a much slower event than the dephosphorylation of E_2P. In the now classic experiments by Post et al. (1972), they measured phosphorylation by ATP of Na^+-K^+-ATPase enzyme that had just been dephosphorylated from E_2P, and found a slow rate of phosphorylation when a K^+ congener (Rb^+) had been present during E_2P hydrolysis. It followed that the hydrolysis product must have continued to bind Rb^+ (in the E_2 form which cannot phosphorylate from ATP) for some time after the cleavage of the acyl phosphate bond in E_2P. In this period the ion apparently was not free to dissociate. Time-resolved direct binding experiments using radioactive isotopes later confirmed the hypothesis that there is an intermediate occluded state in the transport pathway, in which the cation being translocated is neither free to dissociate to the cytoplasmic nor to the extracytoplasmic membrane side (Forbush, 1987; Glynn and Karlish, 1990; Glynn, 1993; see Figures 7 and 8).

Ion occlusion does not necessarily mean that the binding site in the occluded state has an unusually high affinity for the bound ion, but in the occluded state high activation barriers may prevent rapid equilibration between the binding site and the medium on either side of the membrane (Figure 9). The transition between the enzyme forms exposing the binding sites at opposing sides of the membrane may thus be viewed as a successive closing and opening of two gates corresponding to the two energy barriers. The existence of the occluded state with both gates closed is probably one of nature's means of minimizing slippage to maintain a fixed stoichiometry and a high efficiency of pumping. The present understanding of ion occlusion can be formulated in a third essential coupling rule: the translocated species bound in the phosphorylated E_1P form (a-type ion) cannot escape from the binding site, until either a slow conformational change has occurred in the forward direction of the reaction cycle, or the enzyme has donated the phosphoryl group back to ADP forming ATP (in the latter situation the escape is to the cytoplasmic side with no net translocation). For countertransport systems (exemplified by the Na^+-K^+-ATPase), a corollary states that the E_2 dephospho-form likewise releases the countertransported c-type ion at the cytoplasmic side, only after a slow conformational change, and at the extracytoplasmic side, only after phosphorylation by inorganic phosphate.

There is no rule without an exception, and passive downhill movements of K^+ and Ca^{2+} in the wrong direction, through the respective Na^+-K^+- and Ca^{2+}-pumps, have been demonstrated under certain conditions, even without phosphorylating substrates present. These leakage fluxes have been explained by conformational

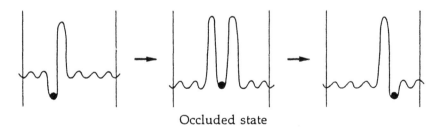

Occluded state

Figure 9. Energy diagram for the cation binding sites during ion translocation. In the occluded state, cytoplasmic and extracytoplasmic gates are closed corresponding to the existence of high energy barriers separating the occluded state from conformations with ready access of the binding sites to the aqueous phase. Reproduced from Läuger, 1991 with permission from Sinauer Associates, Inc.

equilibria that allow a small population of E_2 forms to expose their ion binding sites at the extracytoplasmic surface in the absence of phosphorylation, so that the pump can work as a passive carrier oscillating between unphosphorylated E_1 and E_2 forms with alternate orientation of the ion binding sites. Although under physiological conditions the leakage fluxes are believed to be too small to affect the overall efficiency of the pumps to an appreciable extent, it cannot be excluded that they are regulated and play a role in a physiological mechanism for varying the coupling ratio. The leakage flux through Ca^{2+}-ATPase (De Meis, 1991) has been found to be greatly increased by certain drugs, notably phenothiazines and local anesthetics. It has also been reported that the Na^+-K^+-ATPase can be converted reversibly into a massive univalent cation leak by toxins. This type of channel–like behavior suggests a close relationship between ion pumps and channels. Thus, it may not be too far-fetched to consider ion pumps as channels equipped with locks and gate controls.

Tanford et al. (1987) suggested that the first and third of the coupling rules described above may have a common structural basis. They proposed a hypothetical model in which the nucleotide-binding domain and the phosphorylation site of the pump protein are linked to opposite sides of an ion-binding cavity with negatively charged walls, so that ion binding draws the walls of the cavity together, thereby bringing the ATP molecule in position for transfer of the γ-phosphoryl group to the protein simultaneously with the occlusion of the ion (Figure 8).

A consecutive pattern of cation binding has been demonstrated not only for the two calcium ions that bind from the cytoplasmic side to form Ca_2E_1 in the SR Ca^{2+}-ATPase (cf., Figure 8), but also for the two potassium ions that bind from the extracytoplasmic side to form K_2E_2P in Na^+-K^+-ATPase (Forbush, 1987). Isotope exchange studies have shown that one of the two bound ions (the one binding last) prevents the other from dissociating at the uptake side (see Figure 10). Hence, in the Ca^{2+}-ATPase the calcium ion that is first to bind becomes occluded in the E_1 form as a consequence of the binding of the second calcium ion, even before phosphorylation from ATP has occurred, and in Na^+-K^+-ATPase, the potassium ion that is first to bind likewise becomes occluded in the E_2P form as a result of the binding of the second potassium ion, before any dephosphorylation has occurred. This constitutes the basis for the hypothesis that one of the ion-binding sites is located deeper in the protein structure than the other, so that steric hindrance in a channel-like binding pocket is involved as shown in Figures 8 and 10B. The phosphorylation (dephosphorylation in case of K^+-occlusion) appears to stabilize the conformation containing both ions in the occluded state. It is also possible that the superficial ion bound last spends part of the time in the occluded state, even before any change in the phosphorylation state has occurred, as envisioned in a flickering gate model proposed by Forbush (1987). The role played by the change in the phosphorylation state would then be analogous to the locking of a door that has already been closed. Evidence supporting this conclusion is that the occluded

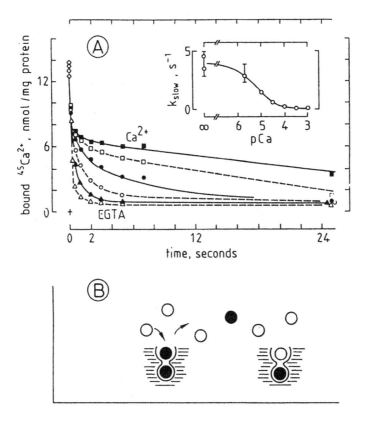

Figure 10. Experiment demonstrating consecutive dissociation of two calcium ions from the uptake sites of the SR Ca^{2+}-ATPase in the unphosphorylated E_1 state. $^{45}Ca^{2+}$ bound from the cytoplasmic side at the two high-affinity sites on the enzyme can be seen to dissociate back to the medium on the cytoplasmic side in two distinct phases upon dilution of the radioactivity (**A**). The first phase is rapid and independent of the concentration of Ca^{2+} in the medium. The second phase is also rapid in the absence of Ca^{2+} (presence of EGTA) in the buffer medium (open triangles) but slow in the presence of nonradioactive Ca^{2+} in the buffer medium (closed triangles, 10 μM Ca^{2+}; open circles, 30 μM Ca^{2+}; closed circles, 100 μM Ca^{2+}; open squares, 300 μM Ca^{2+}; closed squares 1 mM Ca^{2+}). The inset shows the rate constant for dissociation of the second Ca^{2+} as function of the Ca^{2+} concentration in the medium. The lower panel (**B**) explains these observations in terms of steric hindrance of the dissociation of the deeper ion by the presence of a more superficial ion at its binding site (closed circles, radioactive $^{45}Ca^{2+}$; open circles, nonradioactive $^{40}Ca^{2+}$ from the medium). Reproduced from Orlowski and Champeil, 1991a with permission from The American Chemical Society.

19

E_1 form can be stabilized by means other than phosphorylation, for instance by binding the non-phosphorylating substrate analog CrATP (Vilsen et al., 1987; Vilsen and Andersen, 1992b), or by oligomycin binding in the case of Na$^+$-K$^+$-ATPase (Esmann and Skou, 1985).

If the release of the deoccluded ions at the discharge side occurred through a narrow channel–like structure, one might expect a single-file behavior in the release process similar to that described for binding at the uptake sites. Much of the available evidence argues against an ordered release of the two calcium ions from the extracytoplasmic discharge sites in Ca^{2+}-ATPase under physiological conditions, or of the two potassium ions from the cytoplasmic discharge sites in Na$^+$-K$^+$-ATPase. It has thus not been possible to distinguish the two binding sites kinetically during the release process, either because the dissociation occurs too fast (the sites are of low affinity; see Figure 2), or because the sites are no longer located in a file one above the other after the conformational change that confers low affinity (Forbush, 1987; Orlowski and Champeil, 1991b). Studies of voltage effects on Na$^+$-K$^+$-ATPase partial reactions have accordingly demonstrated that the release of the two potassium ions at the cytoplasmic side is unaffected by membrane potential, indicating that any access channel to the aqueous medium on the cytoplasmic side must be of a high conductance (wide).

A somewhat different picture has emerged for the extracytoplasmic Na$^+$ discharge sites in the Na$^+$-K$^+$-ATPase. There is evidence that the extracellular Na$^+$ release involves the consecutive appearance of at least three phosphorylated intermediates with different kinetic properties in the transition between E_1P with three occluded Na$^+$ and E_2P (Nørby and Klodos, 1988). One intermediary form (often designated E*P) accumulates if the cholesterol content of the membrane is high. This phosphoenzyme intermediate can be slowly hydrolyzed, carrying two sodium ions back to the cytoplasmic surface as if Na$^+$ substituted for K$^+$ (Yoda and Yoda, 1987). Digitalis glycosides such as ouabain (Hansen, 1984), and other cardiotonic steroids, bind to E*P and stabilize a phosphorylated state retaining two Na$^+$ in an occluded state. Therefore, one sodium ion seems to be released at the extracellular side before the other two. In addition, the apparent affinities of the respective discharge and uptake sites for extracellular Na$^+$ and K$^+$ have been demonstrated to depend on membrane potential in a way consistent with the presence of a narrow low-conductance access channel (ion well) connecting the ion binding sites to the extracellular aqueous medium in Na$^+$-K$^+$-ATPase (Figure 7).

Current State of the Model

The concepts outlined above are tentatively summarized in the minimal reaction cycles shown in Figure 11. The cycles show not just two conformational states, but multiple states. The E_1/E_2 notation is used primarily to refer to distinct catalytic specificities for ATP-ADP exchange and P_i-H$_2$O exchange, respectively. There is

A

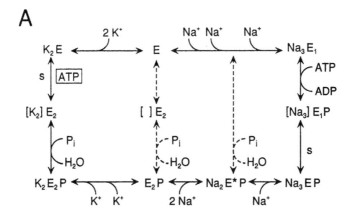

B

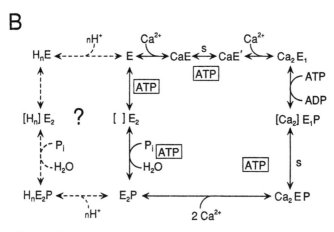

Figure 11. Proposed reaction cycles of the Na$^+$-K$^+$-ATPase (**A**) and the SR Ca^{2+}-AT-Pase (**B**) involving transitions between different conformational states of the enzymes (see text for further explanation). The cytoplasmic side of the membrane is upward and the extracytoplasmic side downward. Brackets indicate that all the cation binding sites reside in an occluded state. A tentative H$^+$-countertransport limb is shown for the SR Ca^{2+}-ATPase (most likely n = 2). s indicates a relatively slow reaction step. ATP boxes indicate steps accelerated by ATP not being hydrolyzed. Mg^{2+} serving as a cofactor in phosphorylation and dephosphorylation is not shown.

preference for a-type ions in E$_1$ and for c-type ions in E$_2$. Each conformational class consists of subconformations. In E$_1$ states the orientation of the cation binding sites can be either cytoplasmic or occluded, the occluded state being stabilized by phosphorylation. In E$_2$ states the orientation can be either extracytoplasmic or occluded, the occluded state being destabilized by phosphorylation. Forms without well defined catalytic specificity have been classified as neither E_1 nor E_2. These include intermediate states with less than full occupancy of the ion-binding sites, as well as transient states in which low affinity sites are occupied with translocated ions that are about to leave (deoccluded forms denoted E and EP in the upper left and lower right corners of the schemes). For Ca^{2+}-ATPase a distinct conformational change (CaE→CaE′) is shown to be involved in the consecutive binding pattern at the cytoplasmic uptake sites in accordance with the model in Figure 8. This type of structural model explains why the second site is not exposed at the cytoplasmic surface until after the binding of the first ion: the second site does not exist at all, until after the first site has been occupied. For Na^+-K^+-ATPase, the events at the extracellular surface, the dissociation of Na^+ from discharge sites and the binding of K^+ at uptake sites occur in a consecutive manner. An intermediate in this sequence is the E*P form with two Na^+ bound, which can be stabilized by ouabain binding. The broken arrows indicate reactions in the Na^+-K^+-ATPase cycle that lead to Na^+ transport without K^+ countertransport. Moreover, based on recent evidence (Levy et al., 1990), the Ca^{2+}-ATPase cycle is tentatively extended with a countertransport limb containing an E_2 form with occluded protons. The steps thought to be rate-limiting for the overall turnover (the occlusion/deocclusion steps) are indicated by s (slow) in Figure 11.

Modulation by ATP

As indicated by the ATP boxes in the schemes in Figure 11, ATP comes into play not only as a phosphorylating substrate, but also as an allosteric effector that activates some of the partial reactions without being hydrolyzed. The potassium-occluded E_2 form of the Na^+-K^+-ATPase binds ATP (or non-hydrolyzable analogs) with a 100–1,000-fold lower affinity than the E_1 form, and deocclusion of K^+ from E_2 is greatly accelerated when ATP is bound at the low-affinity site (Glynn, 1985, 1993). The mutual destabilization of bound K^+ and ATP is usually referred to as K^+-ATP antagonism. In fact, the pump cycle is virtually stopped in the potassium-occluded state unless ATP is present in a millimolar concentration. The concentration of ATP required to saturate the phosphorylation of E_1 is only micromolar, and the Lineweaver-Burk plot for Na^+-K^+-ATPase activity as a function of ATP concentration is curvilinear with one K_m in the micromolar range, reflecting the ATP requirement for phosphorylation, and another K_m in the millimolar range, reflecting the ATP requirement for activation of deocclusion of K^+. A similar biphasic ATP-dependence of turnover exists for the SR Ca^{2+}-ATPase, but in Ca^{2+}-ATPase the E_1P-E_2P transition with associated deocclusion of Ca^{2+} constitutes the slowest step in the cycle, and is actually the step being accelerated by millimolar ATP. In addition, other partial reaction steps in the Ca^{2+}-ATPase cycle are accelerated by ATP (Figure 11B), but these modulations occur with lower K_m values in the micromolar range and do therefore not show up so clearly in the

ATP-dependence of overall turnover. The behavior of the Ca^{2+}-ATPase cycle is in accordance with a labile nature of the putative E_2 form of Ca^{2+}-ATPase, and contrary to the very stable K^+-occluded form of Na^+-K^+-ATPase, the Ca^{2+}-ATPase E_2 form does not possess very low affinity for ATP.

The enhancement of the rate of conversion of one enzyme form into another by binding of ATP may be attributed to a destabilizing effect of ATP binding on the state displaying low nucleotide affinity and/or to creation of an alternative reaction pathway (stabilization of a transition state). There is evidence to support the view that the effector site for ATP is identical to, or is a structurally modified version of, the catalytic site (in the E_2 form). A fluorescent trinitrophenyl derivative of ATP has been shown to bind at the catalytic site on the E_1P phosphoenzyme of Ca^{2+}-ATPase after departure of ADP (Seebregts and McIntosh, 1989), and a corresponding binding of ATP in exchange with ADP can explain the ATP-induced acceleration of the E_1P-E_2P transition in Ca^{2+}-ATPase, if it occurs with low affinity (possibly due to repulsion by the phosphate already present in the site). The existence of an additional (purely allosteric) ATP binding site is, however, not excluded (Coll and Murphy, 1991). For technical reasons it is difficult to determine the presence of a site of low affinity by direct measurement of equilibrium nucleotide binding, since background nonspecific binding is too high at millimolar concentrations of nucleotide.

Role of Mg^{2+}

Magnesium ions play an essential role (ignored in Figure 11) as cofactors in the phosphorylation and dephosphorylation reactions. The MgATP and MgP_i complexes seem to be the true substrates in phosphorylation (Champeil et al., 1985), and Mg^{2+} becomes trapped with the phosphoryl group on the enzyme. It has not been determined whether the enzyme-bound Mg^{2+} remains coordinated by the covalently attached phosphoryl group. There is evidence that Mg^{2+} and the chromium complex of ATP (CrATP) can bind simultaneously, suggesting the existence of a separate Mg^{2+} site on the protein (Vilsen and Andersen, 1987). Other divalent and trivalent cations such as Ca^{2+} and La^{3+} can act as cofactors in phosphorylation, but with these ions substituting for Mg^{2+} the rate of the $E_1P\rightarrow E_2P$ transition, and hence the overall turnover rate, is reduced.

STRUCTURE OF THE P-TYPE PUMPS

Subunit Composition

In most P-type pumps the polypeptide chain that carries out ATP hydrolysis and ion translocation has a size ranging between 900 and 1,200 amino acid residues (although shorter and longer versions exist). In the Ca^{2+}-ATPases no other peptide is required for pump activity. In Na^+-K^+-ATPase and H^+-K^+-ATPase the catalytic peptide (the α-subunit), which is homologous to the peptide of Ca^{2+}-ATPase, occurs together with a glycosylated β-subunit consisting of about 300 amino acid residues plus a sugar moiety of about 10 kDa. Although the β-subunit is bound only noncovalently to the α-chain, the binding is strong enough to require a denaturing detergent such as sodium dodecyl sulfate to separate them. There is evidence that the major function of the β-subunit is related to the structural maturation and correct expression of the α-subunit, i.e., stabilization of a functional folding pattern of the α-subunit and the intracellular transport of the α-subunit to the plasma membrane (Geering, 1991).

The P-type pumps can exist as dimers ($(\alpha\beta)_2$ in case of Na^+-K^+-ATPase and H^+-K^+-ATPase) and higher oligomers in the membrane, and the possible functional consequences of this is a widely debated issue (Glynn, 1985; Andersen, 1989). Oligomerization could be fortuitous, reflecting nonspecific peptide-peptide interactions as a result of the high concentration of peptide present in the enriched membranes used in most studies. It is possible with the aid of suitable nonionic detergents to obtain active preparations of soluble Ca^{2+}-ATPase monomers and Na^+-K^+-ATPase $\alpha\beta$-protomers. The solubilized monomer/protomer is able to hydrolyze ATP coupled with conformational changes similar to those occurring in the membrane-bound enzyme. In particular, the soluble monomer is able to occlude the cations (Vilsen and Andersen, 1987; Vilsen et al., 1987; Andersen, 1989). Therefore, the monomer comprises a fully functional unit containing the proteinaceous cavity that provides the passage for the ions across the membrane. Some aspects of ligand binding in Na^+-K^+-ATPase may, on the other hand, be difficult to interpret without involving functional interaction between catalytic sites on different protomers (Jensen and Ottolenghi, 1983). A regulatory role of oligomerization is thus not excluded.

Gross Structural Features of the Catalytic Peptide

The protein structure is not known at high-resolution for any P-type pump, as crystals of a size and quality suitable for X-ray analysis have not yet been produced. However, the discovery that ordered two-dimensional semicrystalline arrays of ATPase molecules can be formed *in situ* in the membrane by incubation with tightly binding inhibitors such as vanadate (Skriver et al., 1981; Dux and Martonosi, 1983) has been of great help in obtaining a relatively detailed picture of the overall shape of the protein by electron microscopy. Fourier analysis of tilt series of electron micrographs permitted 3D-reconstructions of the Na^+-K^+- and Ca^{2+}-ATPase protein structures to a resolution of 20–25 Å (Figure 12a). Recently, this type of reconstruction analysis has been taken a step further to a resolution of about 14 Å, and detailed structural features have become visible, even for the transmembrane domain (Toyoshima et al., 1993). The model of the Ca^{2+}-ATPase peptide shown in

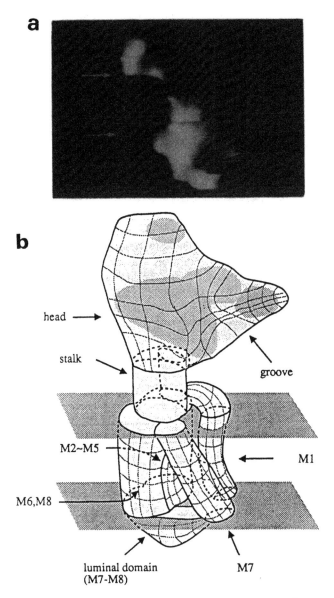

a

b

head →

stalk

groove

M2~M5

M1

M6,M8

luminal domain
(M7-M8)

M7

Figure 12. Three-dimensional structures of Na$^+$-K$^+$-ATPase and Ca^{2+}-ATPase based on image reconstruction analysis of electron microscopic data obtained from two-dimensional membrane crystals. (**a**). Na$^+$-K$^+$-ATPase molecule consisting of one α-subunit and one β-subunit. The horizontal arrows indicate a tentative location of the membrane surfaces (upper arrow: cytoplasmic surface; lower arrow: extracellular surface; suggested membrane thickness: 39 Å). Data from Hebert et al., 1988. (**b**). Ca^{2+}-ATPase molecule consisting only of a catalytic subunit. A tentative location of the transmembrane helices (M1–M10) is indicated. The cytoplasmic part (head) is pointing upwards. In this study, the membrane (sarcoplasmic reticulum) was found to be only 32 Å thick (surfaces indicated by shaded areas). Modified from Toyoshima et al., 1993.

Figure 12b has been constructed from such data. In the Ca^{2+}-ATPase, only 5% of the mass protrudes on the extracytoplasmic (luminal) side, and since the protrusions of Ca^{2+}-ATPase and Na^+-K^+-ATPase on the cytoplasmic side are highly similar, it may be inferred that the extracellular protrusion seen in the case of the Na^+-K^+-ATPase corresponds to the β-subunit (Figure 12a), while most of the cytoplasmic protrusion corresponds to the α-subunit. In the model of Ca^{2+}-ATPase, the cytoplasmic protrusion, in total 70% of the mass of the catalytic peptide, extends 75 Å above the membrane surface, and has a complex structure consisting of several subdomains. The head, which is subdivided into lobes, is joined to the membrane by a 25 Å long stalk. The intramembranal structure consists of three distinguishable segments, the largest of which seems to form a continuation of the stalk. This segment consists of two parts that separate as the extracytoplasmic surface is approached, possibly forming a mouth for delivery of the translocated ions. It should be emphasized that there is no solvent-filled cavity, as is seen for channel-proteins. The amount and disposition of the intramembranal mass is not inconsistent with the proposal of 10 α-helices discussed below.

Amino Acid Sequence and Topology

In 1985 the first amino acid sequences of Na^+-K^+-ATPase and the SR Ca^{2+}-ATPase appeared, deduced from the corresponding sequences of cloned cDNA (Shull et al., 1985; MacLennan et al., 1985). Since then, sequences have been determined for more than 50 members of the P-type family, and it has become clear that the functional and overall structural similarity of the P-type pumps is matched by a considerable sequence homology, which supports the hypothesis of a common evolutionary origin (Serrano, 1988; Jørgensen and Andersen, 1988; Green and MacLennan, 1989; Green, 1989). The Na^+-K^+-ATPases and H^+-K^+-ATPases show as much as 60–70% overall amino acid sequence identity, while the overall identity between the Na^+-K^+-ATPases and the sarco(endo)plasmic reticulum Ca^{2+}-ATPases is 20–30%, and that between the Na^+-K^+-ATPases and the plant H^+-ATPases is 15–20%. It is perhaps surprising that the sarco(endo)plasmic reticulum Ca^{2+}-ATPases do not resemble the plasma membrane Ca^{2+}-ATPases more than they resemble the Na^+-K^+-ATPases, and that the gastric H^+-K^+-ATPase does not resemble the H^+-K^+-ATPase from colon more than it resembles the Na^+-K^+-ATPase subfamily (Crowson and Shull, 1992).

When different P-type ATPases are compared, certain peptide segments, constituting a total of about 300 residues, show homologies much higher than average (Figure 13), and these can be used as starting points to align all the family members (Green, 1989). The pumps also share a common pattern of hydrophobicity, the most N-terminal four hydrophobic segments being particularly well defined (Figure 14). With some caution the presence of transmembrane α-helices (generally containing about 22–28 residues each) can be predicted from the hydrophobicity pattern, as

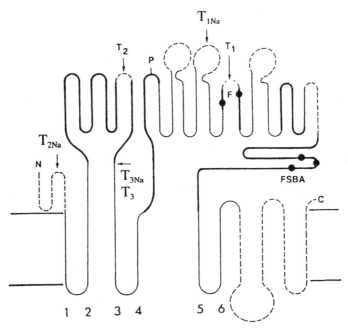

Figure 13. Linear map showing conservation of amino acid sequence homology within the P-type ATPase superfamily. Unbroken lines correspond to segments present in all P-type pumps. Heavy lines highlight those sequences for which different family members show mean pairwise identity > 30% (conserved segments). Broken lines indicate segments with highly variable sequences and segments present only in some P-type ATPases. The data are based on the alignment by Green, 1989. Filled circles indicate residues labeled with TNP-8N$_3$-ATP, FITC, adenosine triphosphopyridoxal, ClR-ATP, and FSBA (see Table 3 for exact position in sequence). T_1, T_2, and T_3 indicate tryptic cleavage sites in the SR Ca^{2+}-ATPase. T_{1Na}, T_{2Na}, and T_{3Na} indicate tryptic cleavage sites in Na$^+$-K$^+$-ATPase (c.f., Figure 19). The chymotryptic cleavage site denoted C_{3Na} in Na$^+$-K$^+$-ATPase (not shown) has approximately the same location as T_{3Na}. Modified from Green and MacLennan, 1989.

suggested by the correlation of prediction with actual structure observed for bacterial photosynthetic reaction center and bacteriorhodopsin, two membrane proteins for which the 3D-structure is known at high resolution. This provides a basis for structural modeling of the P-type pumps (Figures 15–16). Two large cytoplasmic loops connect hydrophobic segment M2 with M3, and M4 with M5. These loops, together with the most C-terminal third of M4, contain all the peptide segments that are highly conserved within the entire family of P-type pumps. The largest cytoplasmic loop between M4 and M5 contains the phosphorylated aspartyl residue (Asp^{351} in the SR Ca^{2+}-ATPase; see Figure 15) and several other highly conserved residues that label covalently with nucleotide-related affinity reagents

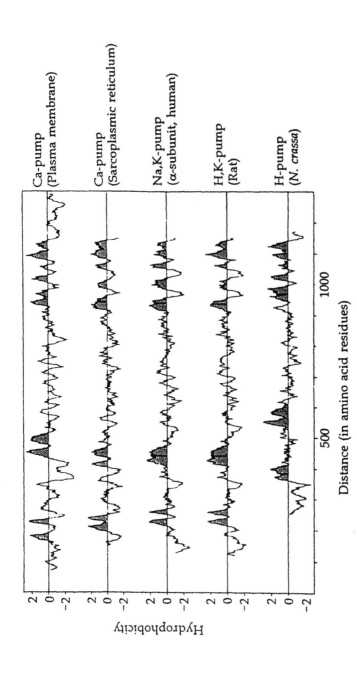

Figure 14. Hydrophobicity analysis of five P-type ATPases according to the Kyte-Doolittle method. A hydrophobicity value between −4.5 and +4.5 is assigned to each type of amino acid residue and mean values are successively calculated along the peptide sequence using a window of 18 residues. Segments corresponding to the transmembrane helices M1–M10 in the structural model in Figure 15 are shaded. Modified from Verma et al., 1988.

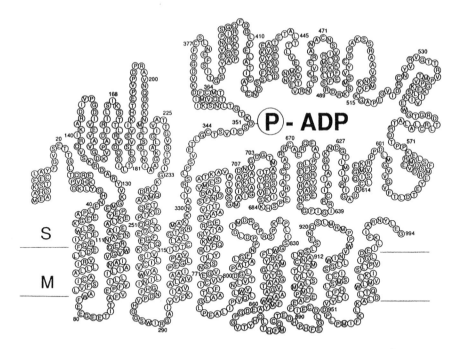

Figure 15. Model of the secondary and tertiary structure of the SR Ca^{2+}-ATPase (MacLennan et al., 1985; Green, 1989). Amino acid residues are indicated by single-letter code. Two types of secondary structure elements are shown; α-helices are represented by diagonal rows consisting of three or four residues as seen for instance in the membrane (M) and stalk (S) sectors; β-strands are indicated by a ladder-type of arrangement of the symbols.

(Figures 13 and 16, and Table 3). The binding of these labels at the ATP site is inferred from the exclusion of ATP binding after their covalent attachment. In several of these cases the prevention of ATP binding is ascribable solely to the presence of the ATP analog and not to the chemical modification of the residue providing the covalent attachment site, since the residue itself is nonessential to function (Table 3 and Figure 16). Labeling with fluorescein isothiocyanate (FITC) or with the trinitrophenyl derivative of ATP does not prevent the Ca^{2+}-ATPase from functioning in Ca^{2+} transport energized with the smaller substrate acetyl phosphate or in backdoor phosphorylation with inorganic phosphate. This suggests a functional subdivision of the M4–M5 loop, the N-terminal part containing the site for receiving the phosphoryl group (phosphorylation domain), and the middle part forming a cleft for binding the adenine ring (nucleotide domain).

The secondary structure of subdomains in the cytoplasmic loops has been predicted from the amino acid sequence by use of standard algorithms for soluble

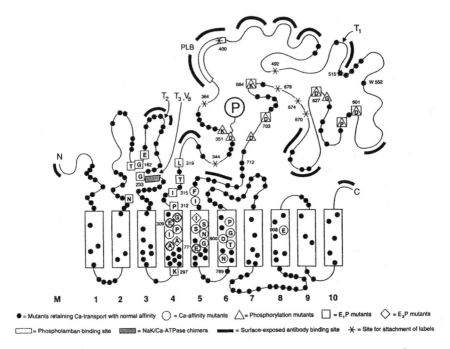

Figure 16. Functional consequences of site-directed mutagenesis of the SR Ca^{2+}-AT-Pase. Single residues were substituted by mutagenesis, and the overall and partial reactions of the mutants were studied by Ca^{2+}-transport and phosphorylation assays. This allowed the functional classification of the residues indicated by filled circles, open circles, triangles, squares, and diamonds. The data were compiled from Maruyama and MacLennan, 1988; Maruyama et al., 1989; Clarke et al., 1989a, 1989b, 1990a,b,d; Andersen and Vilsen, 1992b; Andersen et al., 1989, 1992; Vilsen and Andersen, 1992a,c; Vilsen et al., 1989, 1991a,b, and unpublished results by the same authors. The figure also shows important proteolytic cleavage sites (T$_1$, T$_2$, T$_3$, and V$_8$), the peptide segment involved in interaction with phospholamban, and peptide segments exposed to antibody binding in the absence of detergents or other permeabilizing or denaturing agents (Mata et al., 1992). Residues that are specifically modified by certain chemical labels such as nucleotide-related affinity labels (Table 3) and fluorescent probes are indicated by asterisks. See text for more details.

proteins, giving consensus for the whole family of P-type pumps, even for those parts that are less well conserved (Green, 1989). The M2–M3 loop is predicted to fold into a sandwich of antiparallel β-strands (the β-domain). The predicted structure of the phosphorylation and nucleotide-binding domains of the M4–M5 loop consists of alternating β-strands and α-helices (Figure 15), which would result in parallel β-sheets similar to the findings in several kinases. The kinase analogy may have functional consequences in terms of possible hinge-bending movements

Table 3. Nucleotide-Related Affinity Reagents Labeling P-Type Pumps Covalently

Reagent	Residue[a] Labeled and its Functional Significance[b]	Pumps Labeled
[γ-^{32}P]ATP	Asp^{351} (i)	All P-type pumps
2′,3′-O-(2,4,6-Trinitrophenyl)-8-azido-ATP (TNP-8N$_3$-ATP)	Lys^{492}	SR Ca^{2+}-ATPase
Pyridoxal 5′-phosphate and/or adenosine di- and triphosphopyridoxal	Lys^{492}	SR Ca^{2+}-ATPase (- Ca^{2+}) Na$^+$-K$^+$-ATPase H$^+$-K$^+$-ATPase
Fluorescein 5′-isothiocyanate (FITC)	Lys^{515}	SR and PM Ca^{2+}-ATPases Na$^+$-K$^+$-ATPase H$^+$-K$^+$-ATPase
2-azido-ATP	Asp^{627} (i)	H$^+$-ATPase of yeast
Adenosine triphosphopyridoxal	Lys^{684} (i)	SR Ca^{2+}-ATPase
γ-(4-N-2-chloroethyl-N-methylamino)benzylamide-ATP (ClR-ATP)	Asp^{703} (i)	Na$^+$-K$^+$-ATPase
5′-(p-fluorosulfonyl)benzoyladenosine (FSBA)	Lys^{712}	Na$^+$-K$^+$-ATPase

Notes: [a]The numbering refers to the position of the homologous residue in the SR Ca^{2+}-ATPase (Figure 15).
[b]The labeled enzymes have lost transport function with ATP as substrate, but since the label occupies the ATP site, this does not mean that the residue is essential. The functional role of the residue itself has been tested by mutagenesis (Figure 16) and by modification with other chemical reagents (Stefanova et al., 1993). The (i) for inactive indicates that transport function was lost in all studies, in which the residue was altered.

related to the transfer of the phosphoryl group from ATP bound in one subdomain to the acceptor site in the other subdomain. Such movements could control the gating mechanism in line with the concept outlined in Figure 8. A detailed folding topology similar to that of adenylate kinase, where ATP is bound in a pocket between loops at the carboxy ends of the β-strands, has been suggested (Taylor and Green, 1989). It should be noted, however, that the P-type pumps do not contain a loop with the consensus motif GXXXXGKS/T (Saraste et al., 1990) present in the binding sites of adenylate kinase and other nucleotide-binding proteins including the F- and V-types of ion-motive ATPases. Moreover, some of the conserved residues labeled with ATP analogs, such as Lys^{492} and Lys^{684} of the SR Ca^{2+}-ATPase, are located just outside the fold modeled on the basis of adenylate kinase. This may possibly be taken as evidence that the nucleotide-binding fold of the P-type pumps is unique.

A clue to the folding pattern of the C-terminal part of the M4–M5 loop has been obtained by the observation that one molecule of glutaraldehyde forms a specific cross link between Lys^{492} and Arg^{678} of the SR Ca^{2+}-ATPase, which must mean that at least in one conformation (see below) these two residues are very close in the 3D-structure (McIntosh, 1992). Before entering the membrane, the C-terminal region of the M4–M5 loop (the so-called hinge-domain), which is the most highly conserved part of P-type pumps, probably folds back and comes close to the phosphorylation site (Figure 16). In accordance with this prediction the affinity-labels binding in the hinge domain possess a structure in which the reactive group is located corresponding to the position of the terminal phosphoryl group of ATP. The suggested spatial arrangement is also consistent with studies of energy transfer efficiency between fluorescent probes attached at different positions in the Ca^{2+}-ATPase peptide and at the lipid bilayer (Bigelow and Inesi, 1992). The measured energy transfers permit calculation of approximate distances between the probes, which show that iodoacetamide derivatives attached at Cys^{670} or Cys^{674} and maleimide derivatives attached at Cys^{344} and Cys^{364} (Figure 16) are located about midway between the bilayer and FITC (labeling Lys^{515}). From the calculated distance between the membrane surface and FITC (> 60 Å), the latter probe seems to be located close to the top of the ATPase molecule.

Surface-exposed parts of the protein have been distinguished by their ability to bind specific antibodies and proteolytic enzymes under nondenaturing conditions (see Figure 16). Many of these epitopes are located at positions in the sequence that correspond to the external side of the suggested nucleotide-binding fold (Mate et al., 1992). Antibodies to fluorescein bind only to denatured FITC-labeled Ca^{2+}-ATPase, and not to the native FITC-labeled enzyme. This is consistent with a location of bound FITC in a hydrophobic cleft corresponding to the ATP site.

The 10-Helix Model

The C-terminal quarter of the peptide chain following the largest cytoplasmic loop, is less well conserved among P-type pumps of different ion specificity than most of the remaining part of the sequence (Figure 13), and the number of transmembrane segments in this region is subject to conflicting interpretations. Numbers varying between three and six have been suggested on the basis of the hydrophobicity analysis shown in Figure 14. As mentioned above, the 3D-reconstruction analysis of electron density patterns would be consistent with a 10 helix model for the SR Ca^{2+}-ATPase (M1–M10, i.e., with six helices in the C-terminal quarter of the molecule). This model, first put forward by MacLennan et al. (1985), has been subjected to a number of critical tests, in which proteolytic enzymes, antibodies, and other nonpenetrating reagents have been used to determine the exposure to the medium of defined segments of pump proteins present in the membranes of closed vesicles of defined sidedness. The following clues have been obtained (referring to the model shown in Figures 15 and 16): a) Both NH_2- and COOH-termini are exposed on the cytoplasmic surface. This has been demonstrated with antibodies specific for these termini and with impermeant sulfhydryl reagents reacting with Cys^{12} of the SR Ca^{2+}-ATPase. For the plasma membrane Ca^{2+}-ATPase, the cytoplasmic orientation of the C-terminus can also be deduced from the presence of a C-terminal regulatory domain that binds calmodulin and other effector molecules present in the cytoplasm. b) The M6–M7 loop reacts with antibodies and proteolytic enzymes from the cytoplasmic side. c) The M7–M8 loop does not react with antibodies present on the cytoplasmic side, as long as the vesicles are tight, but reacts with antibodies present on the extracellular side, or when the vesicles are permeabilized, also with antibodies added on the cytoplasmic side. d) Omeprazole, a therapeutically used inhibitor of gastric acid secretion reacting from the extracellular side, labels cysteine residues in loops M5–M6 and M7–M8 of the H^+-K^+-ATPase. These results are consistent with a 10-helix consensus structure (Clarke et al., 1990c; Sachs et al., 1992; Mata et al., 1992; Karlish et al., 1993), but the presence of the last three putative transmembrane segments (M8–M10) is not very well documented, and the possibility that a single transmembrane segment is present instead of M8–M10 is not excluded. From the hydrophobicity analysis, the M9 segment seems to be the most likely candidate in this case. One may also have to consider possible individual variation between P-type pumps (cf., Figure 13).

In addition, the assumptions that all membrane-buried segments are full membrane-spanning α-helices and that the fundamental folding pattern is unaltered during pumping are questionable. Alternative structures such as β-strands, shorter inner helices, β-turns, and other loops, may be present in the membrane and provide the flexibility needed for ion translocation. Hydrophobicity analysis would not be very useful to reveal the membrane-bound nature of such short segments, which do not even have to be hydrophobic, if they are surrounded (insulated) by transmembrane helices of the conventional type. Indeed, there is some evidence that the M5 segment may be shorter than predicted in the original model, since a cytoplasmically exposed tryptic cleavage site has been found in the gastric H^+-K^+-ATPase, at the lysine located corresponding to Ser^{767} in Ca^{2+}-ATPase (Figure 15). On the other hand, this site was found to be protected when the H^+-K^+-ATPase was phosphorylated in the E_2P form, implying varying positions depending on the functional state (Besancon et al., 1992). As we shall see below, M5 has indeed been implicated in ion binding and translocation on the basis of site-directed mutagenesis studies. To appreciate the possible functional significance of the presence of short flexible structures in the membrane, one need only think of the voltage-gated ion channels, in which a partial transmembrane hairpin loop of suggested β-structure occurs between two membrane spanning helices and acts as an ion-selectivity filter, which is part of the channel-lining (Heinemann et al., 1992). The relevance of this consideration is stressed by the functional analogies between pumps and channels and by a recent discovery of a weak but significant homology between the sequence of the third membrane-spanning sequence in some ion-channels and that of M6 of a member of the P-type ATPase family (Krishna et al., 1993).

Location and Structure of the Ion-Binding Sites

The P-type ATPases do not contain sequence elements of distinct homology with cation-binding folds in soluble proteins of known structure. By analogy to

a

Wild type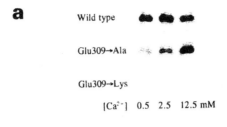

Glu309→Ala

Glu309→Lys

[Ca²⁺] 0.5 2.5 12.5 mM

b

Wild type   Glu309→Ala   Glu309→Lys

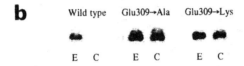

E   C   E   C   E   C

c

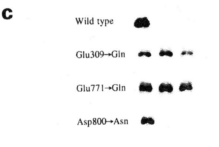

Wild type

Glu309→Gln

Glu771→Gln

Asp800→Asn

0   5s   10s   (continued)

Figure 17. Analysis of phosphoenzyme intermediates of SR Ca²⁺-ATPase mutants with alterations to carboxylate-containing residues in the transmembrane sector. Wild-type or mutant Ca²⁺-ATPases expressed in the endoplasmic reticulum membranes of COS-1 cells were phosphorylated with [γ-³²P]ATP (panel a) or [³²P]Pᵢ (panels b and c). Following acid-quench of the phosphorylated intermediate, the samples were subjected to SDS-polyacrylamide gel electrophoresis under acid pH conditions and the dried gels were autoradiographed to visualize the radioactivity associated with the covalently bound phosphate. Panel a shows the Ca²⁺-concentration dependence of phosphorylation from ATP. The Glu³⁰⁹→Lys mutant is unable to phosphorylate, even at 12.5 mM Ca²⁺. In the wild-type Ca²⁺-ATPase the phosphorylation reaction is fully saturated at 10 μM Ca²⁺. Panel b shows lack of Ca²⁺ inhibition of backdoor phosphorylation from Pᵢ in the mutants. E indicates the presence of EGTA to chelate Ca²⁺ (normally a requirement for phosphorylation by the backdoor route). C indicates the

such folds and to the structure of small cation ionophores, it is, however, expected that the cation-binding sites in ion pumps consist of 6–8 oxygen-donating liganding groups provided by amino acid side chains (carboxylate, carboxamide, hydroxyl) and peptide-backbone carbonyls (Glynn and Karlish, 1990). The liganding groups replace the negative dipoles of the primary hydration shell of the cations, the affinity being directly related to the number of ligands provided by the protein (cf., Figure 2). The finding that the Na-limb of the Na^+-K^+-ATPase cycle is voltage sensitive, whereas the K-limb apparently is not, has led to the speculation that in Na^+-K^+-ATPase, the protein sites occupied by the ions during their movement relative to the electric field contain a total of two negative charges (carboxylates donated by aspartic acid and glutamic acid residues). These charges would be neutralized when the sites translocate two potassium ions, whereas one positive charge would be in excess, when the three sodium ions are translocated (Figure 7). This hypothesis is in accordance with the observation that one of the three Na^+ sites (the one not containing a carboxylate group) has a lower affinity than the other two. The number of charges in the binding sites can, however, not be generalized to other pumps, since in the highly homologous H^+-K^+-ATPase, the H^+-limb, as well as the K^+-limb are voltage sensitive.

The availability of the cDNA clones of P-type ATPases has opened up the possibility for introducing defined point mutations in the proteins by *in vitro* oligonucleotide directed mutagenesis of the cDNA. This approach can be used to search for the cation-binding sites, provided a suitable system is available for functional expression of the mutant cDNA. In addition to the requirement for a high expression level of the exogenous mutant cDNA, the absence of a significant contribution from endogenous pumps is demanded. The sarcoplasmic reticulum Ca^{2+}-ATPase can be expressed in the endoplasmic reticulum of COS-1 cells transfected with Ca^{2+}-ATPase cDNA, to a level which is more than 100-fold higher than the expression level of the endogenous COS-1 cell Ca^{2+}-pump. In the first model of the Ca^{2+}-ATPase peptide put forward on the basis of the amino acid sequence (MacLennan et al., 1985; Brandl et al., 1986), Ca^{2+}-binding sites were placed in the stalk helices, since these contain an excessive number of carboxylate and carboxamide residues (Figure 15), but this assignment had to be discarded after a total of more than 30 Asp, Glu, Asn, and Gln residues in the stalk had been

---

Figure 17. (continued) presence of 100 μM Ca^{2+}, which inhibits backdoor phosphorylation in the wild type Ca^{2+}-ATPase. Panel **c** shows the rate of dephosphorylation of the E_2P phosphoenzyme intermediate formed in the backdoor reaction with P_i. To observe the dephosphorylation, phosphorylation by radioactive P_i was terminated by dilution of the sample in medium containing non-radioactive P_i. The wild type and the $Asp^{800} \rightarrow Asn$ mutant dephosphorylates within 5 seconds, while dephosphorylation from E_2P is blocked in mutants $Glu^{309} \rightarrow Gln$ and $Glu^{771} \rightarrow Gln$. These data were compiled from Vilsen and Andersen, 1992a and Andersen and Vilsen, 1992b.

replaced with Ala by mutagenesis, singly or in clusters, without significant effects on Ca^{2+}-activation of the pump (Clarke et al., 1989b; MacLennan, 1990). Since this first round of mutational analysis, residues have been changed all over the protein (Figure 16). The most significant finding with respect to the location of the Ca^{2+}-transport sites was that alterations to either one of six oxygen-containing residues Glu^{309}, Glu^{771}, Asn^{796}, Thr^{799}, Asp^{800}, and Glu^{908} in the predicted trans-membrane segments M4, M5, M6, and M8 led to loss of the ability to transport Ca^{2+}, and to reduced Ca^{2+} sensitivity of the phosphorylation reaction (Clarke et al., 1989a; Andersen and Vilsen 1992b; Vilsen and Andersen, 1992a). Direct measurement of equilibrium Ca^{2+} binding at the transport sites is not possible with the minute amounts of protein harvested from the transfected cell culture, but an effect of the mutations on Ca^{2+} binding could be deduced indirectly from measurement of phosphorylation (since the phosphoryl group is attached covalently, the $[\gamma\text{-}^{32}P]$-phosphorylated protein can be isolated by gel electrophoresis and measured by autoradiography with less than 0.2 µg enzyme). These mutants did not form significant amounts of phosphoenzyme from ATP at a Ca^{2+} concentration of 100 µM, i.e., 100-fold that was required for half-saturation of the phosphorylation reaction in the non-mutated (wild-type) Ca^{2+}-ATPase. Moreover, micromolar concentrations of Ca^{2+} did not cause inhibition of phosphorylation from inorganic phosphate in the mutants, in contrast to the inhibitory effect of Ca^{2+} on phosphorylation with inorganic phosphate observed for the wild-type Ca^{2+}-ATPase (Ca^{2+} confers E_1 specificity). Figure 17a,b shows examples of this type of analysis of mutants in which the glutamic acid residue in M4 (Glu^{309}) was replaced. Note that the most conspicuous reduction in the apparent Ca^{2+} affinity measured in the phosphorylation reaction with ATP was observed after substitution with a lysine. This may be explained by repulsion of Ca^{2+} by the presence of the positive charge of the lysine in the binding site. These experiments did not exclude the possibility that the effects of the mutations on phosphorylation were caused by a structural change in the phosphorylation site, but the conclusion that it was the function of the Ca^{2+} sites which was defective, has recently been substantiated by the demonstration that the mutants with replacements of Glu^{309}, Glu^{771}, and Asp^{800} were unable to occlude Ca^{2+} in the presence of CrATP (Vilsen and Andersen, 1992c). In the wild type, this ATP analog stabilizes the occluded state, and the enzyme containing occluded radioactively labeled $^{45}Ca^{2+}$ stabilized with CrATP can be separated from non-specifically bound Ca^{2+} and measured by high performance liquid chromatography (Figure 18). The occluded intermediate is stable when CrATP is bound, even though the γ-phosphoryl group of CrATP is not transferred to the protein. Therefore, it was concluded that the inability of mutants to occlude Ca^{2+} observed with this method resulted from a defect in Ca^{2+} binding and not from a defect in phosphorylation.

Although a total of close to 250 amino acid residues of the SR Ca^{2+}-ATPase have now been analyzed by mutagenesis, in addition to those mentioned above,

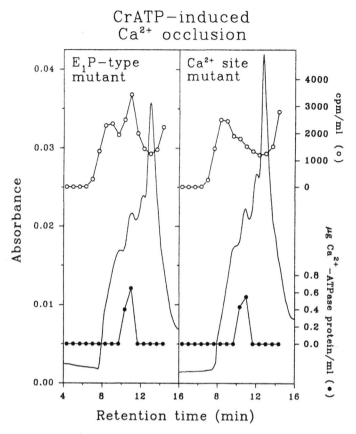

CrATP-induced
Ca²⁺ occlusion

Figure 18. Molecular sieve HPLC examination of CrATP-induced Ca^{2+} occlusion in mutants of the SR Ca^{2+}-ATPase. The Ca^{2+} occluded enzyme was formed by incubation with 10 μM $^{45}Ca^{2+}$ and CrATP. The enzyme was solubilized by non-ionic detergent and injected into an HPLC column. Absorbance of the eluate was read continuously at 226 nm (–), and fractions were collected for analysis of radioactivity (O) or immunoreactivity specific for the Ca^{2+}-ATPase (O). The right panel shows the absence of occluded $^{45}Ca^{2+}$ from the Ca^{2+}-ATPase protein peak corresponding to a mutant with severely reduced apparent Ca^{2+} affinity in the phosphorylation assay (c.f., Figure 17). The left panel shows as control a peak of $^{45}Ca^{2+}$ associated with a mutant that binds Ca^{2+} normally.

only a few other residues, all located in or close to the predicted transmembrane segments M4, M5, and M6, have been implicated in Ca^{2+} binding (Figure 16). Of these, Ser^{766}, Ser^{767}, and Asn^{768} contain side chain oxygens, but the effect on Ca^{2+} sensitivity of replacing these residues with Ala was much less dramatic (2–5-fold reduction of apparent Ca^{2+} affinity with preserved transport capability) than for the critical six residues discussed above. The Ca^{2+} affinity was also found to be reduced after mutation of two prolines (Pro^{308} and Pro^{803}) and three glycines (Gly^{310}, Gly^{770}, and Gly^{801}) located next to the critical carboxylic acid residues in M4, M5, and M6 (Vilsen et al., 1989; Andersen et al., 1992). The extent to which alterations to the glycines affected the apparent Ca^{2+} affinity depended on the size of the side chain used for replacement; the larger the side chain the more reduction in Ca^{2+} affinity. The importance of the proline and glycine residues for Ca^{2+} binding can be attributed to the unique properties of these amino acids. Prolines have the ability to break α-helical structures and may create kinks and flexible loops for ion binding to juxtaposed carboxylate groups and exposed backbone carbonyls. Since a single hydrogen atom constitutes the only side chain of glycine, this amino acid is helpful in exposing backbone carbonyls and in providing flexibility, so the backbone can wrap around the cation. Glycine can function as a D-amino acid, resembling those found in the gramicidin helix pore. Pure cyclic peptides based on proline and glycine have been found to function as selective ionophores for cations.

The results obtained by mutagenesis of single residues point to a mechanism for ion translocation, where all the major cation-binding structures are located in or close to (depending on the presumed hydrophobic segment boundaries) the membrane domain. This view has been confirmed independently in studies with the Na^+-K^+-ATPase, in which proteolytic enzymes have been used to shave off most of the cytoplasmic part of the protein. It was demonstrated that the shaved membranes, which contained the predicted transmembrane segments plus their cut cytoplasmic extensions (10–20 residues in length), were able to occlude the K^+-congener Rb^+ as well as Na^+ (the latter in the presence of oligomycin, which stabilizes the occluded state in the absence of the phosphorylation domain). These membranes also sustained slow Rb^+-Rb^+) exchange in sided membrane vesicles, suggesting retention of the whole transport pathway (Karlish et al., 1990, 1992).

Most of the oxygen-donating residues and the two prolines implicated in Ca^{2+} binding in Ca^{2+}-ATPase by the mutagenesis studies are well conserved in the Na^+-K^+-ATPase subfamily as well. This raises the question of which residues actually determine the difference between the cation specificities of these pumps. Steric factors imposed by the packing of the non-conserved hydrophobic residues in the membrane may of course be important. It is also possible that some of the conserved residues that appear important for cation binding are not directly involved in formation of the ion-binding site, but contribute indirectly by stabilizing the appropriate conformation (E_1 for the binding and occlusion of a-type ions, E_2 for the c-type ions) (Andersen and Vilsen, 1992b).

Studies with Na^+-K^+-ATPase mutants expressed in COS-1 cells have pointed to a direct role of the proline Pro^{328} in formation of the binding site for Na^+ (Vilsen, 1992). To distinguish exogenous expressed rat kidney Na^+-K^+-ATPase from the Na^+-K^+-ATPase endogenously present in COS-1 cells, advantage was taken of the 1,000-fold species difference in sensitivity to inhibition by the cardiac glycoside ouabain. Since ouabain selectively inhibits the endogenous Na^+-K^+-ATPase, experiments with COS-cell membranes expressing both exogenous and endogenous enzymes were carried out in the presence of ouabain, and this permitted titration of the apparent affinities for Na^+, K^+, and ATP in mutants. It was found that the $Pro^{328} \rightarrow Ala$ mutant of the rat kidney enzyme displayed lower affinity for both Na^+ and K^+ relative to wild-type rat kidney enzyme. Since the apparent affinity of the proline mutant for ATP was higher than that of the wild type, the low K^+ affinity may have been caused by a displacement of the E_1-E_2 conformational equilibrium in favor of E_1 (the form with highest affinity for ATP), rather than by a direct distortion of a K^+-site. However, by the same argument, the high ATP affinity of the $Pro^{328} \rightarrow Ala$ Na^+-K^+-ATPase mutant also excluded an indirect effect (through stabilization of the E_2 form) as cause of the low affinity for Na^+ (Vilsen, 1992). This proline is located in M4, and begins the motif PEGL (Figure 15), which is highly conserved among P-type ATPases, and which contains, in addition to the proline, the residues Glu^{309} and Gly^{310} of importance for Ca^{2+} binding in Ca^{2+}-ATPase (Figure 16).

An important problem to solve is whether the ions transported in opposite directions in countertransport systems actually are bound by the same set of liganding groups working consecutively, such as hypothesized for Na^+ and K^+ transport by Na^+-K^+-ATPase. It is well recognized that Na^+ bound at some extracellular site in Na^+-K^+-ATPase influences the affinity of the intracellular Na^+ uptake sites, and it has sometimes been discussed whether the extracellular allosteric site might be identical to one of the extracellular K^+ uptake sites implying coexistence of two sets of sites, each of which might be used simultaneously for transport (Cornelius, 1991). Evidence against simultaneous existence of different sites for Na^+- and K^+-transport is provided by the finding that chemical modification of carboxyl groups in the transmembrane region of Na^+-K^+-ATPase with dicyclohexylcarbodiimide (DCCD) inactivates both Na^+- and K^+-occlusion (Karlish et al., 1992). However, it remains to be determined whether this is due to a functional disturbance of a common site for alternate Na^+- and K^+-occlusion, or the bulky carbodiimide moiety sterically hinders access of Na^+, as well as K^+, at their respective non-identical sites. One of the DCCD-binding carboxylic acid residues in the membrane has been identified as Glu^{953} of M9, which has no homologous counterpart in Ca^{2+}-ATPase. Replacement of this residue by mutagenesis did not affect enzyme function significantly, indicating that it is not the DCCD-modification of oxygen-ligands contributed by this residue that disturbs ion occlusion (Van Huysse et al., 1993).

The mutagenesis studies on the SR Ca^{2+}-ATPase showed that in addition to their apparent role in Ca^{2+} binding at cytoplasmic uptake sites, the residues Glu^{309} and Glu^{771} located in M4 and M5, respectively, were crucial also for the dephosphorylation of E_2P (see Figure 17c). Another carboxylate residue implicated in Ca^{2+} binding, Asp^{800} located in M6, could, on the other hand, be changed without effect on dephosphorylation kinetics (Figures 16 and 17c). Hence, it appears that M4 and M5 are involved in reactions associated with the E_1 state as well as in reactions associated with the E_2 state, while M6 plays a more static role. The data could mean that the residues Glu^{309} and Glu^{771} take part in countertransport of H^+, and therefore in the associated dephosphorylation. The carboxylate groups of Glu^{309} and Glu^{771} might thus accomplish functions as cation ligands in cytoplasmically, as well as extracytoplasmically, orientated sites. Asp^{800} would on the other hand contribute oxygen ligands only at the cytoplasmic uptake sites (Andersen and Vilsen, 1992b).

If one takes the view that the ligands forming the cation-binding structures in the sodium form (E_1) and the potassium form (E_2) of Na^+-K^+-ATPase are at least in part identical, then the question which arises is what determines the different cation specificities of the two conformations? It should be noted in this connection that normally the change in specificity for the ions is accompanied by a change in capacity. The ability of the Na^+-K^+-ATPase enzyme to phosphorylate from ATP is normally conferred only after the binding of three sodium ions, while the E_2 forms bind maximally two ions. The 3/2 stoichiometry is maintained even when extracellular K^+ is removed and replaced by Na^+, resulting in an ATP-driven electrogenic exchange of three cytoplasmic Na^+ for two extracellular Na^+. Under extreme conditions (very low cytoplasmic Na^+ concentration and low pH), however, $2H^+$ or $1Na^+ + 1H^+$ may be extruded from the cytoplasm in exchange for $2K^+$ in each cycle, and there is also evidence that $3H^+$ can substitute for $2K^+$ at the extracellular surface at low pH conditions (Blostein and Polvani, 1992). Therefore, protonation may modify the capacity of the ion-binding sites without corresponding changes in catalytic specificity. At least for the gastric H^+-K^+-ATPase a pH-dependent variation in pump stoichiometry from $2H^+/2K^+$ to $1H^+/1K^+$ may play a physiological role by enabling the gastric pump to work against the steepest possible pH gradient (Rabon and Reuben, 1990). In analogy with the ability of H^+ to serve as substitute for Na^+ in Na^+-K^+-ATPase, Na^+ can substitute for H^+ in H^+-K^+-ATPase (Polvani et al., 1989). In view of these findings, it seems likely that H^+ is transported as a hydronium ion, H_3O^+, which is about the same size as Na^+ in the dehydrated state (Boyer, 1988). A parallel to this exists in the discovery that an F-type ATPase/synthase with specificity for transport of Na^+ as well as H^+ is present in the strictly anaerobic bacteria *Propionigenium modestum*. This has led to the rejection of hypotheses involving proton-specific transport mechanisms such as H^+ transfer in a chain of protonable groups (a proton wire). Of the factors involved in the fine tuning of the differing specificities of the cation sites in the E_1 and E_2 states of Na^+-K^+-ATPase and H^+-K^+-ATPase, the size of the ion may be a major one. The a-type ions (Table 2) form one category of smaller ions (in the dehydrated state), which include the hydronium ion. The K^+-congeners form a category of larger size, which are the most efficient c-type ions, and which do not qualify as a-type ions. Discrimination against the larger ions in the E_1 form may be a result of steric factors on the protein, while discrimination against the smaller ions in the E_2 state may be a result of their higher dehydration energy. Occlusion of K^+ by Na^+-K^+-ATPase has been compared to the binding of K^+ by valinomycin (Glynn and Karlish, 1990). The latter process begins with the formation of a short-lived precomplex with little K^+/Na^+ selectivity and with only some of the waters of hydration replaced. This is followed by a change of conformation that allows replacement of all waters of hydration with peptide carbonyl oxygen ligands in a very stable and highly specific bracelet structure around K^+.

STRUCTURAL CHARACTERIZATION OF CONFORMATIONAL CHANGES ASSOCIATED WITH ION TRANSLOCATION

Understanding how the catalytic site in the cytoplasmic domain communicates with the cation binding sites in the membrane to accomplish energy interconversion and movement of the ions requires that the structural changes associated with the transitions between the various reaction intermediates be described in molecular terms. Conformational changes related to ion transport have been demonstrated by various biochemical, spectroscopic, crystallographic, and protein engineering techniques. Most of the approaches fall into the following categories: (1) studies of the accessibility of defined structural elements in the peptide to modifying agents, such as proteolytic enzymes, antibodies or chemical reagents; (2) studies of spectroscopic signals from reporter groups, either intrinsically present in the protein (such as tryptophan residues) or attached by noncovalent or covalent binding; (3) studies of the functional consequences of modification of the peptide by site-directed mutagenesis and by other means, and (4) studies of the electron density profiles in bidimensional membrane crystals. Some illustrative examples will be discussed below (see also previous reviews by Glynn, 1985; Jørgensen and Andersen, 1988; and Andersen and Vilsen, 1992a). A message to be derived from these studies is that almost every corner of the peptide moves during the transport cycle. The various steps in the cycles shown in Figure 11 are not all revealed by the same technique, but in combination the techniques applied provide evidence for structural differences between most of the intermediates.

Change in Exposure to Modifying Agents

Jørgensen (1975) demonstrated that the sodium form and the potassium form of the Na^+-K^+-ATPase display different patterns of inactivation when exposed to limited trypsin digestion (Figure 19). When digestion was carried out in the presence of Na^+ (i.e., in the Na_3E_1 form, see Figure 11A), the loss of Na^+-K^+-ATPase activity followed a biphasic time course, because trypsin cleaves rapidly at Lys^{30} (T_{2Na} site, see Figure 13) and more slowly at Arg^{262} (T_{3Na} site, See Figure 13). When digested in the presence of K^+ (i.e., in the $[K_2]E_2$ form, see Figure 11A), a monophasic first-order time course was followed. Under these conditions trypsin

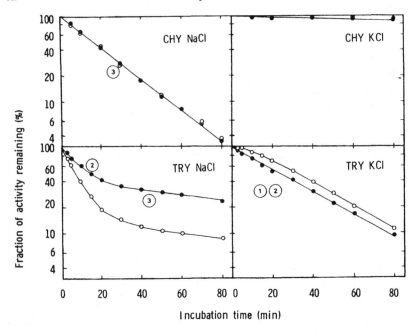

Figure 19. Time course of inactivation of Na$^+$-K$^+$-stimulated ATP hydrolysis (●) or K$^+$-phosphatase (○) activity of the Na$^+$-K$^+$-ATPase enzyme treated with trypsin or chymotrypsin under carefully controlled conditions in NaCl or KCl media. In NaCl, chymotrypsin (CHY) cleaves at Leu^{266} (3), while trypsin (TRY) cleaves at Lys^{30} (2) and Arg^{262} (3). In KCl, trypsin cleaves at Arg^{438} (1) and Lys^{30} (2) in sequence, while there is no cleavage site exposed to chymotrypsin. Data from Jørgensen and Andersen, 1988.

cleaves first at Arg^{438} (T$_{1Na}$ site, see Figure 13), and subsequently at Lys^{30}. Differing inactivation patterns were also demonstrated upon treatment with chymotrypsin (Figure 19). Chymotrypsin cleaves the Na$_3$E$_1$ form at Leu^{266} (C$_{3Na}$ site), but does not cleave the [K$_2$]E$_2$ form under the limited digestion conditions applied. Thus, transition from Na$_3$E$_1$ to [K$_2$]E$_2$ consists of an integrated structural change involving protection of the bonds at Arg^{262} and Leu^{266}, both of which are located in the C-terminal part of the β-domain connecting M2 with M3, and exposure of Arg^{438} in the central domain. It may be suggested that this reflects motion within the segment that connects Arg^{262} with Arg^{438}. This segment contains the phosphorylated aspartyl residue and the transmembrane segments M3 and M4. Likewise in the SR Ca^{2+}-ATPase, cleavage sites for trypsin (T$_2$ at Arg^{198} and T$_3$ at Lys^{234} or Arg^{236}), and for V8-protease (V8 at Glu^{231}) are exposed in the Ca$_2$E$_1$ state, and protected to various degrees in the absence of Ca^{2+}. Complete protection of T$_2$ requires, in addition to the absence of Ca^{2+}, quantitative phosphorylation with P$_i$ to form E_2P, or the binding of the phosphoryl transition state analog vanadate, which associates tightly with the phosphorylation site without forming a covalent bond (Andersen et al., 1986). An accessible cleavage-site exists also in the central cytoplasmic domain of Ca^{2+}-ATPase (T_1 at Arg^{505}), but the rate of cleavage at this site does not depend on ligand binding to the enzyme. The tryptic cleavage sites in Ca^{2+}-ATPase, which are sensitive to ligand binding, are located in the same region as the T_{3Na} and C_{3Na} sites in Na^+-K^+-ATPase, Lys^{234} in Ca^{2+}-ATPase being homologous to Arg^{262} in Na^+-K^+-ATPase. Site-directed mutagenesis experiments have excluded that this region is directly involved in Ca^{2+} binding (Andersen et al., 1989), but it may possibly participate in formation of the catalytic site in E_2P (see below). When digestion is carried out during enzyme turnover with ATP, the rate of cleavage at the T_2-site is high under conditions where the predominant phosphoenzyme intermediate accumulated in steady state is ADP-sensitive (E_1P), and low if the predominant phosphoenzyme intermediate is ADP-insensitive (E_2P) (Figure 20). The change in exposure of the T_2-site associated with the E_1P-E_2P transition must reflect a rearrangement in the protein conformation, since even if steric hindrance to trypsin access imposed by the presence of the phosphoryl group were involved in the protection against tryptic cleavage at the T_2-site in E_2P, a change in the spatial relationship would have to occur for this hindrance to be relieved in E_1P (Andersen et al., 1985, 1986).

When the phosphorylated Na^+-K^+-ATPase conformation containing two bound Na^+ ($E*P$, lower part of Figure 11A) is stabilized with ouabain, the proteolytic cleavage pattern corresponds to that of E_2. This suggests a classification of $E*P$ as a subconformation belonging to the E_2-pool (Jørgensen, 1991), and indicates that the E_1P-E_2P conformational rearrangement reflected in the proteolytic cleavage pattern occurs in association with or immediately after the extracytoplasmic dissociation of the first Na^+, before exchange of the remaining two Na^+ for K^+. ATP turns the unphosphorylated Na^+-K^+-ATPase conformation into one with an E_1-type cleavage pattern, even in the presence of K^+. Therefore, in accordance with the K^+-ATP antagonism described above, the deoccluded form with K^+ bound at cytoplasmic sites (upper left corner of Figure 11A) may be classified as belonging to an E_1-pool of conformations, although E_1 catalytic specificity (ability to phosphorylate from ATP) is not conferred until K^+ has been exchanged for Na^+.

Changes in the reactivity of amino acid side chains can also, in many cases, provide information about conformational changes. The ATP analog adenosine triphosphopyridoxal reacts selectively with lysine residues in the central cytoplasmic loop. Both Lys^{492} and Lys^{684} are labeled in the SR Ca^{2+}-ATPase in the absence of Ca^{2+} at the cytoplasmic uptake sites, while only Lys^{684} is labeled in the presence of Ca^{2+} (Yamamoto et al., 1989). McIntosh (1992) demonstrated the formation by glutaraldehyde of a cross link between Lys^{492} and Arg^{678} in the SR Ca^{2+}-ATPase. It was shown that this reaction is sensitive to the functional state of the phosphorylation site. While Ca^{2+} had no effect on cross linking, phosphorylation to the

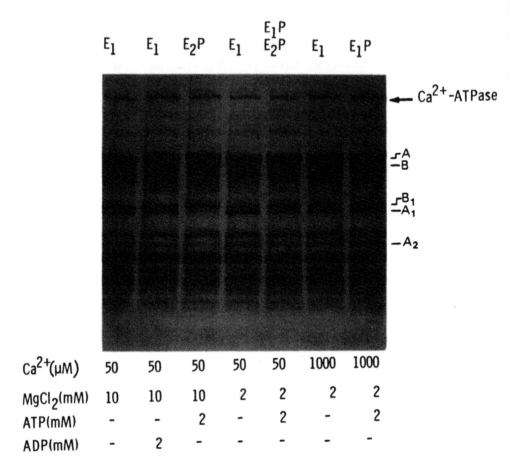

Figure 20. SDS-polyacrylamide gel electrophoresis showing tryptic cleavage patterns of the SR Ca^{2+}-ATPase in various functional states. The predominant enzyme forms present during proteolysis are indicated above the lanes. The concentrations of ligands present to accumulate these forms are indicated below the lanes. Titration of the steady-state distribution of E_1P and E_2P was achieved by varying the Ca^{2+}/Mg^{2+} concentration ratio present during turnover with ATP. The tryptic fragments are indicated by the following nomenclature. A: residues 1–505; B: residues 506–994; A_1: residues 199–505; A_2: residues 1–198; B_1: residues 506–825. It can be seen that E_1 and E_1P forms of the enzyme produce A_1 and A_2 fragments in a much higher yield than does E_2P, indicating protection of the bond at Arg^{198} in E_2P (Andersen et al., 1985; Andersen, 1989).

ADP-sensitive E_1P intermediate enhanced the rate of cross linking 3–4-fold. Further development of the phosphoenzyme intermediate to ADP-insensitive E_2P completely prevented the cross linking. This was interpreted in terms of sequential hinge-bending movements, leading to closure of the active site in E_2P with possible extrusion of water and restriction of the access of glutaraldehyde. It is also possible that the glutaraldehyde did have access, but Lys^{492} and Arg^{678} were no longer close enough to cross link in the rearranged catalytic site in E_2P.

Nonspecific hydrophobic photoactivatable reagents such as trifluoromethyl iodophenyl diazirine (TID) and phospholipid derivatives have been used to label Ca^{2+}-ATPase and Na^+-K^+-ATPase from the membrane phase. The degree of hydrophobic labeling was found to be higher in E_2 (stabilized with vanadate in Ca^{2+}-ATPase and K^+ in Na^+-K^+-ATPase) than in E_1, suggesting an increased exposure of part of the protein to the lipid phase in E_2 (Andersen et al., 1986; Modyanov et al., 1991). This could be explained either by movement of cytoplasmic segments (perhaps the stalk) into the bilayer, or by a local rearrangement of the intramembraneous segments. The results with the Ca^{2+}-ATPase showed that the major part of the preferential labeling of E_2 was located in the tryptic fragment (A_1) spanning between Arg^{198} and Arg^{505}, i.e., the part of the peptide encompassing the phosphorylation domain and transmembrane segments M3 and M4.

The cardiac glycoside ouabain (Hansen, 1984) and the hydrophobic amine SCH28080 (Rabon and Reuben, 1990) are noncovalently binding specific inhibitors of the Na^+-K^+-ATPase and the H^+-K^+-ATPase, respectively. Either of these compounds act from the extracellular side and prevent K^+-stimulated dephosphorylation by binding preferentially to E_2P forms of the enzymes (the subconformation denoted E*P in Figure 11A has been emphasized as being particularly reactive in the Na^+-K^+-ATPase; see above). Residues of importance for the binding seem to be present in the extracellular M1–M2 loop (Lingrel et al., 1990; Sachs et al., 1992). With a fluorescent compound related to SCH28080 it has been demonstrated that the environment of the binding site becomes more hydrophobic with the formation of E_2P. Based on these studies it appears that the M1–M2 loop undergoes a structural rearrangement in relation to formation of E_2P (E*P). Activation of the H^+-K^+-ATPase by extracellular K^+ is also prevented by the binding to the enzyme of a monoclonal antibody whose epitope is located in the central cytoplasmic loop (Bayle et al., 1992), and omeprazole binding to the extracytoplasmic loop M5–M6 in H^+-K^+-ATPase blocks phosphorylation at the catalytic site in the cytoplasmic domain. Moreover, accumulation of the E_2P form of H^+-K^+-ATPase in the presence of ATP and SCH28080 leads to protection of proteolytic cleavage sites normally accessible from the cytoplasmic side (Besancon et al., 1992). These are but a few out of many examples of inhibitor effects illustrating the transmembrane cross-talk between cytoplasmic and extracytoplasmic domains through conformational changes.

The SR Ca^{2+}-ATPases of cardiac and slow skeletal muscle are regulated *in vivo* by cAMP-mediated phosphorylation of another intrinsic membrane protein, phospholamban. It is believed that the phosphorylated form of phospholamban does not bind to the Ca^{2+}-ATPase, whereas the unphosphorylated form of phospholamban binds preferentially to the conformation of Ca^{2+}-ATPase present in the absence of Ca^{2+} (E or E_2, see Figure 11B), thereby displacing the conformational equilibrium so that the apparent Ca^{2+}-affinity of the Ca^{2+}-ATPase displayed during turnover is reduced. A region located in the variable part of the central cytoplasmic domain of the Ca^{2+}-ATPase peptide, between the phosphorylation site and the FITC-binding residue (Figure 16), has been shown to be crucial to the interaction of the Ca^{2+}-pump with phospholamban. It has been demonstrated that in the absence of Ca^{2+}, but not in the presence of Ca^{2+}, phospholamban conjugated to Denny-Jaffe's reagent cross-links to Lys^{400} in the C-terminal part of this region (Figure 16) (James et al., 1989; Toyofuku et al., 1993). It may therefore be presumed that this region of the peptide undergoes structural changes preventing contact with phospholamban when the Ca^{2+}-ATPase binds Ca^{2+}.

Spectroscopic Signals

The binding of Ca^{2+} at the high-affinity cytoplasmic uptake sites in the SR Ca^{2+}-ATPase induces a 3–5% increase in the fluorescence from intrinsic tryptophan residues (Dupont, 1976). The calcium dependence and rapid kinetics of this fluorescence enhancement follows closely that of enzyme activation. In the 10-helix structural model, 12 out of a total of 13 tryptophan residues in Ca^{2+}-ATPase are located in or near the transmembrane segments, while one tryptophan (Trp^{552}, see Figure 16) is in the cytoplasmic domain. The use of hydrophobic fluorescence quenchers has shown that the tryptophan(s) responding to Ca^{2+} is at the protein-lipid interface, and the fluorescence change may therefore reflect local effects of Ca^{2+} binding on the arrangement of the transmembrane segments. The fluorescence also changes in relation to formation of E_2P by backdoor phosphorylation with P_i. This involves opposite changes in the fluorescence quantum yield of tryptophans located in the membrane and in the cytoplasmic region, in line with the view that the phosphorylation not only affects the conformation at the catalytic center, but also reorganizes the membrane portion of the Ca^{2+}-ATPase by long-range action, possibly allowing the Ca^{2+} sites to become accessible from the extracytoplasmic side in E_2P (De Foresta et al., 1990).

In Na^+-K^+-ATPase, the tryptophan fluorescence changes in relation to E_1-E_2 and E_1P-E_2P conformational transitions. The fluorescence quantum yield is highest in E_2 and E_2P forms, opposite Ca^{2+}-ATPase, where the highest fluorescence is associated with the Ca_2E_1 state. This variation can be ascribed to the different locations of the tryptophans in the amino acid sequences of the two proteins.

Certain fluorescent compounds, such as formycin nucleotides and eosin, behave as noncovalently binding ATP analogs, and their fluorescence increases upon association with the ATPases. The transition from the Na^+-form to the K^+-form of Na^+-K^+-ATPase gives rise to a decrease in fluorescence from these probes. The fluorescence of fluorescein isothiocyanate (FITC) bound covalently at the nucleotide site of Na^+-K^+-ATPase (Table 3) also decreases in relation to transition from the Na^+-form to the K^+-form. With the noncovalent as well as with the covalently-binding probes the fluorescence decrease probably reflects the structural change in the nucleotide site associated with the reduced nucleotide affinity in the E_2 form (see above). By contrast, when FITC is attached to the SR Ca^{2+}-ATPase (at the residue homologous to the FITC-binding residue in Na^+-K^+-ATPase) the fluorescence increases upon removal of Ca^{2+}. Most likely this relates to the above discussed difference between Na^+-K^+-ATPase and Ca^{2+}-ATPase with respect to the existence of a stable E_2 form with low affinity for nucleotide.

Trinitrophenyl-nucleotides constitute a unique class of fluorescent ATP-analogs, since they bind at the catalytic site of the phosphoenzyme of the Ca^{2+}-ATPase after departure of ADP and give off a tremendous fluorescence signal upon conversion of the phosphoenzyme intermediate from ADP-sensitive to ADP-insensitive (c.f., Figure 6), possibly reflecting a hydrophilic-hydrophobic transition related to closure of the catalytic site (Andersen et al., 1985; Seebregts and McIntosh, 1989).

Fluorescent sulfhydryl reagents like iodoacetamido fluorescein (IAF) (reacting with Cys^{457} in Na^+-K^+-ATPase), benzimidazolylphenyl maleimide (BIPM) (reacting with Cys^{964} in Na^+-K^+-ATPase), chloro-nitrobenzyl oxadiazole (NBD-Cl) (reacting with Cys^{344} in Ca^{2+}-ATPase), and iodoacetyl sulfonaphtyl ethylenediamine (I-EDANS) (reacting with Cys^{674} in Ca^{2+}-ATPase) allow ATP binding and phosphorylation after their specific modification of the enzyme, and these probes all seem to be sensitive to E_1P-E_2P transitions of the phosphoenzyme formed with ATP (Taniguchi et al., 1984; Steinberg and Karlish, 1989; Suzuki et al., 1989; Wakabayashi et al., 1990). In Ca^{2+}-ATPase labeled with NBD-Cl and Na^+-K^+-ATPase labeled with IAF, the fluorescence changes accompanying E_1-E_2 and E_1P-E_2P transitions are large and of a similar magnitude, while almost no changes in fluorescence occur upon the phosphorylating E_1-E_1P and E_2-E_2P transitions, in accordance with the concept of two families of conformations corresponding to E_1/E_1P and E_2/E_2P pools (Figure 21). However, in studies with other probes the fluorescent properties of phospho- and dephosphoforms have been shown to differ too much to justify a pooling of subconformations according to the two-state E_1-E_2 model. There is also evidence for the existence of subconformations of E_1 and E_1P. The I-EDANS and other probes reacting with Cys^{674} in the SR Ca^{2+}-ATPase appear to be more sensitive to ATP-induced transitions between subconformations of E_1 than to E_1-E_2 or E_1P-E_2P transitions (Suzuki et al., 1989). Multiple conformational states of E_1P have been detected with BIPM in Na^+-K^+-ATPase and have been

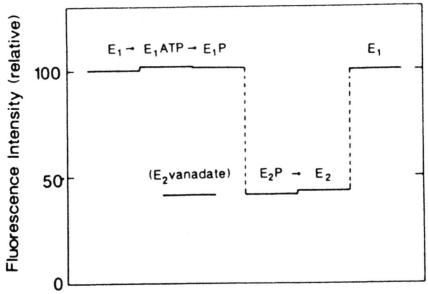

Figure 21. Diagram showing the relative fluorescence levels of various functional states of the SR Ca^{2+}-ATPase labeled with 4-nitrobenzo-2-oxa-1,3-diazole (NBD) at Cys^{344}. Modified from Wakabayashi et al., 1990. The pH was 6.0 to stabilize the unphosphorylated E_2 form in the absence of Ca^{2+}.

ascribed to consecutive steps in the Na^+ translocation (Sasaki et al., 1991). The site of attachment for BIPM (Cys^{964}) is in the predicted transmembrane segment M9 in the 10-helix model, and it is therefore reasonable to assume that BIPM registers conformational changes associated with transit of Na^+ through the membrane domain. Kinetic studies with doubly labeled enzyme preparations show that the changes registered by probes present in different microenvironments of the enzyme (e.g., FITC and BIPM) do not occur simultaneously. This suggests that conformational changes are transmitted between domains in a sequential rather than a concerted manner (Taniguchi et al., 1988).

Functional Consequences of Modification of the Peptide

The use of site-directed mutagenesis and expression of mutant cDNA provides a direct way to examine the functional roles of individual amino acid residues, not only in cation binding, but also in the conformational changes associated with ion translocation. A series of studies of point mutants of the SR Ca^{2+}-ATPase, applying a panel of assays comprising Ca^{2+}-transport, ATP-hydrolysis, phosphoenzyme formation from ATP and inorganic phosphate, ADP-sensitivity, and dephosphorylation kinetics, has led to a functional classification of the amino acid residues as indicated by the various symbols in Figure 16. Profound functional changes can be induced by substitution of single amino acid residues, and it is impressive how specific the effects are. The fact that a great number of residues can be replaced without any significant functional changes (closed circles in Figure 16) highlights the interesting effects observed upon replacement of certain residues. The open circles in Figure 16 have been discussed above; they signify residues whose substitutions lead to reduced Ca^{2+} sensitivity. The diamonds located in M4 and M5 indicate residues whose substitutions lead to inhibition of the dephosphorylation of E_2P. Some of these residues are bifunctional, since they are important also for the response to Ca^{2+} (Ala^{305}, Glu^{309}, Gly^{310}, and Glu^{771}, doubly labeled with diamonds and circles in Figure 16). As mentioned above, their dual functional roles can be explained by alternate participation in transport of Ca^{2+} and protons. Either these residues take part directly in the binding of the hydronium ions at the extracytoplasmic side, or they are involved through participation in the structural rearrangements leading to formation of a dephosphorylated (occluded) E_2 form.

The squares in Figure 16 indicate amino acid substitutions that lead to a block of the E_1P-E_2P conformational change. The first residue to be implicated in this transition on the basis of mutagenesis studies was the proline in the upper part of M4 (Pro^{312}). The role played by this proline is different from the role in cation binding at the uptake sites played by the prolines further down in the transmembrane sectors described above. It was discovered that replacement of Pro^{312} with Ala resulted in a complete block of Ca^{2+} uptake, although the ability to phosphorylate from ATP in a normal Ca^{2+}-dependent manner was preserved (Vilsen et al., 1989). Figure 22 shows how the properties of the phosphoenzyme intermediates of Pro^{312} mutants were analyzed. The rapid disappearance of the incorporated radioactivity from [γ-^{32}P]ATP upon addition of ADP demonstrated that the major phosphoenzyme intermediate accumulated in steady state was ADP-sensitive (E_1P). This was also the case for the wild type under the prevailing ionic conditions. However, when EGTA was added to chelate free Ca^{2+} and thereby terminated the new formation of phosphoenzyme, there was only a very slow decay of the E_1P phosphoenzyme intermediate in the $Pro^{312}{\to}$Ala mutant, whereas the wild-type phosphoenzyme intermediate decayed rapidly through conversion of E_1P to E_2P and hydrolysis of the latter intermediate. The mutant phosphoenzyme was not only stable, it also retained the ability to transfer the phosphoryl group back to ADP for several minutes after the addition of EGTA (not shown in Figure 22, but see Vilsen et al., 1989). This indicates that in the mutant phosphoenzyme Ca^{2+} remained bound in an occluded state inaccessible to chelation. On the other hand, the $Pro^{312}{\to}$Ala mutant was not defective in the dephosphorylation from the E_2P intermediate formed in the backdoor reaction with P_i (Figure 22, right panel). Thus, it could be concluded that the lack of dephosphorylation from E_1P was due to a block of the E_1P-E_2P interconversion. Replacement of Pro^{312} in Ca^{2+}-ATPase by a leucine, which is the residue present at the homologous position in Na^+-K^+-ATPase, also produced an enzyme with defective E_1P-E_2P transition. In this case, however, the

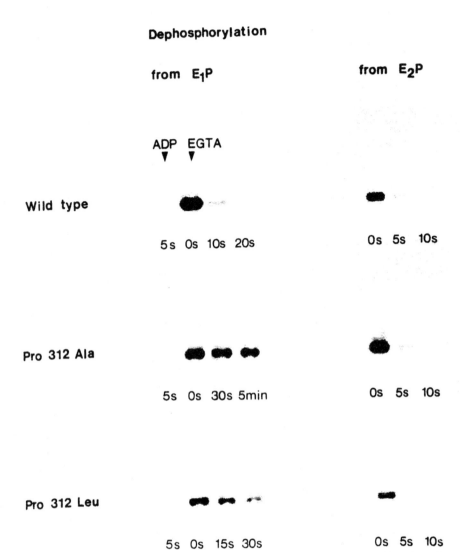

Figure 22. Analysis of the rates of dephosphorylation of phosphoenzyme intermediates of wild-type and Pro[312] mutants of the SR Ca^{2+}-ATPase. Left panel: Phosphorylation was performed with [γ-^{32}P]ATP in the presence of Ca^{2+}. The buffer conditions were adjusted to obtain predominantly E$_1$P in steady state as indicated by the complete disappearance of the phosphoenzyme upon addition of ADP. EGTA was added to terminate phosphorylation by chelation of Ca^{2+} and permit observation of the dephosphorylation of E$_1$P through conversion to E$_2$P and hydrolysis of the latter intermediate. A rapid dephosphorylation is observed with the wild type, while in the mutants the dephosphorylation of E$_1$P is inhibited. More than 80% of the phosphoenzyme remains five minutes after addition of EGTA in the Pro[312]→Ala mutant. Right panel: Phosphorylation was performed with [^{32}P]P$_i$ by the backdoor reaction in the absence of Ca^{2+}

inhibition was less severe than with alanine (Figure 22). There was also some indication that the replacement of Pro^{312} with either Ala or Leu led to higher apparent Ca^{2+} affinity of the Ca^{2+}-ATPase, i.e., to a shift in the equilibrium of the dephosphoenzyme in favor of E_1 (Vilsen et al., 1989). Parallel mutagenesis studies of Na^+-K^+-ATPase have demonstrated that substitution of the leucine (Leu^{332}) homologous to Pro^{312} in Ca^{2+}-ATPase with Ala leads to a displacement of the conformational equilibrium of Na^+-K^+-ATPase in favor of E_1 (Vilsen, 1992). The crucial role of Pro^{312} in the conformational change of Ca^{2+}-ATPase may be ascribed to its ability to kink the M4 helix at the top and thereby introduce a defect in helix packing in the transmembrane sector. It is possible that to some extent the bulky side chain of the leucine present in Na^+-K^+-ATPase can mimic the effect of the kink dictated by proline. As shown in Figure 16, additional residues at the top of the M4 helix, in the fourth stalk segment linking this helix with the phosphorylation site, have been implicated in the E_1P-E_2P transition of Ca^{2+}-ATPase by mutagenesis experiments (Vilsen et al., 1991b). The demonstration of the importance of M4 for the E_1P-E_2P transition, in conjunction with the data indicating that M4 and M5 are involved in cation binding and dephosphorylation of E_2P, suggest that M4 and M5 play pivotal roles in the long-range communication between the catalytic site and the ion binding domain. This can be seen as a natural consequence of the direct physical link between these two transmembrane segments and the central cytoplasmic domain. E_1P-E_2P conformational changes originating in the catalytic site may be transmitted through the stalk to M4, leading to a rotation or tilting of this helix, which disrupts the Ca^{2+}-binding domain, so that the previously occluded calcium ions become available to the extracytoplasmic surface. Conversely, conformational changes elicited at the ion binding sites upon ion binding or dissociation may be transmitted from M4 and M5 back through the stalk to the catalytic center to alter its reactivity.

A glycine (Gly^{233}) in the C-terminal part of the β-domain close to M3 was also among the first residues to be shown by mutagenesis to be crucial to the E_1P-E_2P transition in the SR Ca^{2+}-ATPase (Andersen et al., 1989). Replacement of Gly^{233} led not only to an enzyme that was unable to form E_2P in the forward direction from E_1P, but also the backdoor phosphorylation with inorganic phosphate was defective, contrary to the situation with the $Pro^{312} \rightarrow$ Ala mutant, in which P_i-phosphorylation was unaffected. On the basis of this finding, it was proposed that while the proline residue participates in the transmission of signals between the phosphory-

---

Figure 22. (continued) to obtain E_2P. Dephosphorylation is observed upon dilution of the sample in medium containing non-radioactive P_i. The high rate of dephosphorylation of E_2P in the mutants indicates that the reason for the inhibition of E_1P dephosphorylation in the Pro^{312} mutants must have been a block of the E_1P-E_2P interconversion rather than a block of the ensuing E_2P hydrolysis (Vilsen et al., 1989).

lation site and the cation sites, the glycine residue is involved in the E_1P-E_2P transition of the phosphorylation site itself (Andersen and Vilsen, 1990). The triplet ThrGlyGlu[183] in the β-domain is also crucial to the E_1P-E_2P transition in Ca^{2+}-ATPase (Figure 16). These residues as well as Gly^{233} are highly conserved within the family of P-type ATPases, and replacement of some of their homologous counterparts in the yeast plasma membrane H^+-ATPase has been shown to lead to insensitivity of this enzyme to inhibition by vanadate (Ghislain et al., 1987; Portillo and Serrano, 1989). As turnover was completely blocked in the Ca^{2+}-ATPase mutants, their vanadate sensitivity could not be tested, but recently vanadate insensitivity has also been demonstrated in a Ca^{2+}-ATPase mutant, in which seven non-conserved residues adjacent to Gly^{233} on the C-terminal side were replaced by the corresponding 7-residue segment present at the homologous position in Na^+-K^+-ATPase without complete loss of activity (Figure 16, Na^+-K^+/Ca^{2+}-ATPase chimera). Considering that vanadate is an analog of the pentacoordinated transition state of the phosphoryl group, the findings with the yeast H^+-ATPase mutants and the more recent finding with the chimeric Ca^{2+}-ATPase mutant might be taken as evidence that the β-domain functions as a phosphatase domain crucial to the hydrolysis of the aspartyl phosphate bond (Serrano, 1988). Although this hypothesis differs from the suggestion of a role for the β-domain in the E_1P-E_2P transition, the dual roles are not mutually exclusive. If the β-domain is assumed to pack against the largest cytoplasmic loop, thereby forming the third wall of the catalytic cleft, it might be able to participate in phosphoenzyme hydrolysis. The E_1P-E_2P transition might possibly serve to move the β-domain into this position, and therefore the phosphatase reaction would also depend on the ability of the enzyme to undergo a proper E_1P-E_2P transition. It is not safe to guess why the substitution of the small 7-residue segment in Ca^{2+}-ATPase with the corresponding Na^+-K^+-ATPase peptide segment in the chimera would interfere with the phosphatase function, which is common to all P-type ATPases. A clue might lie in the need for K^+ binding at the ion occlusion sites to activate dephosphorylation of E_2P in Na^+-K^+-ATPase. The Na^+-K^+-ATPase segment used for substitution may be catalytically defective in the Ca^{2+}-ATPase environment, because of the absence of a conformational coupling with K^+-occlusion sites.

Gly[233] and the 7-residue segment swapped in the chimera discussed above are located near proteolytic cleavage sites (T_3 and V8 in Ca^{2+}-ATPase, T_{3Na} and C_{3Na} in Na^+-K^+-ATPase) that are hidden in the E_2 state. Cleavage of the peptide chain at these sites modifies enzyme function in a way very similar to the mutation of the glycine. The effect of the chymotryptic split C_{3Na} in Na^+-K^+-ATPase has been examined in some detail (Jørgensen and Andersen, 1988). After cleavage, noncovalent forces hold the peptide fragments together. The cleaved enzyme still phosphorylates from ATP, but the phosphorylated intermediate is locked up in the ADP-sensitive state holding three Na^+ in a stable occluded state. Charge transfer and fluorescence changes normally found to be linked with the E_1P-E_2P transition are blocked in the cleaved enzyme. However, while some properties of the cleaved enzyme are consistent with a stabilization of the E_1/E_1P states, the ability to bind and occlude K^+-congeners (an E_2-property) is retained, and the usual K^+-ATP antagonism is lost in the cleaved enzyme. A straightforward explanation for this could be that the C_{3Na} split interrupts the communication between the catalytic site and the cation sites.

Mutagenesis analysis of the central cytoplasmic domain of the SR Ca^{2+}-ATPase has shown that several of the residues located here are crucial to formation of the phosphoenzyme intermediate (triangles in Figure 16). Not unexpectedly, the phosphorylated aspartic acid residue, Asp^{351}, itself, is one of these residues. Generally, residues belonging to this category are highly conserved within the P-type ATPase family, and some have been implicated in ATP binding by the above mentioned structure predictions and affinity labeling (Table 3). So far, the studies of mutants have not pinpointed residues with a specific role in binding of the adenosine part of the ATP molecule, since phosphorylation was lost in the mutants symbolized by triangles, irrespective of whether ATP or P_i was used as substrate. A defective phosphorylation with either substrate points to a direct interaction of the mutated residue with the phosphoryl group or with the catalytic Mg^{2+}, or to a role in the local conformational changes associated with such interaction. Some of the residues in this category can be replaced with certain amino acids without loss of phosphorylation. Such replacements lead to the blocking of E_1P-E_2P transition (residues doubly labeled with squares and triangles in Figure 16). One example is Lys^{684}. If this residue is replaced with Ala, His, or Gln, phosphorylation is defective, but replacement with Arg, retaining the positive charge, allows phosphorylation from ATP, but not E_1P-E_2P transition or backdoor phosphorylation from P_i. It was suggested that the positive charge is required for stabilization of a transition state in the phosphoryl transfer reaction (Vilsen et al., 1991a). The bulky guanidinium group of arginine may interfere with E_1P-E_2P transition by preventing an interdomain movement that normally closes the active site cleft in E_2P. Interestingly, the functional properties of the Ca^{2+}-ATPase enzyme with a glutaraldehyde-induced cross-link between Arg^{678} and Lys^{492} (McIntosh, 1992) were similar to those of the $Lys^{684} \rightarrow Arg$ mutant. After formation of the cross link, the enzyme could still be phosphorylated by ATP, but the transformation of the E_1P phosphoenzyme intermediate to E_2P was completely blocked. No extracytoplasmic Ca^{2+} release from the discharge sites occurred in the cross-linked enzyme (McIntosh et al., 1991), supporting the hypothesis that normally the deocclusion of Ca^{2+} at the extracytoplasmic side (and therefore the formation of low-affinity extracytoplasmically orientated Ca^{2+} discharge sites) depends on propagation of the movements in the cytoplasmic domain down to the Ca^{2+}-binding domain in the membrane. Cross linking also prevented formation of E_2P in the backdoor reaction with P_i suggesting that similar active site movements are executed during the loss of ADP sensitivity in the forward direction and upon P_i phosphorylation in the reverse. Chemical modifications of cysteine sulfhydryls have likewise, in some cases, given rise to blocks of E_1P-E_2P transition in Ca^{2+}-ATPase and Na^+-K^+-ATPase. The cysteine involved in Ca^{2+}-ATPase has been identified as Cys^{344} near the phosphorylation site (Figure 16).

How Large Are the Conformational Changes?

Transport models in which the ion-binding sites move the whole way across the membrane have generally been abandoned and replaced by the concept of access channels in which the ions move part of the way without being bound. In principle, this allows transport to be accomplished without a large conformational rearrangement (Figure 7). Nevertheless, the fact that residues critical to conformational changes associated with cation translocation are so widely distributed over the pump protein, in regions encompassing extracytoplasmic as well as cytoplasmic and membranous domains, naturally leads one to ask whether extensive rearrangements might be involved in the translocation process. One indication that this might be the case comes from the electron microscopic studies of bidimensional membrane crystals. Unit cell size and symmetry varies, and the protomers show different shapes and points of contact with each other in the crystalline arrays, depending on which ligand is used for stabilization of the crystalline state. Even the distribution of protein mass between the membrane and the extramembranous projections may vary (Maunsbach et al., 1991). Time-resolved X-ray diffraction studies on oriented multilayers containing SR Ca^{2+}-ATPase have provided evidence that the cylindrically averaged ATPase profile changes upon utilization of ATP, and the data were interpreted in terms of a net redistribution of approximately 8% of the protein mass from the cytoplasmic region to the membrane region (Blasie et al., 1990). Spectroscopic measurements of the relative content of α-helix and β-structure have given rise to conflicting answers to the question whether the functional cycle involves changes in secondary structure, but most data seem to indicate that such changes are relatively small, if they do occur. Therefore, any extensive conformational rearrangements must comprise alteration of tertiary structure, i.e., movement of whole peptide segments and subdomains. On the other hand, the distances measured by fluorescence energy transfer between probes located in different cytoplasmic subdomains (see above) do not change measurably during enzyme cycling (Bigelow and Inesi, 1992).

REGULATION

Principles in Regulation of Pump Activity

The prime regulators of pump activity are the concentrations of the transported ions at their uptake sites. In the normal resting state of the cell, the intracellular Na^+

and Ca^{2+} concentrations are such that the Na^+-K^+- and Ca^{2+}-pumps operate at a fraction of their maximal turnover numbers, and there is a great reserve potential for increasing turnover as a response to temporary increases in the cation concentrations. Apart from this direct regulation by the transported cations, many ion pumps are regulated by a multitude of hormonal and environmental factors. Table 4 shows examples of some well documented hormonal influences on the Na^+-K^+- and Ca^{2+}-pumps (see also the reviews by Lingrel et al., 1990; Missiaen et al., 1991; and Clausen, 1986). Hormonal regulation can be divided into two general categories. Corticosteroids and thyroid hormones alter the level of gene expression, and their manifestation requires hours to days. Control of expression is thought to be mediated through hormone-receptor DNA-binding complexes that alter gene transcription by influencing control elements upstream to the coding gene sequences. Processing and stability of the mRNA transcripts, as well as their translation, may

Table 4. Examples of Hormones That Stimulate Na^+,K^+- and Ca^{2+}-Pumps

Hormone	Property Affected	Suggested Mechanism(s)
Na^+-K^+-ATPase		
Catecholamines	Molecular turnover	a) Protein kinase mediated phosphorylation of the catalytic peptide $\rightarrow$ alteration of kinetic parameter(s)
		b) Change in Na_i^+ (?)
Insulin	Molecular turnover	a) Change in Na_i^+ (only in some cell types)
	Number of active pumps	b) Altered interaction with unknown regulator $\rightarrow$ change in kinetic parameters (Na^+ affinity?)
		c) Increased transport of latent pumps from intracellular store to surface membrane
Thyroid hormone	Synthesis of pumps	a) Increased gene transcription (through direct interaction of nuclear T3-receptor complex with *cis*-regulatory upstream gene element, or secondary to T3 induction of other proteins, or to altered ion fluxes)
		b) Posttranscriptional effects on mRNA levels
		c) Translational effects
Aldosterone	Synthesis of pumps	Increased gene transcription (through direct interaction of nuclear aldosterone-receptor complex with *cis*-regulatory upstream gene element, and/or secondary to elevated Na_i^+ caused by induction of channel proteins)
Ca^{2+}-ATPase		
Catecholamines	Molecular turnover	Protein kinase mediated phosphorylation of phospholamban (SR Ca^{2+}-ATPase) or autoinhibitory domain (PM Ca^{2+}-ATPase) relieving negative influence on kinetic parameters
Thyroid hormones	Synthesis of pumps	Same as for Na^+-K^+-ATPase

also be subject to regulation. There is evidence that in some cases the long-term effects are mediated through a permanent increase in the concentration of the cation to be transported, such as Na^+ by Na^+-K^+-ATPase, affecting the expression level by an undetermined mechanism. The availability of β-subunits to be assembled with the α-subunits before translocation to the plasma membrane is often a limiting factor in the expression of active Na^+-K^+-pumps, and under these conditions control can be exerted through regulation of β-subunit synthesis.

Peptide hormones and catecholamines modulate the activity of preexisting pumps, and the response is rapid, occurring within seconds or minutes. This is mediated through plasma membrane receptors and intracellular second messengers, which activate regulatory proteins such as protein kinases and calmodulin. In principle, one can think of three major routes for further transmission of the signal to the pumps. (1) The regulatory protein may modulate the activity of ion channels or secondary ion pumps such as Na^+/H^+ and Na^+/Ca^{2+} antiports, thereby altering the cation substrate concentration at the uptake sites of the primary pumps and thus the pump activity, as described above. (2) The regulatory protein may modulate the pump activity by binding directly to the pump or by executing a covalent modification (e.g., serine or threonine phosphorylation) of the pump. Alternatively, a regulator may relieve an inhibition of the pump exerted by another protein. The prime example is the cAMP-mediated phosphorylation of phospholamban, which relieves the inhibitory influence of phospholamban on SR Ca^{2+}-pumps (James et al., 1989). The plasma membrane Ca^{2+}-pumps contain an autoinhibitory domain at the C-terminus with a remarkable structural and functional homology to phospholamban, and the self-inhibition of the pump by this domain is relieved by calmodulin binding and by kinase-mediated phosphorylation. (3) The third possible route for rapid up-regulation of pump activity is by translocation of preexisting pump molecules from a nonfunctional intracellular store to a functional location such as the plasma membrane. This type of regulated membrane traffic is well known from vacuolar type H^+-ATPases in kidney distal tubule and for glucose transporters, and has been suggested as an explanation of the rapid insulin stimulation of Na^+-K^+-pump activity (Hundal et al., 1992).

Isoform Diversity

Molecular cloning has revealed that many organisms contain a multiplicity of genes encoding different P-type ATPases with the same cation specificity and with similar subcellular localization. Hence, three genes located on different chromosomes encode isoforms of the Na^+-K^+-ATPase α-subunit (α_1, α_2, and α_3), and another three genes encode isoforms of the β-subunit (β_1, β_2, and β_3). Sarco(endo)plasmic reticulum Ca^{2+}-ATPases are likewise products of three genes (Table 5), while there are at least four different genes that encode the plasma membrane Ca^{2+}-ATPase (Carafoli, 1992). In general, isoforms show amino acid

Table 5. Characteristics of Sarco(endo)plasmic Reticulum Ca^{2+}-ATPase (SERCA) Isoforms and Tissue Distribution of Isoform mRNAs

	Isoform				
	$1a^a$	$1b^b$	2a	2b	3
Number of amino acid residues present[c]	993 + 1	993 + 8	993 + 4	993 + 49	999
Homology[d]		100%		84%	75%
Relative abundance of mRNA in tissues					
Uterine smooth muscle	–		–+	+++	+
Skeletal muscle	+++++		+++	+	+
Heart	–		+++++	+++	+
Brain	–		–+	+++	+
Lung	–		–+	+	+++
Liver	–		–	+	–
Kidney	–		–	+	–+
Spleen	–		–	+	+++
Testes	–		–	+	–
Stomach	–		–+	+	+
Small intestine	–		–+	+	+++
Large intestine	–		–+	+	++++
Pancreas	–		–	–	–+
Platelets[e]	–		–	+	+

Notes: [a]Found only in adult fast twitch muscle.
[b]Found only in neonatal muscle.
[c]Each of the SERCA 1 and 2 genes encodes 993 residues plus a C-terminus that can be alternatively spliced. Alternative splicing has not been described for the SERCA 3 gene.
[d]The fast twitch muscle gene is used for reference. The alternatively spliced C-terminal parts are ignored in the calculation of homology.
[e]Data for platelets are based on Papp et al. (1992). Apart from this, all data are from Burk et al. (1989).

sequence identities higher than 70%, and the differences have a tendency to cluster in those regions that also show the least homology among pumps of differing cation specificity. Analysis of the sequence relationships between isoforms in different species suggests that the isoform diversity originated early in evolution and was maintained rigidly through the past 200 million years, presumably to serve a physiological purpose. Isoform diversity results not only from the gene multiplicity, but also from alternative splicing of the transcribed mRNA, commonly at the C-terminus and at other domains involved in interaction with short-term regulators (Table 5, see also Carafoli, 1992 and Keeton et al., 1993 for plasma membrane Ca^{2+}-ATPase isoforms). The isoforms are expressed in a tissue- and cell-specific manner (Tables 5 and 6), and their relative abundancy is often subject to developmental and hormonal regulatory influences. Some isoforms (like α_1 of Na^+-K^+-AT-Pase and SERCA2b of SR Ca^{2+}-ATPase) deserve the status of housekeeping enzymes, since they are more widely distributed than the others.

Table 6. Relative Abundance of α Isoform mRNAs of Na^+-K^+-ATPase

Tissue	α Isoform mRNA		
	α_1	α_2	α_3
Rat Kidney			
Fetal	+++	−	−
Neonatal	++++	−	−
Adult	+++++	−+	−
Lung			
Fetal	+	−	−
Neonatal	++++	−	−
Adult	+	−+	−
Muscle			
Fetal	+	−+	−
Neonatal	+	+	+
Adult	+	++++	+
Brain			
Fetal	+	+	+
Neonatal	+	+	+++
Adult	+++	+++	+++
Heart			
Fetal	+++	−	+
Neonatal	+++	+	+
Adult	+++	++	−
Human Heart			
Fetal	+		+
Neonatal			
Adult	+++	+++	+++

Note: Based on data in Lingrel, 1992.

The physiological significance of the presence of the isoforms is still far from being fully understood. As a consequence of their different amino acid sequences, the isoforms could differ functionally and/or respond differently to short-term control by hormones and other effector molecules. A cell which temporarily has to cope with sudden increases in passive ion influxes might thus find it advantageous to have a high-capacity low-affinity pump in addition to the high-affinity house-keeping pump, or it might need an extra pump which could be turned on by a particular hormonal signal. Indeed, there is evidence of functional heterogeneity among isoforms, both with respect to their cation affinities, maximum turnover rates, and ability to interact with regulatory molecules (Sweadner, 1989; Jewell and Lingrel, 1991; Lytton et al., 1992; Toyofuko et al., 1993). The lower affinity of the α_3 isoform of Na^+-K^+-ATPase for Na^+ and of SERCA3 for Ca^{2+} is especially noteworthy. For Na^+-K^+-ATPase and H^+-K^+-ATPase the existence of multiple isoforms of the β-subunit adds to the possible functional diversities of the assembled αβ-complexes, since the properties displayed by the catalytic subunit may depend on the type of β, with which it is assembled.

In many instances, however, the functional differences between two isoforms appear too subtle to justify the existence of both forms. Hence, SERCA1a and SERCA2a Ca^{2+}-pumps are functionally indistinguishable in the assays available, and both are susceptible to inhibition by phospholamban. Since phospholamban is expressed mainly in the tissues, where SERCA2a is present (i.e., in cardiac and slow twitch muscle), and not where SERCA1a is present (fast twitch muscle), the regulation by phospholamban appears much more isoform specific than it actually is.

An obvious advantage of having different genes encoding similar pumps is that their expression can be regulated independently. To achieve this, the gene products do not need to differ functionally. Thus, rather than overlaying a very complicated gene regulation mechanism on a single gene, organisms may have divided the regulatory requirements among several genes with different regulatory response elements. Indeed, comparisons of the nucleotide sequences of the 5'-flanking regions upstream to the coding sequences of the genes encoding Na^+-K^+- and Ca^{2+}-ATPases have shown that each gene has its own distinct set of regulatory elements, which appear to be more conserved between the same isoforms of different species than between the various isoforms of the same species.

Isoform diversity could also play a role in differential sorting and targeting to the various subcellular structures. This is pertinent to the SERCA type of pumps, whose distribution among various postulated intracellular compartments (IP_3-sensitive and -insensitive sarco/endoplasmic reticulum subtypes and calciosomes) is controversial. Interaction between Na^+-K^+-ATPase and the membrane cytoskeleton components has recently been demonstrated and may be isoform-specific. To understand the sorting signals hidden in the amino acid sequences is one of the future challenges. One need only think of the Na^+-K^+- and H^+-K^+-pumps of the gastric parietal cells. Why does the first go to the basolateral membrane, while the second goes to the apical membrane?

SUMMARY

The P-type ATPases constitute a major group of primary ion pumps of crucial importance for cell function. These pumps are characterized by the formation during their functional cycle of an aspartyl phosphorylated intermediate. The catalytic site alternates between states with differing catalytic specificities for ATP-ADP exchange and H_2O-P_i exchange, respectively, and energy transduction depends on the coupling of the change in catalytic specificity with a rearrangement of the cation binding sites. Ion translocation involves changes in the orientation and the affinity of the cation binding sites and occurs through intermediary forms in which the cations are occluded. A single polypeptide chain of around 1,000 amino acid residues is responsible for the pumping. This peptide chain has a cytoplasmic head piece containing the binding sites for ATP and phosphate, and a transmembrane part (consisting of at least 8–10 transmembrane helices) containing the cation-binding sites. These two parts are coupled through conformational changes that involve residues located over most of the protein. Site-directed mutagenesis has demonstrated a pivotal role of the two transmembrane segments M4 and M5 in the signal transmission. These segments link the membrane region to the head-piece through stalk helices. The ion pumps are regulated by the internal concentrations of the ions being pumped and by various hormonal factors influencing the kinetics and availability of the pumps, as well as their expression levels. The existence of various isoforms of pumps with the same ion specificity is a central aspect of this regulation.

REFERENCES

Andersen, J.P. (1989). Monomer-oligomer equilibrium of sarcoplasmic reticulum Ca-ATPase and the role of subunit interaction in the Ca^{2+} pump mechanism. Biochim. Biophys. Acta 988, 47–72.

Andersen, J.P., Jørgensen, P.L., & Møller, J.V. (1985). Direct demonstration of structural changes in soluble, monomeric Ca^{2+}-ATPase associated with Ca^{2+}-release during the transport cycle. Proc. Natl. Acad. Sci. USA 82, 4573–4577.

Andersen, J.P., Vilsen, B., Collins, J.H., & Jørgensen, P.L. (1986). Localization of E_1-E_2 conformational transitions of sarcoplasmic reticulum Ca-ATPase by tryptic cleavage and hydrophobic labeling. J. Membr. Biol. 93, 85–92.

Andersen, J.P. & Vilsen, B. (1990). Primary ion pumps. Current Op. Cell Biol. 2, 722–730.

Andersen, J.P. & Vilsen, B. (1992a). Structural basis for the E_1/E_1P-E_2/E_2P conformation changes in the sarcoplasmic reticulum Ca^{2+}-ATPase studied by site-specific mutagenesis. Acta Physiol. Scand. 146, 151–159.

Andersen, J.P. & Vilsen, B. (1992b). Functional consequences of alterations to Glu^{309}, Glu^{771}, and Asp^{800} in the Ca^{2+}-ATPase of sarcoplasmic reticulum. J. Biol. Chem. 267, 19383–19387.

Andersen, J.P., Vilsen, B., Leberer, E., & MacLennan, D.H. (1989). Functional consequences of mutations in the β-strand sector of the Ca^{2+}-ATPase of sarcoplasmic reticulum. J. Biol. Chem. 264, 21018–21023.

Andersen, J.P., Vilsen, B., & MacLennan, D.H. (1992). Functional consequences of alterations to Gly^{310}, Gly^{770}, and Gly^{801} located in the transmembrane domain of the Ca^{2+}-ATPase of sarcoplasmic reticulum. J. Biol Chem. 267, 2767–2774.

Bayle, D., Robert, J.C., Bamberg, K., Benkouka, F., Cheret, A.M., Lewin, M.J.M., Sachs, G., & Soumarmon, A. (1992). Location of the cytoplasmic epitope for a K^+-competitive antibody of the (H^+,K^+)-ATPase. J. Biol. Chem. 267, 19060–19065.

Besancon, M., Shin, J.M., Mercier, F., Munson, K., Rabon, E., Hersey, S., & Sachs, G. (1992). Chemomechanical coupling in the gastric H,K ATPase. Acta Physiol. Scand. 146, 77–88.

Bigelow, D.J. & Inesi, G. (1992). Contributions of chemical derivatization and spectroscopic studies to the characterization of the Ca^+ transport ATPase of sarcoplasmic reticulum. Biochim. Biophys. Acta 1113, 323–338.

Blasie, J.K., Pascolini, D., Asturias, F., Herbette, L.G., Pierce, D., & Scarpa, A. (1990). Large-scale structural changes in the sarcoplasmic reticulum ATPase appear essential for calcium transport. Biophys. J. 58, 687–693.

Blostein, R. & Polvani, C. (1992). Altered stoichiometry of the Na,K-ATPase. Acta Physiol. Scand. 146, 105–110.

Boyer, P.D. (1988). Bioenergetic coupling to protonmotive force: should we be considering hydronium ion coordination and not group protonation? TIBS 13, 5–7.

Brandl, C.J., Green, N.M., Korczak, B., & MacLennan, D.H. (1986). Two Ca^{2+} ATPase genes: Homologies and mechanistic implications of deduced amino acid sequences. Cell 44, 597–607.

Burk, S.E., Lytton, J., MacLennan, D.H., & Shull, G.E. (1989). cDNA cloning, functional expression, and mRNA tissue distribution of a third organellar Ca^{2+} pump. J. Biol. Chem. 264, 18561–18568.

Carafoli, E. (1991). Calcium pump of the plasma membrane. Physiol. Rev. 71, 129–153.

Carafoli, E. (1992). The Ca^{2+} pump of the plasma membrane. J. Biol. Chem. 267, 2115–2118.

Champeil, P., Guillain, F., Vénien, C., & Gingold, M.P. (1985). Interaction of magnesium and inorganic phosphate with calcium-deprived sarcoplasmic reticulum adenosinetriphosphatase as reflected by organic solvent induced perturbation. Biochemistry 24, 69–81.

Clarke, D.M., Loo, T.W., Inesi, G., & MacLennan, D.H. (1989a). Location of high affinity Ca^{2+}-binding sites within the predicted transmembrane domain of the sarcoplasmic reticulum Ca^{2+}-ATPase. Nature (London) 339, 476–478.

Clarke, D.M., Maruyama, K., Loo, T.W., Leberer, E., Inesi, G., & MacLennan, D.H. (1989b). Functional consequences of glutamate, aspartate, glutamine, and asparagine mutations in the stalk sector of the Ca^{2+}-ATPase of sarcoplasmic reticulum. J. Biol. Chem. 264, 11246–11251.

Clarke, D.M., Loo, T.W., & MacLennan, D.H. (1990a). Functional consequences of alterations to polar amino acids located in the transmembrane domain of the Ca^{2+}-ATPase of sarcoplasmic reticulum. J. Biol. Chem. 265, 6262–6267.

Clarke, D.M., Loo, T.W., & MacLennan, D.H. (1990b). Functional consequences of mutations of conserved amino acids in the β-strand domain of the sarcoplasmic reticulum. J. Biol. Chem. 265, 14088–14092.

Clarke, D.M., Loo, T.W., & MacLennan, D.H. (1990c). The epitope for monoclonal antibody A20 (amino acids 870–890) is located on the luminal surface of the Ca^{2+}-ATPase of sarcoplasmic reticulum. J. Biol. Chem. 265, 17405–17408.

Clarke, D.M., Loo, T.W., & MacLennan, D.H. (1990d). Functional consequences of alterations to amino acids located in the nucleotide binding domain of the Ca^{2+}-ATPase of sarcoplasmic reticulum. J. Biol. Chem. 265, 22223–22227.

Clausen, T. (1986). Regulation of active Na^+-K^+ transport in skeletal muscle. Physiol. Rev. 66, 542–580.

Coll, R.J. & Murphy, A.J. (1991). Kinetic evidence for two nucleotide binding sites on the CaATPase of sarcoplasmic reticulum. Biochemistry 30, 1456–1461.

Cornelius, F. (1991). Functional reconstitution of the sodium pump. Kinetics of exchange reactions performed by reconstituted Na/K-ATPase. Biochim. Biophys. Acta 1071, 19–66.

Crowson, M.S. & Shull, G.E. (1992). Isolation and characterization of a cDNA encoding the putative distal colon H^+,K^+-ATPase. J. Biol. Chem. 267, 13740–13748.

De Foresta, B., Champeil, P., & Le Maire, M. (1990). Different classes of tryptophan residues involved in the conformational changes characteristic of the sarcoplasmic reticulum Ca^{2+}-ATPase cycle. Eur. J. Biochem. 194, 383–388.

De Meis, L. & Vianna, A.L. (1979). Energy interconversion by the Ca^{2+}-dependent ATPase of the sarcoplasmic reticulum. Ann. Rev. Biochem. 48, 275–292.

De Meis, L. (1981). The sarcoplasmic reticulum. In: Transport and Energy Transduction (Bitter, E.E., ed.), Vol. 2, pp. 1-163, John Wiley & Sons, New York.

De Meis, L. (1989). Role of water in the energy of hydrolysis of phosphate compounds—energy transduction in biological membranes. Biochim. Biophys. Acta 973, 333–349.

De Meis, L. (1991). Fast effux of Ca^{2+} mediated by the sarcoplasmic reticulum Ca^{2+}-ATPase. J. Biol. Chem. 266, 5736–5742.

Dupont, Y. (1976). Fluorescence studies of the sarcoplasmic reticulum calcium pump. Biochem. Biophys. Res. Comm. 71, 544–550.

Dux, L. & Martonosi, A. (1983). Two-dimensional arrays of proteins in sarcoplasmic reticulum and purified Ca^{2+}-ATPase vesicles treated with vanadate. J. Biol. Chem. 258, 2599–2603.

Esmann, M. & Skou, J.C. (1985). Occlusion of Na by Na,K-ATPase in the presence of oligomycin. Biochem. Biophys. Res. Commun. 127, 857–863.

Forbush III, B. (1987). Rapid release of ^{42}K or ^{86}Rb from two distinct transport sites on the Na,K-pump in the presence of P_i or vanadate. J. Biol. Chem. 262, 11116–11127.

Forgac, M. (1989). Structure and function of vacuolar class of ATP-driven proton pumps. Physiol. Rev. 69, 765–796.

Garrahan, P.J. & Rega, A.F. (1990). Plasma Membrane Calcium Pump In: Intracellular Calcium Regulation (Bronner, F., ed.), pp. 271–303, Alan R. Liss, Inc., New York.

Geering, K. (1991). The functional role of the β-subunit in the maturation and intracellular transport of Na,K-ATPase. FEBS Lett. 285, 189–193.

Ghislain, M., Schlesser, A., & Goffeau, A. (1987). Mutation of a conserved glycine residue modifies the vanadate sensitivity of the plasma membrane H^+-ATPase from the schizosaccharomyces pombe. J. Biol. Chem. 262, 17549–17555.

Glynn, I.M. (1985). The Na^+,K^+-transporting adenosine triphosphatase. In: The Enzymes of Biological Membranes (Martonosi, A.N., ed.), Vol. 3, pp. 35–114, Plenum Press, New York.

Glynn, I.M. (1993). All hands to the sodium pump. J. Physiol. 462, 1–30.

Glynn, I.M. & Karlish, S.J.D. (1990). Occluded cations in active transport. Ann. Rev. Biochem. 59, 171–205.

Green, N.M. (1989). ATP-driven cation pumps: Alignment of sequences. Biochem. Soc. Trans. 17, 970–972.

Green, N.M. & MacLennan, D.H. (1989). ATP driven ion pumps: An evolutionary mosaic. Biochem. Soc. Trans. 17, 819–822.

Hansen. O. (1984). Interaction of cardiac glycosides with $(Na^+ + K^+)$-activated ATPase. A biochemical link to digitalis-induced inotropy. Pharmacol. Rev. 36, 143–163.

Hebert, H., Skriver, E., Söderholm, M., & Maunsbach, A.B. (1988). Three-dimensional structure of renal Na,K-ATPase determined from two-dimensional membrane crystals of the p1 form. J. Ultrastruct. Mol. Struct. Res. 100, 86–93.

Heinemann, S.H., Terlau, H., Stühmer, W., Imoto, K., & Numa, S. (1992). Calcium channel characteristics conferred on the sodium channel by single mutations. Nature (London) 356, 441–443.

Hundal, H.S., Marette, A., Mitsumoto, Y., Ramlal, T., Blostein, R., & Klip, A. (1992). Insulin induces translocation of the α2 and β1 subunits of the Na^+/K^+-ATPase from intracellular compartments to the plasma membrane in mammalian skeletal muscle. J. Biol. Chem. 267, 5040–5043.

Inesi, G. (1985). Mechanism of calcium transport. Annu. Rev. Physiol. 47, 573–601.

James, P., Inui, M., Tada, M., Chiesi, M., & Carafoli, E. (1989). Nature and site of phospholamban regulation of the Ca^{2+} pump of sarcoplasmic reticulum. Nature (London) 342, 90–92.

Jencks, W.P. (1989a). Utilization of binding energy and coupling rules for active transport and other vectorial processes. Methods in Enzymol. 171, 145–164.

Jencks, W.P. (1989b). How does a calcium pump pump calcium? J. Biol. Chem. 264, 18855–18858.

Jensen, J. & Ottolenghi, P. (1983). ATP binding to solubilized $(Na^+ + K^+)$-ATPase. The abolition of subunit-subunit interaction and the maximum weight of the nucleotide-binding unit. Biochim. Biophys. Acta 731, 282–289.

Jewell, E.A. & Lingrel, J.B. (1991). Comparison of the substrate dependence properties of the rat Na,K-ATPase α1, α2, and α3 isoforms expressed in HeLa cells. J. Biol. Chem. 266, 16925–16930.

Jørgensen, P.L. (1975). Purification and characterization of (Na^+K^+)-ATPase. V. Conformational changes in the enzyme. Transitions between the Na-form and the K-form studied with tryptic digestion as a tool. Biochim. Biophys. Acta 401, 399–415.

Jørgensen, P.L. (1991). Conformational transitions in the α-subunit and ion occlusion. In: The Sodium Pump: Structure, Mechanism, and Regulation (Kaplan, J.H. & De Weer, P., eds.), pp. 189–200, The Rockefeller University Press, New York.

Jørgensen, P.L. & Andersen, J.P. (1988). Structural basis for E_1-E_2 conformational transitions in Na,K-pump and Ca-pump proteins. J. Membr. Biol. 103, 95–120.

Karlish, S.J.D., Goldshleger, R., & Stein, W.D. (1990). A 19 kDa C-terminal tryptic fragment of the α-chain of Na,K-ATPase is essential for occlusion and transport of cations. Proc. Natl. Acad. Sci. USA. 87, 4566–4570.

Karlish, S.J.D., Goldshleger, R., Tal, D.M., Capasso, J.M., Hoving, S., & Stein, W.D. (1992). Identification of the cation binding domain of Na/K-ATPase. Acta Physiol. Scand. 146, 69–76.

Karlish, S.J.D., Goldshleger, R., & Jørgensen, P.L. (1993). Location of Asn[831] of the α chain of Na/K-ATPase at the cytoplasmic surface. J. Biol. Chem. 268, 3471–3478.

Keeton, T.P., Burk, S.E., & Shull, G.E. (1993). Alternative splicing of exons encoding the calmodulin-binding domains and C termini of plasma membrane Ca^{2+}-ATPase isoforms 1, 2, 3, and 4. J. Biol. Chem. 268, 2740–2748.

Krishna, S., Cowan, G., Meade, J.C., Wells, R.A., Stringer, J.R., & Robson, K.J. (1993). A family of cation ATPase-like molecules from *Plasmodium falciparum*. J. Cell Biol. 120, 385–398.

Läuger, P. (1991). Electrogenic Ion Pumps. Sinauer Associates, Inc., Sunderland, Mass.

Levy, D., Seigneuret, M., Bluzat, A., & Rigaud, J.-L. (1990). Evidence for proton countertransport by the sarcoplasmic reticulum Ca^{2+}-ATPase during calcium transport in reconstituted proteoliposomes with low ionic permeability. J. Biol. Chem. 265, 19524–19534.

Lingrel, J.B., Orlowski, J., Shull, M.M., & Price, E.M. (1990). Molecular genetics of Na,K-ATPase. Progr. Nucl. Acid Res. 38, 37–89.

Lingrel, J.B. (1992). Na,K-ATPase: Isoform structure, function, and expression. J. Bioenerg. Biomem. 24, 263–270.

Lytton, J., Westlin, M., Burk, S.E., Shull, G E., & MacLennan, D.H. (1992). Functional comparisons between isoforms of the sarcoplasmic or endoplasmic reticulum family of calcium pumps. J. Biol. Chem. 267, 14483–14489.

MacLennan, D.H. (1990). Molecular tools to elucidate problems in excitation-contraction coupling. Biophys. J. 58, 1355–1365.

MacLennan, D.H., Brandl, C.J., Korczak, B., & Green, N.M. (1985). Amino-acid sequence of a Ca^{2+} + Mg^{2+}-dependent ATPase from rabbit muscle sarcoplasmic reticulum, deduced from its complementary DNA sequence. Nature (London) 316, 696–700.

Maruyama, K. & MacLennan, D.H. (1988). Mutation of aspartic acid-351, lysine-352, and lysine-515 alters the Ca^{2+} transport activity of the Ca^{2+}-ATPase expressed in COS-1 cells. Proc. Natl. Acad. Sci. USA 85, 3314–3318.

Maruyama, K., Clarke, D.M., Fujii, J., Inesi, G., Loo, T.W., & MacLennan, D.H. (1989). Functional consequences of alterations to amino acids located in the catalytic center (isoleucine 348 to threonine 357) and nucleotide-binding domain of the Ca^{2+}-ATPase of sarcoplasmic reticulum. J. Biol. Chem. 264, 13038–13042.

Mata, A.M., Matthews, I., Tunwell, R.E.A., Sharma, R.P., Lee, A.G., & East, J.M. (1992). Definition of surface-exposed and trans-membranous regions of the $(Ca^{2+}\text{-}Mg^{2+})$-ATPase of sarcoplasmic reticulum using anti-peptide antibodies. Biochem. J. 286, 567–580.

Maunsbach, A.B., Skriver, E., & Hebert, H. (1991). Two-dimensional crystals and three-dimensional structure of Na,K-ATPase analyzed by electron microscopy. In: The Sodium Pump: Structure, Mechanism, and Regulation (Kaplan, J.H. & De Weer, P., eds.), pp. 159–172, The Rockefeller University Press, New York.

McIntosh, D.B. (1992). Glutaraldehyde cross-links Lys-492 and Arg-678 at the active site of sarcoplasmic reticulum Ca^{2+}-ATPase. J. Biol. Chem. 267, 22328–22335.

McIntosh, D.B., Ross, D.C., Champeil, P., & Guillain, F. (1991). Crosslinking the active site of sarcoplasmic reticulum Ca^{2+}-ATPase completely blocks Ca^{2+} release to the vesicle lumen. Proc. Natl. Acad. Sci. USA 88, 6437–6441.

Missiaen, L., Wuytack, F., Raeymaekers, L., De Smedt, H., Droogmans, G., Declerck, I., & Casteels, R. (1991). Ca^{2+} extrusion across plasma membrane and Ca^{2+} uptake by intracellular stores. Pharmac. Ther. 50, 191–232.

Modyanov, N., Lutsenko, S., Chertova, E., & Efremov, R. (1991). Architecture of the sodium pump molecule: Probing the folding of the hydrophobic domain. In: The Sodium Pump: Structure, Mechanism, and Regulation (Kaplan, J.H. & De Weer, P., eds.), pp. 99–115, The Rockefeller University Press, New York.

Nelson, N. (1992). Evolution of organellar proton-ATPases. Biochim. Biophys. Acta 1100, 109–124.

Nørby, J.G. & Klodos, I. (1988). The phosphointermediates of Na,K-ATPase. Progr. Clin. Biol. Res. 268A, 249–270.

Orlowski, S. & Champeil, P. (1991a). Kinetics of calcium dissociation from its high-affinity transport sites on sarcoplasmic reticulum ATPase. Biochemistry 30, 352–361.

Orlowski, S. & Champeil, P. (1991b). The two calcium ions initially bound to nonphosphorylated sarcoplasmic reticulum Ca^{2+}-ATPase can no longer be kinetically distinguished when they dissociate from phosphorylated ATPase toward the lumen. Biochemistry 30, 11331–11342.

Papp, B., Enyedi, A., Pászty, K., Kovács, T., Sarkadi, B., Gárdos, G., Magnier, C., Wuytack, F., & Enouf, J. (1992). Simultaneous presence of two distinct endoplasmic-reticulum-type calcium-pump isoforms in human cells. Biochem. J. 288, 297–302.

Pedersen, P.L. & Carafoli, E. (1987). Ion motive ATPases. I. Ubiquity, properties, and significance to cell function. TIBS 12, 146–150.

Polvani, C., Sachs, G., & Blostein, R. (1989). Sodium ions as substitutes for protons in the gastric H,K-ATPase. J. Biol. Chem. 264, 17854–17859.

Portillo, F. & Serrano, R. (1989). Growth control strength and active site of yeast plasma membrane ATPase studied by site-directed mutagenesis. Eur. J. Biochem. 186, 501–507.

Post, R.L., Hegyvary, C., & Kume, S. (1972). Activation by adenosine triphosphate in the phosphorylation kinetics of sodium and potassium ion transport adenosine triphosphatase. J. Biol. Chem. 247, 6530–6540.

Rabon, E.C. & Reuben, M.A. (1990). The mechanism and structure of the gastric H,K-ATPase. Ann. Rev. Physiol. 52, 321–344.

Sachs, G., Shin, J.M., Besancon, M., Munson, K., & Hersey, S. (1992). Topology and sites in H,K-ATPase. Ann. N.Y. Acad. Sci. 671, 204–216.

Saraste, M., Sibbald, P.R., & Wittinghofer, A. (1990). The P-loop—a common motif in ATP- and GTP-binding proteins. TIBS 15, 430–434.

Sasaki, T., Shinoguchi, E., Kamo, Y., Ito, E., & Taniguchi, K. (1991). The change in Na^+ binding states before and after accumulation of ADP-sensitive phosphoenzyme in Na^+,K^+-ATPase. In: The Sodium Pump: Recent Developments (Kaplan, J.H. & De Weer, P., eds.), pp. 413–417, The Rockefeller University Press, New York.

Seebregts, C.J. & McIntosh, D.B. (1989). 2′,3′-0-(2,4,6-trinitrophenyl)-8-azido-adenosine mono-, di-, and triphosphates as photoaffinity probes of the Ca^{2+}-ATPase of sarcoplasmic reticulum. J. Biol. Chem. 264, 2043–2052.

Serrano, R. (1988). Structure and function of proton translocating ATPase in plasma membranes of plants and fungi. Biochim. Biophys. Acta 947, 1–28.

Shull, G.E., Schwartz, A., & Lingrel, J.B. (1985). Amino-acid sequence of the catalytic subunit of the $(Na^+ + K^+)$ATPase deduced from a complementary DNA. Nature (London) 316, 691–695.

Skou, J.C. & Esmann, M. (1992). The Na,K-ATPase. J. Bioenerg. Biomembr. 24, 249–261.

Skriver, E., Maunsbach, A.B., & Jørgensen, P.L. (1981). Formation of two-dimensional crystals in pure membrane-bound Na^+,K^+-ATPase. FEBS Lett. 131, 219–222.

Stefanova, H.I., Mata, A.M., East, J.M., Gore, M.G., & Lee, A.G. (1993). Reactivity of lysyl residues on the $(Ca^{2+}$-$Mg^{2+})$-ATPase to 7-amino-4-methylcoumarin-3-acetic acid succinimidyl ester. Biochemistry 32, 356–362.

Steinberg, M. & Karlish, S.J.D. (1989). Studies on conformational changes in Na,K-ATPase labeled with 5-iodoacetamidofluorescein. J. Biol. Chem. 264, 2726–2734.

Suzuki, H., Obara, M., Kubo, K., & Kanazawa, T. (1989). Changes in the steady-state fluorescence anisotropy of N-iodoacetyl-N′-(5-sulfo-1-naphthyl)ethylenediamine attached to the specific thiol of sarcoplasmic reticulum Ca^{2+}-ATPase throughout the catalytic cycle. J. Biol. Chem. 264, 920–927.

Sweadner, K.J. (1989). Isozymes of the Na^+/K^+-ATPase. Biochim. Biophys. Acta 988, 185–220.

Tanford, C. (1982a). Simple model for the chemical potential change of a transported ion in active transport. Proc. Natl. Acad. Sci. USA 79, 2882–2884.

Tanford, C. (1982b). Steady state of an ATP-driven calcium pump: Limitations on kinetic and thermodynamic parameters. Proc. Natl. Acad. Sci. USA 79, 6161–6165.

Tanford, C. (1983). Translocation pathway in the catalysis of active transport. Proc. Natl. Acad. Sci. USA 80, 3701–3705.

Tanford, C. (1984). Twenty questions concerning the reaction cycle of the sarcoplasmic reticulum calcium pump. Crit. Rev. Biochem. 17, 123–151.

Tanford, C., Reynolds, J.A., & Johnson, E.A. (1987). Sarcoplasmic reticulum calcium pump: A model for Ca^{2+} binding and Ca^{2+}-coupled phosphorylation. Proc. Natl. Acad. Sci. USA 84, 7094–7098.

Taniguchi, K., Suzuki, K., Kai, D., Matsuoka, I., Tomita, K., & Ilida, S. (1984). Conformational change of sodium- and potassium-dependent adenosine triphosphatase. J. Biol. Chem. 259, 15228–15233.

Taniguchi, K., Tosa, H., Suzuki, K., & Kamo, Y. (1988). Microenvironment of two different extrinsic fluorescence probes in Na^+,K^+-ATPase changes out of phase during sequential appearance of reaction intermediates. J. Biol. Chem. 263, 12943–12947.

Taylor, W.R. & Green, N.M. (1989). The homologous secondary structures of the nucleotide-binding sites of six cation-transporting ATPases lead to a probable tertiary fold. Eur. J. Biochem. 179, 241–248.

Toyofuku, T., Kurzydlowski, K., Tada, M., & MacLennan, D.H. (1993). Identification of regions in the Ca^{2+}-ATPase of sarcoplasmic reticulum that affect functional association with phospholamban. J. Biol. Chem. 268, 2809–2815.

Toyoshima, C., Sasabe, H., & Stokes, D.L. (1993). Three-dimensional cryo-electron microscopy of the calcium ion pump in the sarcoplasmic reticulum membrane. Nature (London) 362, 469–471.

Van Huysse, J.W., Jewell, E.A., & Lingrel, J.B. (1993). Site-directed mutagenesis of a predicted cation binding site of Na,K-ATPase. Biochemistry 32, 819–826.

Verma, A.K., Filoteo, A.G., Stanford, D.R., Wieben, E.D., Penniston, J.T., Strehler, E.E., Fischer, R., Heim, R., Vogel, G., Mathews, S., Strehler-Page, M.-A., James, P., Vorherr, T., Krebs, J., & Carafoli, E. (1988). Complete primary structure of a human plasma membrane Ca^{2+} pump. J. Biol. Chem. 263, 14152–14159.

Vilsen, B. (1992). Functional consequences of alterations to Pro[328] and Leu[332] located in the 4th transmembrane segment of the α-subunit of the rat kidney Na^+,K^+-ATPase. FEBS Lett. 314, 301–307.

Vilsen, B. & Andersen, J.P. (1987). Characterization of CrATP-induced calcium occlusion in membrane-bound and soluble monomeric sarcoplasmic reticulum Ca^{2+}-ATPase. Biochim. Biophys. Acta 898, 313–322.

Vilsen, B., Andersen, J.P., Petersen, J., & Jørgensen, P.L. (1987). Occlusion of $^{22}Na^+$ and $^{86}Rb^+$ in membrane-bound and soluble protomeric αβ-units of Na,K-ATPase. J. Biol. Chem. 262, 10511–10517.

Vilsen, B., Andersen, J.P., Clarke, D.M., & MacLennan, D.H. (1989). Functional consequences of proline mutations in the cytoplasmic and transmembrane sectors of the Ca^{2+}-ATPase of sarcoplasmic reticulum. J. Biol. Chem. 264, 21024–21030.

Vilsen, B., Andersen, J.P., & MacLennan, D.H. (1991a). Functional consequences of alterations to amino acids located in the hinge domain of the Ca^{2+}-ATPase of sarcoplasmic reticulum. J. Biol. Chem. 266, 16157–16164.

Vilsen, B., Andersen, J.P., & MacLennan, D.H. (1991b). Functional consequences of alterations to hydrophobic amino acids located at the M_4S_4 boundary of the Ca^{2+}-ATPase of sarcoplasmic reticulum. J. Biol. Chem. 266, 18839–18845.

Vilsen, B. & Andersen, J.P. (1992a). Mutational analysis of the role of Glu^{309} in the sarcoplasmic reticulum Ca^{2+}-ATPase of frog skeletal muscle. FEBS Lett. 306, 247–250.

Vilsen, B. & Andersen, J.P. (1992b). Interdependence of Ca^{2+} occlusion sites in the unphosphorylated sarcoplasmic reticulum Ca^{2+}-ATPase complex with CrATP. J. Biol. Chem. 267, 3539–3550.

Vilsen, B. & Andersen, J.P. (1992c). CrATP-induced Ca^{2+} occlusion in mutants of the Ca^{2+}-ATPase of sarcoplasmic reticulum. J. Biol. Chem. 267, 25739–25743.

Wakabayashi, S., Imagawa, T., & Shigekawa, M. (1990). Does fluorescence of 4-nitrobenzo-2-oxa-1,3-diazole incorporated into sarcoplasmic reticulum ATPase monitor putative E_1-E_2 conformational transition? J. Biochem. 107, 563–571.

Yamamoto, H., Imamura, Y., Tagaya, M., Fukui, T., & Kawakita, M. (1989). Ca^{2+}-dependent conformational change of the ATP-binding site of Ca^{2+}-transporting ATPase of sarcoplasmic reticulum as revealed by an alteration of the target-site specificity of adenosine triphosphopyridoxal. J. Biochem. 106, 1121–1125.

Yoda, A. & Yoda, S. (1987). Two different phosphorylation-dephosphorylation cycles of Na,K-ATPase proteoliposomes accompanying Na^+ transport in the absence of K^+. J. Biol. Chem. 262, 110–115.

RECOMMENDED READINGS

Bamberg, E. & Schoner, W. (eds.) (1994). The Sodium Pump. Structure, Mechanism, Hormonal Control and its Role in Disease. Steinkopff, Darmstadt.

Glynn, I.M. (1993). All hands to the sodium pump. J. Physiol. 462, 1–30.

Glynn, I.M. & Karlish, S.J.D. (1990). Occluded cations in active transport. Annu. Rev. Biochem. 59, 171–205.

De Meis, L. (1981). The sarcoplasmic reticulum. In: Transport and Energy Transduction (Bittar, E., ed.), Vol. 2, pp. 1–163, John Wiley & Sons, New York.

Kaplan, J.H. & De Weer, P. (eds.) (1991). The Sodium Pump: Structure, Mechanism, and Regulation. The Rockefeller University Press, New York.

Läuger, P. (1991). Electrogenic ion pumps. Sinauer Associates, Inc., Sunderland, Mass.

Lingrel, J.B., Orlowski, J., Shull, M.M., & Price, E.M. (1990). Molecular genetics of Na,K-ATPase. Progr. Nucl. Acid Res. 38, 37–89.

MacLennan, D.H. (1990). Molecular tools to elucidate problems in excitation-contraction coupling. Biophys. J. 58, 1355–1365.

Pedersen, P.L. & Carafoli, E. (1987). Ion motive ATPases. I. Ubiquity, properties, and significance to cell function. TIBS 12, 146–150.

Scarpa, A., Carafoli, E., & Papa, S. (eds.) (1992). Ion-motive ATPases: Structure, function, and regulation. Ann. N.Y. Acad. Sci. Vol. 671.

Skou, J.C., Nørby, J.G., Maunsbach, B., & Esmann, M. (eds.) (1988). The Na^+,K^+-Pump, Part A: Molecular Aspects. Part B: Cellular Aspects. Progr. Clin. Biol. Res. Vol. 268. Alan R. Liss, Inc., New York.

Tanford, C. (1984). Twenty questions concerning the reaction cycle of the sarcoplasmic reticulum calcium pump. Crit. Rev. Biochem. 17, 123–151.

Wallmark, B., Sachs, G., Karlish, S., & Kaplan, J. (eds.) (1992). Ion pumps—Structure and Mechanism. Acta Physiol. Scand. 146, Suppl. 607.

Chapter 2

Facilitative Glucose Transport

CHARLES F. BURANT

Principles of Medical Biology, Volume 4
Cell Chemistry and Physiology: Part III, pages 67–86.
Copyright © 1996 by JAI Press Inc.
All rights of reproduction in any form reserved.
ISBN: 1-55938-807-2

INTRODUCTION

Transport across cell membranes is the first step in the metabolism of glucose and other simple sugars by mammalian cells. After uptake, glucose is utilized by the cell in a number of ways. Glucose provides ATP through both aerobic and anaerobic metabolism, it can be stored either as glycogen or converted to fat for later use, or it can be converted into structural moieties for the synthesis of macromolecules such as glycoproteins, proteoglycans, glycolipids, and nucleic acids. Insulin action in muscle and adipose tissue involves stimulation of glucose transport. Disruption of the normal pathways of glucose metabolism which occurs in diabetes mellitus can result in significant alterations in cellular function. The complications of diabetes mellitus, such as damage to the retina, kidney, cardiovascular and nervous systems, are likely due, in part, to the alterations in glucose uptake and metabolism.

Glucose transport occurs by two distinct mechanisms, one which requires energy in the form of an ion gradient and one which is facilitative and nonenergy requiring. Facilitative transport is the mechanism by which most tissues take up glucose. When facilitative glucose transport was compared in different tissues, distinct kinetics for glucose uptake and inhibition by various agents were found. Although the liver was not dependent upon insulin for glucose uptake, fat and skeletal and cardiac muscle were dependent upon the action of insulin for stimulation of glucose uptake, implying that more than one facilitative transporter existed which had tissue specificity. In addition, fructose transport in the small intestine appeared to be mediated by a distinct transporter. With the cloning of the facilitative glucose transporter of erythrocytes in 1986 and subsequently, additional facilitative transporters from other tissues, the basis for these differences were clarified. In this chapter, the structure of glucose transporters, their kinetics, substrate specificity, and cellular and organismal regulation of the individual facilitative glucose transporters will be examined.

TRANSCELLULAR GLUCOSE TRANSPORT

Glucose traverses the plasma membranes through two distinct families of glucose carrier proteins. The first is the Sodium/Glucose Transporter (SGLT). In the SGLT family, glucose and sodium are transported across the membrane together. Galactose, another major dietary sugar, shares this transport mechanism while the third major dietary sugar, fructose, is not a substrate for SGLT. In the small intestine and kidney tubules an electrochemical gradient of sodium generated by the Na^+-K^+-ATPase is used to transport glucose against its concentration gradient. Details of this transporter system are described in the nephrobiology module.

The second family of carrier proteins are the facilitative Glucose Transporters (GLUT). These proteins, which are structurally unrelated to the SGLT family of proteins (Figure 1), allow for the rapid, bidirectional, nonenergy requiring transport of glucose and closely related sugars across cell membranes. The net transport of

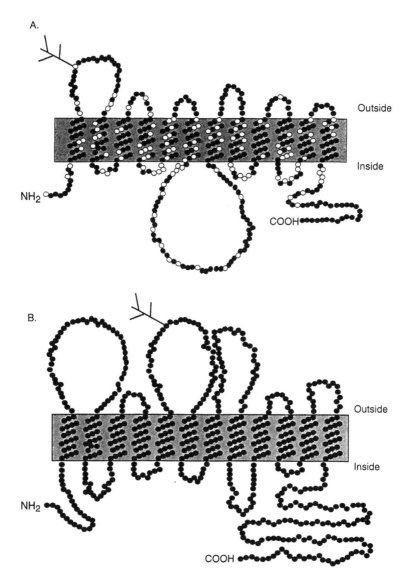

Figure 1. Models for the orientation of: **A.**) Members of the facilitative glucose transporter family (GLUT1 to GLUT7), and **B.**) the sodium-dependent glucose transporter (SGLT1). The branched structure is at the site of glycosylation for both transporters. In **A**, the open residues represent amino acids which are identical in GLUT1 through GLUT5.

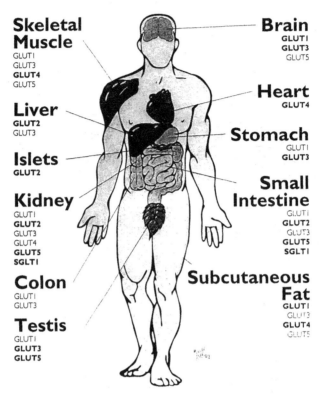

Figure 2. Sites of distribution of glucose transporters expressed in human tissue. Bold type signifies that high levels of transporter mRNA are expressed in the indicated tissue.

glucose by these carriers is always from high to low glucose concentrations and is not coupled with the transit of any other molecule. To date, seven cDNAs encoding facilitative glucose transporters have been cloned. Designated GLUT1 through GLUT7 (numbered in order in which each was cloned) six encode functional proteins and one is a pseudogene. These transporters have distinct and overlapping tissue distributions (Figure 2). In addition to differences in tissue distribution, each appears to have a distinct subcellular location as well (Figure 3). The kinetic properties of the individual transporters allow for the precise regulation of glucose delivery to the individual tissues under a constantly changing metabolic milieu.

STRUCTURE OF FACILITATIVE GLUCOSE TRANSPORTERS

Mueckler and colleagues achieved the molecular cloning of the GLUT1 isoform of facilitative glucose transporters by using an expression library. In this technique, cDNA was prepared from a human hepatoma cell line known to express large

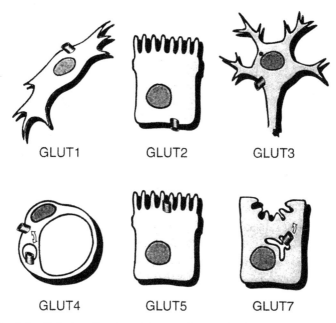

GLUT1   GLUT2   GLUT3

GLUT4   GLUT5   GLUT7

Figure 3. Subcellular localization of the members of the family of facilitative glucose transporters. Shown are the primary sites of facilitative transporter expression in the cell types where these transporters are highly expressed.

quantities of facilitative glucose transporters that were immunologically identical to the glucose transporter of erythrocytes, a well studied protein to which specific antibodies had been prepared. The proteins encoded by this cDNA library were expressed *in vitro* and reacted with the antibodies against the erythrocyte glucose transporter. The clone which produced immunological protein was isolated and sequenced. This molecular cloning study showed that GLUT1 was a 492 amino acid protein with distinct structural features which were predicted by computer modeling of the amino acid sequence. The modeling showed GLUT1 to have 12 membrane spanning domains with both the carboxy and amino termini located intracellularly (Figure 1). Between membrane spanning domains 1 and 2 (M1 and M2) is a large extracellular loop which contains a site for glycosylation at Asn-45. There is an additional large loop, predicted to be intracellular, between M6 and M7. Short loops of 7–14 amino acids connect the membrane spanning domains and these short loops place a severe constraint on possible tertiary structures of the protein. Many of these predictions for the structure of GLUT1 have been confirmed by either limited proteolytic digestion or by the use of antibodies directed at specific regions of the transporter.

The observation that liver and skeletal muscle, the two major sites of glucose uptake, express very little GLUT1 mRNA and protein led other investigators to search for additional facilitative glucose transporters. These investigators used radioactively labeled GLUT1 cDNA to screen cDNA libraries prepared from these tissues. Using this approach six additional cDNAs were isolated from various tissues, five of which encoded functional proteins. Each of these facilitative transport proteins have identical predicted secondary structures. The amino acid sequences between the individual transporters are 39–65% identical and 50–76% similar (conservative amino acid changes) (Figure 1). The highest degree of homology between the transporters are in the transmembrane segments and the short loops at the cytoplasmic side of the membrane. In contrast, the intracellular carboxy and amino termini and the extracellular and intracellular loops are the most divergent, suggesting that these domains may be responsible for determining the specific characteristics of the individual transporters. Each of the transporter cDNAs have a consensus sequence for Asn-linked glycosylation in the extracellular loop connecting M1 and M2 (Figure 1).

Among the facilitative glucose transporters, there is a strong conservation of sequences between species with greater than 85% identity at the amino acid level from rodent to man. This suggests that an array of glucose transporters with different characteristics are important to proper maintenance of glucose homeostasis in mammals.

CATALYSIS OF GLUCOSE TRANSPORT BY FACILITATIVE GLUCOSE TRANSPORTERS

The transport of glucose into erythrocytes has been one of the most extensively studied transport processes. From these studies much is known about the properties of facilitative sugar transport both kinetically and biochemically. Experiments in erythrocytes (which contain only GLUT1) and other tissues, as well as in cells expressing the cloned glucose transporter genes, have shown that the uptake of glucose has several definable properties. 1.) Transport occurs from high substrate concentration to low concentration without the need of cellular energy or other cotransported substrate. 2.) Transport follows simple Michaelis-Menten kinetics. 3.) Transport is stereospecific, i.e., only the D isomer of glucose is transported. 4.) Transport of glucose is inhibited by cytochalasin B, a fungal toxin. Thus, it appears that glucose transport is much like any enzymatic reaction with the substrate being glucose outside the cell and the product being glucose inside the cell.

Mechanism of Glucose Transport: The Alternating Conformer Model

The generally accepted model of glucose transport is the alternating conformer model. In this model, the glucose transporter molecule has binding sites for substrate at both the extracellular and intracellular faces of the membrane (Figure 4A). These binding sites are mutually exclusive and the transporter oscillates between conformations. Evidence for structural separation of these binding sites

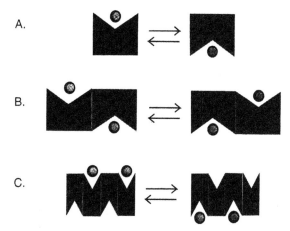

Figure 4. Models for glucose transporter catalyzed glucose movement. **A.**) is the alternating conformer model. **B.**) is an extension of **A** with more than one pore which would allow for accelerated exchange. **C.**) demonstrates the possible interaction of two or more transporter molecules providing for the possibility of allosteric interactions.

have been provided by kinetic studies and by specific ligands which bind to either the outer face or inner face of the transporter. For instance, 3-iodo-4-azidophenethylamido-7-*O*-succinyldeacetylforskolin (IAPS-forskolin) and the bis-mannose compound 2-N-4-(1-azi-2,2,2-trifluoroethyl)benzoyl-1,3(D-mannose-4-yloxy)-2-propylamine (ATB-BMPA) bind at or near the outer glucose binding site and inhibit glucose influx. Cytochalasin B appears to bind to the inner face of the transporter and is a competitive inhibitor of sugar efflux. Binding of cytochalasin B appears to lock the transporter in an inward orientation and inhibits binding of ligands which bind to the outer face of the transporter.

This model predicts that the rate of glucose uptake or turnover number is defined by a number of different steps in the interaction of substrate with the transporter. First is the binding of sugar to the outward binding site of the transporter. Higher affinity of the transporter for substrate will increase the rate of this step. Second is the conformational change of the substrate-occupied transporter from an outward to an inward facing conformation. It has been calculated for the erythrocyte glucose transporter that there are about 100 oscillations per second in the unoccupied transporter and the oscillations increase 10-fold in the presence of sugar due to a lowering of energy required to change between the inward and outward conformations. Intrinsic properties of the different transporter isoforms and the energy required for this conformational change will determine the rate of transit of substrate from the extracellular to the intracellular side of the membrane. Third is the release of the sugar from the transporter which will be determined by the affinity of the inward face of the transporter for sugar. And fourth is the return of the unoccupied transporter from the inward to the outward facing conformation.

Other Models of Transporter Structure

Other models of facilitative glucose transporters have been proposed which have similar properties to the alternating conformer design and have been advanced to provide a better explanation of some of the subtle properties of glucose transport, such as accelerated exchange where the presence of glucose on one side of the membrane accelerates the uptake of glucose from the opposite side of the membrane. One such model suggests that there are two pores in the protein, one open to the extracellular and one to the intracellular compartment (Figure 4B). Both pores cannot be open to the same compartment simultaneously but occupation of one pore will increase the rate of conformational change to the opposite face of the membrane. This model provides a means of increasing glucose transport from one side of the membrane in the presence of glucose on the other side of the membrane. An extension of this model, proposed by Carruthers, suggests that the glucose transporter exists as a dimer or tetramer in the membrane. In this model, the dimer acts as a simple carrier while the tetramer demonstrates cooperation in ligand binding (Figure 4C).

Structural Requirements of Glucose Binding to Facilitative Glucose Transporters

It is thought that glucose forms transient hydrogen bonds with amino acids that line the channel or pore through which it moves. By using a series of glucose analogues with different substitutions along the alcohol ring, LeFevre, and later Barnett, noted the structural features of glucose which enabled it to bind and be transported by erythrocytes (Figure 5). There is an absolute specificity for D versus L-glucose. The β-conformation at C1 and a hydroxyl at the C3 position appear to be most important in hydrogen binding to the outer pocket of the glucose transporter, or important in the movement of glucose across the pore itself. Substitutions at the C1 or C3 position with fluorine molecules which are hydrogen acceptors have similar affinities of the parent molecules; however, replacement of C1 and C3 with hydrogen reduces the affinity 10-fold or more. Substitutions at C6 are tolerated for binding to the external site of the erythrocyte glucose transporter but large substitutions at C6 decrease affinity for the internal binding site of the transporter, thus demonstrating a difference in structure of the internal and external binding sites. By contrast, when transport of glucose and analogues were studied in adipocytes, substitutions at the C6 position were well tolerated, whereas substitutions at C1 and C4 gave identical results as seen with analogue transport in erythrocytes.

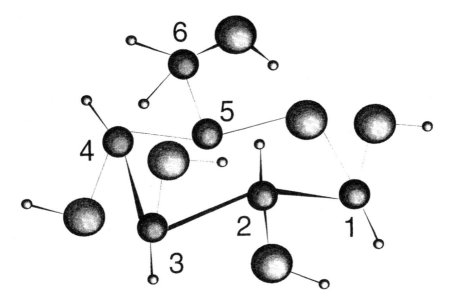

Figure 5. Space filling model of β-ᴅ-glucopyranose in the chair conformation which is thought to be the preferred conformation for transport. The carbon atoms numbered 1, 3, and 6 are thought to be most important for proper interaction with the facilitative glucose transporters.

Transporter Domains Which Interact With Glucose

As stated above, the facilitative glucose transporters appear to have a similar predicted secondary structure with 12 transmembrane domains which are presumed to form the pore with which glucose interacts. The transmembrane segments 3, 5, 7, 8, and 11, which are approximately 20–21 amino acids long, are predicted to form amphipathic α-helices. These segments contain many polar amino acids such as serine, threonine, asparagine, and glutamine which can potentially form hydrogen bonds with glucose, and these amino acids may interact to form a hydrophilic pore through which glucose moves (Figure 6).

The importance of some of these and other amino acids in glucose and inhibitor binding has come to light by mutational analysis of GLUT1. In these experiments, the cDNA which encodes GLUT1 is mutated so that individual nucleotide codons code for alternative amino acids. The mutated cDNA is expressed in *Xenopus* oocytes (frog eggs) or tissue culture cells and the transport properties are then compared with those of the native transporter. For example, mutation of glutamine 282, which resides in the seventh transmembrane domain to leucine, did not change the ability of the transporter to mediate glucose uptake but markedly diminished

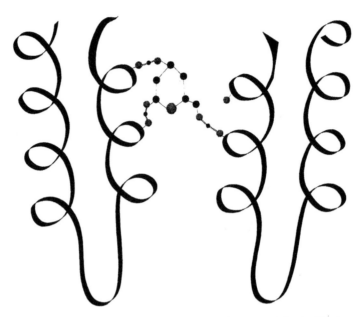

Figure 6. Schematic representation of glucose in the pore of the facilitative glucose transporter. The amphipathic α-helices of the membrane spanning domains are thought to form a hydrophilic pore through which glucose moves via hydrogen bonding to amino acids. The importance of the hydrogen bonds at positions 1, 3, and 6 of the glucose molecule for efficient transport (see text) is shown.

its interaction with ATB-BMPA. In contrast, mutation of glutamine 288 to isoleucine had no effect on transport of glucose or ATB-BMPA binding, suggesting that glutamine 282 is important in binding of the inhibitor but neither residue is necessary for binding or translocation of glucose. Other studies have shown that deletion of the last 33 C-terminal amino acids eliminates transport activity. In addition, the ability of ATB-BMPA to bind to the transporter was impaired, but binding of cytochalasin B (a fungal toxin which binds with high affinity to facilitative glucose transporters) was unaffected, suggesting that the transporter was locked in an inward facing conformation. However, deletion of the last 12 amino acids of GLUT1 had no effect on function, implying that amino acids from within the region of the last 33 and the last 12 are important for the proper conformational changes which occur during the steps of glucose transport.

Mutation of the glycosylation site of GLUT1, Asn-45 to Asp, Tyr, or Gln resulted in a transporter with a twofold lower affinity for glucose than the parental transporter. Since the large extracellular loop which contains the glycosylation site is not thought to be part of the pore structure of the transporter, it may be that glycosylation is important to maintaining overall tertiary or higher structure of the transporter.

Kinetic Differences Among Facilitative Glucose Transporters

Although the uptake of glucose into erythrocytes and other tissues follows Michaelis-Menten kinetics, the kinetic properties of glucose transport differ from tissue to tissue. For example, the uptake of glucose into liver showed a significantly higher K_m (lower affinity) than for other tissues for glucose and various glucose analogues. It is now clear that many of the kinetic differences are due to the expression of different transporters in the tissue studied. Affinity of glucose and other substrates for transporters can be assessed by a number of different methods, the most widely used being the zero-trans and equilibrium exchange. Zero-trans is performed by transporting substrate from the outside of the membrane into a compartment without substrate. In these experiments, different concentrations of radioactively labeled substrate are transported over very short time periods such that the accumulation of substrate in the intracellular compartment is negligible. In equilibrium exchange, the concentration of sugar is varied but identical on both sides of the membrane. The cells are then placed in a solution of substrate at the same concentration to which has been added a radiolabeled sugar. Uptake of the radiolabel is then assessed over different periods of time and the data plotted to assess affinity of the transporter for substrate. While equilibrium exchange is an accurate way to assess relative affinity, it is less physiological in most systems. Any accelerated exchange which may occur due to the presence of substrate on both sides of the membrane will influence the calculated affinity. However, most sugars are metabolized after entry into the cell and this phenomena is seen in only specialized cell types (see below). Zero-trans may be more physiologically relevant but suffers from the fact that there is always some exit of substrate from the cell after any substrate transport, thus true zero-trans conditions are not met unless the subsequent metabolism of substrate is rapid and not rate-limiting.

Both zero-trans and equilibrium exchange have been used to assess the affinity of substrates for the facilitative glucose transporters. Table 1 shows the different apparent affinities for the individual transporters for different substrates. GLUT3 has the highest affinity for glucose and other substrates while GLUT2 has a much lower affinity. GLUT1 and GLUT4 have intermediate affinities of about 5 mM for 2-deoxyglucose, near the physiological concentration of glucose. The capacity of the individual transporters, the amount of substrate transported per unit time, is much more difficult to determine because of the lack of accuracy in determining the number of transporters on the cell surface. In general, GLUT2 and GLUT4 have a higher capacity than GLUT1 and GLUT3. The significance of the kinetic properties of the individual transporters will be discussed later.

The structural features of the glucose transporter that determine its affinity for various sugars are poorly understood but arise from the differences in amino acid sequence which may impart differences in secondary or higher structure. One recent study has suggested that the intracellular C-terminus may determine the affinity for

Table 1. Affinities (K_m) of Mammalian Facilitative Glucose Transporters

Isoform	D-glucose	2-Deoxyglucose	D-galactose	D-fructose
GLUT1	5–30 mM	7 mM	17 mM	Not transported
GLUT2	15–66 mM	11–16 mM	36–86 mM	67 mM
GLUT3	1–3 mM	1–2 mM	6–8 mM	Not transported
GLUT4	2.5–5 mM	5 mM	ND	Not transported
GLUT5	Not transported	Not transported	Not transported	6 mM
GLUT7	ND	ND	ND	ND

Note: These values are taken from various literature values. GLUT6 is a pseudogene and does not encode a functional transporter. ND = no data.

the substrate. Oka and coworkers produced a chimeric glucose transporter in which the last 45 amino acids of GLUT1 were replaced with the corresponding region of GLUT2. This chimera had the kinetic properties of GLUT2 with a markedly decreased affinity for glucose and an increase in glucose transport capacity (turnover number) towards that of GLUT2. In contrast, the affinity of the transporter for cytochalasin B was unchanged, suggesting that the carboxy-terminus of the transporter can confer some kinetic properties to the protein.

PHYSIOLOGICAL PROPERTIES OF INDIVIDUAL FACILITATIVE TRANSPORTER ISOFORMS AND ALTERATIONS IN PATHOLOGICAL STATES

GLUT1

GLUT1 is the glucose transporter of erythrocytes and thus much of the pioneering work in the area of glucose transport kinetics was performed on this isoform. In addition to erythrocytes, GLUT1 is expressed in nearly every tissue examined. It appears to play a housekeeping role and is responsible for the basal glucose uptake and is expressed on the plasma membrane of the cell (Figure 2). The K_m of GLUT1 is around 5 mM which makes it ideal for this task in that it is functioning efficiently at physiological glucose concentrations. GLUT1 is also highly expressed in placenta and in brain microvessels, as well as in most other endothelial cells. Thus, GLUT1 also provides a way in which glucose can efficiently traverse blood vessels and the fetal-placental barrier.

It appears that nearly every cultured cell line expresses GLUT1 and in most it is either by far the predominant or only transporter expressed again, attesting to its housekeeping function. The level of GLUT1 in hepatocytes and adipocytes and other cell types placed in primary culture rises markedly while other isoforms decline rapidly, suggesting that GLUT1 is the transporter of undifferentiated cells. When cells are stimulated to divide by growth factors and oncogenes the levels of GLUT1 rise. When lymphocytes are stimulated to divide by cytokines, antibodies to surface receptors or microbial stimuli, glycolysis rises as does the level of GLUT1. Thus, undifferentiation and proliferation are associated with increased GLUT1 levels, both of which are characteristic of tumor cells. Indeed, the levels of GLUT1 are elevated in most tumors and tumor cell lines and may play a role in the so-called Warburg effect in which transformed cells utilize a disproportionate amount of glucose for glycolytic metabolism.

GLUT1 is also a stress induced protein and proteins in this class are preferentially transcribed and translated in response to cellular stress, such as heat shock, essential metabolite deprivation, and other cellular injury from a variety of causes. GLUT1 levels of cells in culture are modulated by the ambient glucose concentrations. When cells are starved for glucose, the levels of GLUT1 increase, while a relative abundance of glucose in the culture medium decreases the amount of GLUT1. The mechanism for this rise appears to be due to both increased transcription of GLUT1 gene and a posttranscriptional increase in protein levels. *In vivo*, GLUT1 is expressed in the liver only in a few cells around the portal vein. When animals are starved, GLUT1 expression is seen in additional cells many layers deep around the central vein. In the brain, hyperglycemia decreases the expression of GLUT1 in the blood-brain barrier which is reversible by lowering the glucose concentration. Hypoxemia and metabolic poisons also increase the level of GLUT1 expression in cultured cells. Thus, GLUT1 plays an important role in the maintenance of cellular function under basal conditions and may be integral to the alterations in glucose metabolism seen in metabolic stress and oncogenic transformation.

GLUT2

The tissue distribution of GLUT2 is limited to the liver, kidney, small intestine, and pancreatic β-cells. GLUT2 is targeted to the basolateral membrane of each of these cells (Figure 2). In hepatocytes, this is the membrane that faces the sinusoids, and in the intestines and kidney, it is the basolateral membrane which faces the antiluminal side of the cell. In the pancreatic β-cells, GLUT2 has been localized to sites of cell-to-cell contact, is coexpressed in every cell which expresses insulin, and is excluded from other hormone producing cells of the islet. By protein blotting studies, the size of GLUT2 protein is different in each one of the tissues which likely reflects differences in the size and composition of the carbohydrate side chain at the glycosylation site between M1 and M2.

GLUT2 has a relatively high K_m (≈ 20 mM) for both glucose influx and efflux, which is ideal given the function of the cells in which it is expressed. The initial movement of glucose from the small intestine lumen into the absorptive epithelial cell is an active process mediated by the Na^+-dependent glucose transporter, SGLT1. The uptake of glucose by SGLT1 results in a relatively high intracellular glucose concentration in the K_m range of GLUT2. GLUT2 in the basolateral membrane then provides the means for a high capacity, unidirectional flux of glucose out of the cell into the interstitial space and subsequently into the portal blood. The low affinity of GLUT2 will limit the amount of glucose which can enter the enterocyte at ambient glucose concentrations.

The portal blood, which is relatively enriched in glucose in this post-prandial state, is drained directly to the liver. GLUT2 on the surface of the hepatocyte then mediates the uptake of glucose. Again, because of its relatively low affinity, the uptake into the hepatocyte will be in direct proportion to the ambient concentration of glucose in the portal blood. Phosphorylation by glucokinase is the subsequent and rate-limiting step to glucose disposal in the hepatocyte. The glucose is converted to glycogen and stored, or converted primarily to fatty acids for subsequent export and storage as fat. In the fasting state, little glucose is taken up by the liver because the portal blood has essentially the same glucose concentration as the peripheral circulation, about 5.5 mM, which is below the concentration for effective uptake by GLUT2. During fasting, glucose is mobilized from glycogen or made *de novo* by conversion of amino acids and lactate. The intracellular glucose concentration rises to a level above circulating levels and exits the hepatocyte via GLUT2.

A number of studies have been carried out to examine the regulation of GLUT2 expression in altered metabolic states. In the liver of rats fasted for 24 hours, the level of GLUT2 mRNA falls about 50% without a change in protein levels. Refeeding results in a fivefold increase in mRNA and a twofold increase in protein levels in these animals. In rats rendered insulinopenic by the β-cell toxin, streptozotocin, there has been reported either no change or a small increase in mRNA levels. Similarly, minimal changes were observed in the hepatocytes of rats infused with insulin for up to 12 days. In the small intestine, basolateral glucose transport and GLUT2 levels are minimally elevated by 72 hours of fasting. In streptozotocin-induced diabetes, there is a marked rise in the basolateral transport of glucose. In parallel, there is a marked increase in the amount of GLUT2 protein expressed in enterocytes. This is due to a rise in GLUT2 mRNA levels and to a precocious activation of GLUT2 transcription. Normally, GLUT2 transcription occurs in enterocytes which have migrated to the mid to upper villus; in streptozotocin diabetes GLUT2 mRNA is observed in the crypt enterocytes. This premature activation of transcription allows for the accumulation of greater amounts of GLUT2 mRNA in the maturing cells which results in increased GLUT2 protein levels. Analogous changes are observed for GLUT5 and SGLT1 mRNAs and proteins in streptozotocin diabetes.

In the kidney, filtered solutes such as glucose are recovered from the forming urine primarily by active transport mechanisms in the renal tubules. As in the small intestine, glucose is removed from the tubule lumen by SGLT1 and exits across the basolateral membrane via GLUT2. The low affinity of GLUT2 makes flow from the blood into the tubule epithelial cells minimal at normal blood glucose concentrations. To date, little information exists on the regulation of GLUT2 in the kidney in pathological states.

One of the more contentious areas of glucose transporter research surrounds the role of GLUT2 in noninsulin dependent diabetes mellitus (NIDDM). In NIDDM there is a disruption in the normal secretion of insulin from the β-cells of the islets of Langerhans as well as a defect in insulin action. The presence of the high-K_m GLUT2 allows the β-cells to precisely regulate insulin secretion in response to glucose, since the rate of glucose uptake changes in direct proportion to the extracellular glucose concentration over the range of normal glucose concentrations (5–10 mM) that stimulate insulin secretion. Unger and colleagues have demonstrated reduced levels of GLUT2 mRNA and protein in several animal models of acquired and genetic NIDDM, as well as in early stages of diabetes in the BB/Wor rats, a model for insulin-dependent diabetes. They have also reported antibodies to GLUT2 in the blood of individuals with insulin-dependent diabetes mellitus. In animal models, the loss of GLUT2 precedes the onset of overt diabetes. They propose that the loss of GLUT2 renders the β-cells less responsive to glucose-stimulated insulin release since the decrease in GLUT2 will result in decreased glucose uptake but will not affect the response of the β-cells nonglucose secretagogue such as arginine. These results have been challenged since GLUT2 is not limiting to glucose metabolism in β-cells. Rather, as in the liver, glucokinase is the rate limiting step in glucose metabolism, and in some alternative animal models, GLUT2 can be diminished without changes in glucose stimulated insulin release. Any alterations in GLUT2 in human NIDDM have not been examined.

GLUT3

The distribution of GLUT3 mRNA is much more ubiquitous in human tissues than in other species such as monkey, rabbits, rats, and mice where it is confined primarily to neural tissue and testis. GLUT3 appears to be localized to the surface of neurons and other cells in which it is expressed (Figure 3). As stated earlier, GLUT1 appears to mediate the flux of glucose across the blood-brain barrier. Since the concentration of glucose in the cerebral spinal fluid is only about 60% of that seen in the peripheral circulation, the high affinity kinetics displayed by GLUT3 ($K_m \approx 1$ mM, Table 1) mediates the efficient transport of glucose into neural tissue. GLUT3 is also expressed in glial tissue of the brain.

Sperm also express very high levels of GLUT3. In humans, it is thought that sperm, while in seminal fluid, utilize fructose and lactate as their primary energy sources. One sperm enters the female reproductive tract where glucose becomes the primary substrate to supply ATP required for mobility and other metabolic events related to fertilization of the egg. Since the glucose concentration in female reproductive tract secretions is circa 1 mM, the high affinity of GLUT3 would allow for the efficient uptake of glucose after entry into the female reproductive tract.

GLUT3 mRNA or protein has also been detected in white blood cells and platelets and the levels of GLUT3 rise in response to lymphocyte mitogens in a manner similar to GLUT1. Since the glucose concentration at sites of inflammation is lowered due to utilization by inflammatory cells, it is beneficial for these blood components to have a high affinity glucose transport system to ensure sufficient uptake of glucose.

GLUT4

Insulin stimulates glucose transport by causing a rapid and reversible increase in the amount of functional glucose transporters on the plasma membrane of skeletal muscle, heart muscle, and white and brown fat. Cloning of GLUT4 and the subsequent development of antibodies to GLUT4 protein have demonstrated the predominance of GLUT4 expression in these tissues. There is a correlation between the amount of GLUT4 expressed in a cell and the amount of insulin-stimulated glucose uptake. Adipose tissue and heart muscle have a higher level of GLUT4 than red or white skeletal muscle (per gram of tissue) and also have a higher level of insulin-stimulated glucose transport.

A model for the stimulation of glucose transport by insulin was proposed independently by Cushman and Kono. In this model, insulin stimulates the translocation of glucose transporter (GLUT4) containing vesicles from an intracellular site to the plasma membrane of the cell (Figure 2). In fat cells, the translocation event is rapid, occurring within seconds of insulin binding to its receptor, and reversible with the decay occurring within minutes of insulin disassociating from its receptor. There is an approximately 20–40-fold increase in the amount of GLUT4 on the plasma membrane in insulin stimulated cells. Once on the cell surface, GLUT4 can mediate the uptake of large quantities of glucose due to its relatively low K_m and high capacity for transport (Table 1). The signalling pathway and specific components involved in insulin stimulation of this translocation event are not known. There is evidence that the GLUT4 containing vesicles are continuously cycling from the cytoplasm to the cell surface. The effect of insulin is to decrease the internalization rate while not affecting the externalization event. Other experiments have suggested the presence of a unique "small GTP" binding protein in insulin-responsive tissues which participates in the fusion of the intracellular GLUT4 containing vesicles with the plasma membrane in much the same way as other well described vesicular trafficking activities. In addition to this translocation mechanism, there is evidence that some of the increase in glucose transport seen after insulin stimulation may be due to an intrinsic activation of glucose transporters which reside on the cell surface.

When GLUT4 is expressed in cell types which do not usually express this protein, there is no evidence of translocation of GLUT4 to the cell surface, despite the fact that such cells have an adequate number of insulin receptors and can otherwise respond appropriately to insulin. Thus, it is not the mere presence of GLUT4 within the cell which confers insulin sensitivity, rather other parts of the signal transduction pathway are needed in addition to GLUT4.

Both GLUT4 and GLUT1 are expressed in adipocytes but their subcellular distribution differs importantly. In the basal state, the majority (60–70%) of GLUT1 resides on the surface of the cell, while GLUT4 is almost entirely intracellular. That fraction of GLUT1 which is intracellular likely resides in a vesicular population which is distinct from GLUT4 containing vesicles. A number of investigators have tried to understand the structural basis for this alternative targeting. James and colleagues have proposed that a sequence within the first 14 amino acids of GLUT4 confers intracellular sequestration of GLUT4 when expressed in fibroblast cell lines. However, other investigators have suggested sequences in the amino terminus are important for intracellular sequestration. Despite these differences, it is evident that sequence information within GLUT4 itself determines its proper subcellular localization. The identification of the targeting sequence is important since it may provide clues as to the other protein components of the GLUT4 containing vesicles and molecular mechanism whereby insulin stimulates glucose uptake.

Resistance to the action of insulin is a hallmark of many pathophysiological states such as fasting, diabetes, and skeletal muscle denervation. In fasted rats, both adipose tissue and skeletal muscle become resistant to insulin stimulation of glucose transport. In adipose tissue, this may be in part due to a decrease in GLUT4 mRNA and protein content. However, in skeletal muscle there is an increase in GLUT4 protein even though there is insulin resistance. In this case there is a disruption of coupling of the insulin receptor to intracellular events. Likewise, in streptozotocin diabetes there is a marked decline in the levels of GLUT4 in adipose tissue but GLUT4 levels only decline in skeletal muscle after prolonged diabetes and long after there is profound resistance to stimulation of glucose transport in this tissue.

Denervation of skeletal muscle also results in profound defects in glucose transport and glycogen synthesis which occurs within 24 hr of sectioning the nerve. As in the skeletal muscle of starved and diabetic rats, the insulin resistance precedes any detectable changes in GLUT4 levels. On the other hand, exercise has a marked ability to increase glucose transport in the absence of insulin. Chronic exercise training has a minimal effect on the levels of GLUT4 but there is evidence that training may increase the intrinsic activity of the GLUT4 protein in the plasmalemma.

In adult human subjects with NIDDM, adipose tissue is insulin resistant and there is a striking decrease in GLUT4 levels. As in rat models of NIDDM, human skeletal muscle is also insulin resistant. However, there is no change in the amount of GLUT4. These results show that the content of GLUT4 in a tissue itself is not a predictor of the ability of the tissue to respond to insulin by increasing glucose transport. Since normally up to 80% of glucose disposal occurs in skeletal muscle tissue, altered expression of GLUT4 cannot account for the insulin resistance seen in diabetes and other insulin resistant states.

GLUT5

Early studies with GLUT5 showed that it transported glucose either poorly or not at all. GLUT5 is expressed at high levels on the brush border membrane of small intestinal enterocytes, suggesting that it is involved in the uptake of sugars of some sort (Figure 2). Of the major dietary sugars, glucose, galactose and fructose, fructose was the best candidate since uptake of this sugar was not impaired in subjects with glucose-galactose malabsorption syndrome, a recessive disorder due to mutations in SGLT1. Subsequently, it was found that GLUT5 expressed in *Xenopus* oocytes could mediate the transport of fructose with high affinity. GLUT2 also has fructose transport activity but with lower affinity than GLUT5 and likely mediates the exit of fructose from the enterocyte.

Fructose is utilized by human sperm in the male reproductive tract and the high levels of GLUT5 in maturing spermatids and sperm mediate this uptake. GLUT5 in the kidney likely plays the same function as it does in the small intestine, mediating the uptake of fructose from the lumen of the renal tubule. There are also high levels of GLUT5 in the brain where its function is unknown, since the brain utilizes little fructose. This raises the possibility of additional substrates for GLUT5.

In rats, it has been observed that feeding diets enriched in fructose results in a rise in the fructose transport activity of the small intestine with only a minimal change in glucose transport activity. Rats fed a diet with 60% of calories as fructose show a marked up-regulation of GLUT5 levels in the small intestine within 24 hours of introduction of the diet. Interestingly, GLUT5 levels also rise in the kidney. In both cases there is either no or only a minimal increase in the amount of GLUT5 mRNA, suggesting regulation at the level of translation or protein stability.

GLUT6

GLUT6 is a sequence found in a ubiquitously expressed human mRNA. The sequence is 80% homologous with human GLUT3. The coding region of GLUT6 has numerous stop codons and frame shifts and cannot be translated into a functional protein. It likely arose from an insertion of a reverse transcribed pseudogene into the non-coding region of the carrier mRNA.

GLUT7

This transporter is the newest member of the family of glucose transporters. Its amino half is highly homologous to GLUT2 while the carboxy half is more divergent. It is proposed that this transporter is expressed in the endoplasmic reticulum in association with glucose-6 phosphatase and mediates the exit of glucose from the endoplasmic reticulum to the cytoplasm for subsequent export (Figure 2). The kinetic properties or alteration in pathological states has not been elucidated.

SUMMARY

The precise regulation of absorption and distribution of glucose and other sugars in mammals is carried out by two families of proteins, the Na^+-dependent glucose transporter and the family of facilitative glucose transporters. The six functional members of the facilitative transporters, GLUT1 through GLUT5 and GLUT7 have distinct but overlapping tissue distributions and are targeted to distinct subcellular locales in the cell types in which they are expressed. The distinct kinetic properties and substrate selectivity of each of the transporters have evolved such that efficient fuel delivery to appropriate tissues is ensured under a wide variety of metabolic conditions.

The primary, secondary, and presumably tertiary, and higher, structure of the family of facilitative glucose transporters are similar. It is beginning to be understood which domains of these transporters are important for substrate and inhibitor binding and for subcellular localization. Future work will undoubtably uncover the molecular mechanism by which the facilitative transporters catalyze the translocation of sugars across the membrane.

Also becoming clear is the mechanism by which the levels and activity of the glucose transporters change in response to altered homeostasis, such as insulin-dependent and noninsulin-dependent diabetes mellitus and other insulin resistant states. It will be appreciated which changes in transporter levels are a proximal event in altering normal physiology, such as changes in GLUT2 in the β-cell of the islets of Langerhans or in the small intestine with alternative diets, and which are due to the altered metabolic states such as diabetes or uremia but could exacerbate the pathophysiology seen.

ACKNOWLEDGMENTS

I would like to thank Dr. Graeme Bell and Janet Chen for helpful comments on this manuscript and Will Chutkow for the figures.

RECOMMENDED READINGS

Bell, G.I., Kayano, T., Buse, J.B., Burant, C.F., Takeda, J., Lin, D., Fukimoto, H., & Seino, S. (1990). Molecular biology of mammalian glucose transporters. Diabetes Care 13, 198–208.

Burant, C.F., Sivitz, W.I., Fukumoto, H., Kayano, T., Nagamatsu, S., Seino, S., Pessin, J.E., & Bell, G.I. (1992). Mammalian glucose transporters: Structure and molecular regulation. Recent Prog. Hormone Res. 47, 349–387.

Carruthers, A. (1990). Facilitative diffusion of glucose. Physiol. Rev. 70, 1135–1176.

Kahn, B.B. & Flier, J.S. (1990). Regulation of glucose-transporter gene expression in vitro and in vivo. Diabetes Care 13, 198–208.

Kahn, B.B. (1992). Facilitative glucose transporters: Regulatory mechanisms and dysregulation in diabetes. J. Clin Invest. 89, 1367–1374.

Mueckler, M. (1990). Family of glucose-transporter genes: Implications for glucose homeostasis and diabetes. Diabetes 39, 6–11.

Simpson, I.A. & Cushman, S.W. (1989). Hormonal regulation of mammalian glucose transport. Ann. Rev. Biochem. 55, 1059–1089.

Unger, R.H. (1991). Diabetic hyperglycemia: Link to impaired glucose transport in pancreatic β cells. Science 251, 1200–1205.

Wright, E.W., Turk, E., Zabel, B., Mundios, S., & Dyer, J. (1991). Molecular genetics of intestinal glucose transport. J. Clin. Invest. 88, 1435–1440.

Chapter 3

Cation-Coupled Transport

ROSE M. JOHNSTONE and JOHN I. McCORMICK

Principles of Medical Biology, Volume 4
Cell Chemistry and Physiology: Part III, pages 87–127.
Copyright © 1996 by JAI Press Inc.
All rights of reproduction in any form reserved.
ISBN: 1-55938-807-2

HISTORICAL INTRODUCTION

Background

It is now eight decades since experimental evidence was obtained that mammalian cells are capable of extensive accumulation of amino acids (Van Slyke and Meyer, 1913). Shortly thereafter it became apparent that many other low molecular weight solutes were also accumulated in a variety of tissues. However, another 50 years were to pass before the mechanism of energy transduction between primary metabolic events and osmotic work was recognized and accepted.

In the early studies on monosaccharide transport (Wilbrandt and Laszt, 1933; Verzár and Sullmann, 1937) across the small intestine, energy transduction appeared to be associated with substrate phosphorylation. Since metabolizable sugars, capable of undergoing phosphorylation, were tested, it is not surprising that the internalized monosaccharides were found as phosphorylated derivatives. When it was established that non-metabolizable monosaccharides were accumulated against a gradient (Crane and Krane, 1956; Crane, 1960), compulsory phosphorylation ceased to be a viable mechanism for energy dependent accumulation of sugars.

That metabolic changes may influence the interpretation of data from studies on the transport of readily metabolizable compounds highlights the importance of using "gratuitous substrates," which undergo translocation but are resistant to (or incapable of) metabolic change. The use of gratuitous substrates has now become well accepted, although clearly not always feasible experimentally.

Similarly, it was the introduction of synthetic, nonmetabolizable amino acids for transport studies (Noall et al., 1957; Berlinguet et al., 1962) which eliminated the question of formation of Schiff base "transport intermediates" with pyridoxal phosphate.

By the early 1960s it was accepted that energy dependent transport and accumulation could occur independently of metabolic alteration of substrate.

Na$^+$-Dependent Transport: A General Mechanism for Energization of Transport

Like many seminal observations, the special place of Na$^+$ in energy dependent accumulation of solutes by mammalian cells had been observed long before its significance was recognized. Both Riggs et al. (1958) and Ricklis and Quastel (1958) had studied the effects of interchanging K$^+$ for Na$^+$ and vice versa on amino acid and monosaccharide transport, respectively, in different tissues. Because attention was then focused on K$^+$ (whose biological effects were much better understood), the response to the absence of Na$^+$ was initially interpreted as due to excess K$^+$ (Riggs et al., 1958; Ricklis and Quastel, 1958).

The demonstration by Crane (1960, 1965) that Na^+ ions were essential for the translocation of monosaccharides by segments of the intestine brought in a new era of understanding of the central role of ion coupled transport, particularly in higher organisms. While Na^+ is clearly the predominant cation involved in cation driven solute accumulation in mammalian systems, current work has provided examples of H^+ driven solute transport in intestine and kidney (Jessen et al., 1989; Ganapathy and Leibach, 1986). Conversely, in yeast and bacteria, H^+ driven mechanisms are in the majority (Seaston et al., 1973; Hirata et al., 1973), but examples of Na^+-coupled fluxes exist, e.g., proline transport (Dibrov, 1991).

The obligatory Na^+ requirement to accomplish osmotic work in intestinal epithelium led Crane (1965) to propose that the Na^+ gradient across mammalian cell membranes is the actual driving force pulling the solute uphill as Na^+ moved downhill into the cell. Any solute flow tightly coupled to the downhill, exergonic flow of the Na^+ would be moved uphill against its own gradient. The maximum accumulation of the solute at steady state would be the inverse of the Na^+ gradient at steady state, if the degree of coupling and efficiency of transduction were 100%. That is, for a coupling ratio of 1:1 between Na^+ and solute:

$$RT \ln \frac{[Na^+]_o}{[Na^+]_i} \geq RT \ln \frac{[Solute]_i}{[Solute]_o}$$

where R is the universal gas constant and T is absolute temperature.

In general the experimental work has substantiated this hypothesis, yet initial attempts to obtain quantitative support for the proposal fell short of the mark. In many cells, including Ehrlich cells and chicken intestinal mucosa, which were capable of accumulating the driven solute 10 times or more over the medium concentration, the estimates of the driving force fell short of the energy required (Reid and Eddy, 1971; Schafer and Heinz, 1971; Johnstone, 1972).

Although a number of explanations were offered to account for the shortfall in energy, Vidaver (1964a,b) was one of the first to recognize that the membrane potential was a key element in assessing the true magnitude of the available driving force in the Na^+ gradient. If the ternary carrier complex (protein–Na^+-solute) bore a net positive charge, then the driving force $\Delta \tilde{\mu}_{Na^+}$ of the sodium gradient to which solute flow responded should be expressed as the sum of the electrical and chemical potentials of Na^+. This is given by

$$RT \ln \frac{[S]_i}{[S]_0} \leq \left(RT \ln \frac{[Na]_0}{[Na]_i} + VF \right)^n Na^+$$

where n represents the coupling ratio of Na^+/solute, R is the gas constant [8.314 joules degree^{-1}mole^{-1}], T is the absolute temperature, V is the potential difference in volts across the membrane and F is the Faraday constant (96,500 coulombs per mole). For example, in a 1:1 coupling of Na^+/solute, a Na^+ concentration gradient

Table 1. Diversity of Na⁺-Coupled Transport in Animal Cells and Tissues

Substances Transported by Na⁺ Dependent Routes

Substance	Reference
I. General Reviews on Na⁺-Coupled Solute Transport: Sugars and Amino Acids	
Amino acids	Barker & Ellory (1990). Experimental Physiol. 75, 3–26.
	Lerner (1987). Comp. Biochem. Physiol. 87B, 443–457.
	Saier et al. (1988). J. Membr. Biol. 104, 1–20.
Sugars	Kimmich (1990). J. Membr. Biol. 114, 1–27.
	Wright (1993). Ann. Rev. Physiol. 55, 575–589.
General	Christensen (1990). Ann. Rev. Physiol. 70, 43–77.
	Crane (1977). Rev. Physiol. Biochem. and Pharmacol. 78, 99–159.
	Heinz (1978). In: Mechanisms and Energetics of Biological Transport, pp. 120–140, Springer Verlag, Berlin.
	Johnstone (1990). Curr. Opin. in Cell. Biol. 235, 735–741.
	Schultz (1986). In: Physiology of Membrane Disorders (Andreoli, T.E., Hoffman, J.F., Fanestil, D.D., & Schultz, S.G., eds.), pp. 283–294, Plenum Press, New York.
	Stein (1986). In: Transport and Diffusion Across Cell Membranes, Academic Press, New York.

i. Atypical Amino Acids: Emphasis on Nonneural Tissue

	Tissue/System	Reference
β-alanine	Ehrlich cells	Christensen et al. (1954). Cancer Res. 14, 124–127;
		Lambert and Hoffman (1993). J. Membr. Biol. 131, 67–79.
	BBV kidney	Hammerman and Sacktor (1978). Biochim. Biophys. Acta 509, 338–347.
		Turner (1986). J. Biol. Chem. 261, 16060–16066.
	BBV intestine	Miyamoto et al. (1990). Am. J. Physiol. 259, G372–G379.
Betaine	Hamster small intestine	Hagihara et al. (1962). Am. J. Physiol. 203, 637–640.
	Rat kidney	Sung & Johnstone (1969). Biochim. Biophys. Acta 173, 548–553.
Diaminobutyrate	Ehrlich cells	Christensen (1992). J. Membr. Biol. 127, 1–7.
γ-aminobutyrate	Several tissues	Erdo & Wolff (1990). J. Neurochem. 54, 363–372.
	Rat kidney	Goodyer et al. (1985). Biochim. Biophys. Acta 818, 45–54.
p-aminohippurate	Rabbit kidney (C & V)	Sheikh & Molle (1992). Biochem. J. 208, 243–246.
Taurine	Ehrlich cells	Christensen et al. (1954). Cancer Res. 14, 124–127;
		Lambert & Hoffman (1993). J. Membr. Biol. 131, 67–79.
	Kidney	Jessen & Sheikh (1991). Biochim. Biophys. Acta 1064, 189–198.
	Human placenta BBV	Miyamoto et al. (1988). FEBS Lett. 231, 263–267.

(continued)

Table 1. (continued)

Substance	Tissue/System	Reference
ii. Vitamins and Related Substances		
Ascorbate	Adrenal cortex; anterior pituitary	Clayman et al. (1970). Biochem. J. 118, 283–289.
		Sharma et al. (1964). Biochem. J. 92, 564–573.
	Rat adrenal chromaffin cells	Diliberto et al. (1983). J. Biol. Chem. 258, 12886–12894.
	Pancreas	Zhou et al. (1991). Biochem. J. 274, 739–744.
Biotin	Liver BLV	Said et al. (1992). Gastroenter. 102, 2120–2125.
Folate	Jejunum	Strum (1979). Biochim. Biophys. Acta 554, 249–257.
Inositol	Chicken reticulocytes	Isaacks et al. (1989). Arch. Biochem. Biophys. 274, 564–573.
	Ehrlich cells	Johnstone et al. (1967). Biochim. Biophys. Acta 135, 1052–1055.
	Glomerular cells	Whiteside et al. (1991). Am. J. Physiol. 260, F138–F144.
Pantothenate	BBV	Grassl (1992). J. Biol. Chem. 267, 22902–22906.
Pyridoxine	Kidney	Bowman et al. (1990). Ann. NY Acad. Sci. 585, 106–109.
Riboflavin	Intestine	Middleton (1990). J. Nutr. 120, 588–593.

IIA. Reviews on Biogenic Amines and Neurotransmitters

Reference
Amara & Kuhar (1993). Ann. Rev. Neurosci. 16, 73–93.
Graham & Langer (1992). Life Sci. 51, 631–645.
Horne (1990). Progress in Neurobiol. 34, 387–400.
Kanner (1983). Biochim. Biophys. Acta 726, 293–316.
Kanner & Schuldiner (1987). CRC Crit. Revs. Biochem. 22, 1–38.

IIB. Amines—Emphasis on Nonneural Tissues (see IIA for Reviews in Neural Systems)

	Tissue/System	Reference
Carnitine	Heart	Bohmer et al. (1977). Biochim. Biophys. Acta 465, 627–633.
	Kidney cortex	Huth & Shug (1980). Biochim. Biophys. Acta 602, 621–634.
Choline	V. electric organ plasma membrane	Ducis & Whittaker (1985). Biochim. Biophys. Acta 815, 109–127.
	V. lung derived cells	Fisher et al. (1992). Amer. J. Physiol. 263, C1250–1267.
	Kidney cortex	Sung & Johnstone (1965). Can. J. Biochem. 48, 1111–1118.
Creatine	Human cell cultures	Daly & Seifter (1980). Arch. Biochem. Biophys. 203, 317–324.
	Brain & leukemia cells	Guimbal & Kilimann (1993). J. Biol. Chem. 268, 8418–8421.
	Muscle cells	Loike et al. (1988). Proc. Natl. Acad. Sci. USA 85, 807–811.
Dopamine	BBV	Ramamoorthy et al. (1992). Am. J. Physiol. 262, C1189–C1196.
Hydroxytryptamine	Astrocytes	Kimelberg & Katz (1985). Science 228, 889–891.

(continued)

Hydroxytryptamine Astrocytes Kimelberg & Katz (1985). Science 228, 889–891.

Table 1. (continued)

Substance	Tissue/System	Reference
	V. cerebral cortex	O'Reilly & Reith (1988). J. Biol. Chem. 263, 6115–6121.
	V. platelets	Rudick et al. (1989). J. Biol. Chem. 264, 14865–14868.
	BBV	Sugawara et al. (1992). Biochim. Biophys. Acta 1111, 145–150.
Norepinephrine	Astrocytes	Kimelberg & Godine (1993). Brain Research 602, 41–44.
Polyamines (spermadine, etc.)	BALB 3T3 cells	Khan et al. (1990). Cell. Molec. Biology 36, 345–348.
	B16 melanoma	Minchin et al. (1991). Eur. J. Biochem. 200, 457–462.
	LLC-PK cells	Van den Bosch et al. (1990). Biochem. J. 265, 609–612.
III. Inorganic Ions		
Chloride (NaKCl)	Various	O'Grady et al. (1987). Am. J. Physiol. 253, C177–C192.
	Mucous cell acini	Zhang et al. (1993). Am. J. Physiol. 264, C54–C62.
Iodide	Thyroid cells	Bagchi & Fawcett (1973). Biochim. Biophys. Acta 318, 235–251.
	V. thyroid gland	Goldstein et al. (1992). Am. J. Physiol. 263, C590–C597.
		O'Neill et al. (1987). Biochim. Biophys. Acta 896, 263–274.
	Follicle cells	Nilsson et al. (1990). Eur. J. Cell Biol. 52, 270–281.
	Expression cloning	Vilijn & Carrasco (1989). J. Biol. Chem. 264, 11901–11903.
Phosphate	V. sarcolemma	Jack et al. (1987). J. Biol. Chem. 264, 3904–3908.
	BBV kidney	Peerce (1989). Am. J. Physiol. 256, G645–G652.
		Pratt & Petersen (1989). Arch. Biochem. Biophys. 268, 9–19.
	Kidney	Quamme et al. (1989). Am. J. Physiol. 257, F967–F973.
Sulfate	BBV kidney	Tenenhouse et al. (1991). Am. J. Physiol. 261, F420–F426.
		Turner (1984). Am. J. Physiol. 247, F793–F798.
	Liver: cloning	Palacin et al. (1990). J. Biol. Chem. 265, 7142–7144.
IV. Organic Acids		
Bile acids	Hepatocytes	Ananthanarayanan et al. (1988). J. Biol. Chem. 263, 8338–8343.
	Liver	Anwer (1990). Hepatology 12, 1248–1249.
		Hagenbuch et al. (1991). Proc. Natl. Acad. Sci. USA 88, 10629–10633.
Citrate & TCA cycle intermediates	BBV kidney	Wright (1985). Ann. Rev. Physiol. 47, 127–141.

(continued)

Table 1. (continued)

Substance	Tissue/System	Reference
	Chick intestine	Kimmich et al. (1991). Am. J. Physiol. 260, C1151–C1157.
Fatty acids	BLV liver	Stremmel (1987). J. Biol. Chem. 262, 6284–6289.
Lactate	BBV kidney	Stein (1986). Transport and Diffusion Across Cell Membranes. Academic Press, New York.

V. Nucleosides

Adenosine	Liver	Betcher et al. (1990). Am. J. Physiol. 259, G504–G510.
Nucleosides	L1210 cells	Crawford et al. (1990). J. Biol. Chem. 265, 13730–13734.
		Dagnino et al. (1991). J. Biol. Chem. 266, 6308–6311.
	Lymphoma	Plagemann (1991). J. Cell Biochem. 46, 54–59.
	Splenocytes	Darnowski et al. (1987). Cancer Res. 47, 2614–2619.
Purines and pyrimidines	BBV kidney	Williams and Jarvis (1990). Biochem. Soc. Trans. 18, 684–685.

VI. H$^+$-Coupled Transport in Higher Organisms

β alanine	V. kidney	Jessen et al. (1991). Journal of Physiology 436, 149–167.
L-leucine	Chang cells	Mitsumoto & Mohri (1991). Biochim. Biophys. Acta 1061, 171–174.
Peptides	Several	Webb et al. (1992). J. Anim. Sci. 70, 3248–3257.
a:(Dipeptides)	Chick erythrocytes	Calonge et al. (1990). Am. J. Physiol. 259, G775–G780.
	Caco cells	Thwaites et al. (1993). J. Biol. Chem. 268, 7640–7642.
	BBV kidney	Daniel et al. (1991). J. Biol. Chem. 266, 19917–19924.
		Skopicki et al. (1991). Am. J. Physiol. 261, F670–F678.
		Tiruppathi et al. (1990). J. Biol. Chem. 265, 14870–14874.
	BBV rabbit intestine	Okano et al. (1986). J. Biol. Chem. 261, 14130–14134.
b:Tripeptides	BBV kidney	Tiruppathi et al. (1990). Biochem. J. 268, 27–33.
5-Methyl tetrahydro-folate	BLV liver	Horne et al. (1992). Amer. J. Physiol. 262, G150–G158.
	Hepatocytes	Horne (1990). Biochim. Biophys. Acta 1023, 47–55.
Urea	Frog skin	Rapoport et al. (1989). Am. J. Physiol. 256, F830–F835.

Note: Abbreviations: BBV, brush-border vesicles; BLV, basolateral membrane vesicles; C, cells; V, vesicles.

of 5 ($[Na^+]_o = 150\,mM$, $[Na^+]_i = 30\,mM$) in addition to a potential difference of 40 mV (inside negative) would provide a driving force equivalent to ~80 mV, sufficient to establish a maximum solute gradient of ~25-fold. Since the available energy in the Na^+ gradient rises exponentially when the number of cotransported Na^+ ions is greater than one, very large accumulations of transported solutes become energetically feasible. For most systems, the inclusion of the estimated membrane potential provided a sufficient driving force to account for the experimentally observed accumulations. Na^+-coupled transport is now accepted as an example of secondary active transport, i.e., transport coupled to ion gradients, established by a primary metabolic pump, such as Na^+/K^+-ATPase, whose activity in turn depends on a primary metabolic energy source.

Although pumps have been described which utilize ATP directly and bring about ATP hydrolysis in moving specific organic solutes across the membrane (Higgins, 1992), the vast majority of solutes in mammals which are actively transported (i.e., accumulated free in the cell against a concentration gradient), appear to function by transport coupled to Na^+ (or H^+). In these systems, investigators agree that the transporter forms a ternary complex with Na^+ and solute and that the two substrates simultaneously translocate. Table 1 presents the variety of solutes, both organic and inorganic, which are cotransported with Na^+ (or H^+) in animal tissues of many species.

This list is not exhaustive or complete, the intention being to summarize the variety of substances and systems which have been studied. Because practically every natural amino acid has been shown to undergo Na^+-coupled transport in multiple tissues, the common amino acids have not been included in the table. A few reviews are cited dealing with Na^+-coupled general amino acid transport, sugar transport, and energization. It is not an exaggeration to state that most water soluble, organic solutes are transported by this type of mechanism.

The objective of this chapter is to highlight some of the diversification and characteristics of Na^+- (or H^+-) coupled transporters. Additionally the current status of work in transporter reconstitution, purification, and cloning is summarized to gain insight into the transporters' structures at the molecular level.

Na^+-COUPLED TRANSPORTERS

Na^+ Specificity of Coupled Transport

Most cation coupled carriers recognize a class of substances, rather than a unique substance, i.e., incomplete specificity. While some transporters may translocate only a single species (e.g., glycine transporter of pigeon red cells) (Vidaver, 1964a) the majority of transporters show limited substrate specificity. For example, at present it seems that a single amino acid transporter may recognize a dozen or so α amino α carboxylic acids (Christensen et al., 1965). In contrast, the cation specificity for energized coupling is very high. Among the alkali cations, only Li^+

has been found to substitute for Na⁺, with significantly lower efficiency (Christensen et al., 1967).

In mammalian systems, some examples of solute transport coupled to either H^+ or Na^+ flow are known (Jessen et al., 1989), yet most Na^+-dependent systems do not appear to substitute H^+ for Na^+. Until characteristic consensus sequences for Na^+ and H^+ binding are known, the difference between highly specific Na^+ coupling and preferential Na^+ coupling remain unclear.

Two hypotheses have been proposed (Eisenman, 1962; Mullins, 1975) to account for the high Na^+/K^+ selectivity in Na^+ dependent transport systems. Mullins emphasizes a geometrical fit of the cation. Eisenman emphasizes the electrical field strength of the putative binding sites. Many of the predictions made by Eisenman have been borne out experimentally in the ion-coupled systems and are complemented by the construction of ion-specific glass electrodes (Eisenman, 1962).

Characteristics of Na⁺-Coupled Transporters

All models for Na^+-coupled transport contain the following concepts:

1. Both Na^+ and solute combine with a common membrane component (carrier, C) to form a ternary complex with simultaneous translocation of both Na^+ and the coupled solute (S).
2. This combination changes either or both the maximum velocity, V_{max}, and Michaelis constant, K_s, for both solutes.
3. The mobility of the ternary complex across the membrane is greater than that of either binary complex (i.e., $V_{1\rightarrow2}$ of CNaS > CNa ≡ CS < C).
4. While under physiological conditions, the driver molecule is usually Na^+ (in Na^+-coupled flows), the driven solute should enhance the flow of the "normal" driver and bring about its uphill movement, if the experimental conditions are appropriately modified. For example, if the energy from the Na^+ electrochemical gradient, i.e., $\Delta\tilde{\mu}_{Na^+} = 0$, but $\Delta\tilde{\mu}_{solute\,i\rightarrow0}$ is large the outward (downhill) movement of solute from the cell will drive cell Na^+ uphill into the medium.

The diagram shown in Figure 1 summarizes the basic precepts of secondary active transport systems including: (a) binding at cis side, (b) translocation, and (c) release at trans side and relocation of unloaded carrier. Only the fully loaded and the empty carriers are presumed to translocate rapidly. Since the system is reversible, the same characteristics apply at the cis and trans sides of the membrane. Although the numerical values of the K_S and V_{max} may differ at the two surfaces (asymmetric system) even in the absence of electrical and chemical gradients for either solute, the ratios of V_{max}/K_s at both sides of the membrane will be equivalent.

It is evident that the experimentally observed level of accumulation of the driven molecule, when net flow is zero (static head), will depend on several factors,

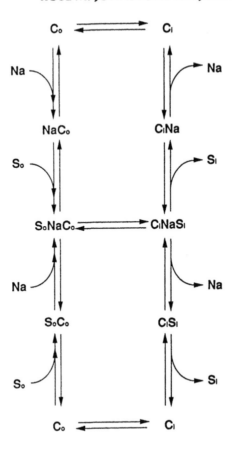

Figure 1. A theoretical model for Na^+-coupled solute transport. The model assumes a reversible system where only two forms of the carrier are mobile, the empty carrier, C, and the ternary complex, $CSNa^+$. Neither CS nor CNa^+ is mobile in either direction so that in absence of Na^+ there is no translocation of S. In this model there is random binding of Na^+ and S. The net direction of movement will be dictated by the direction of the driving forces. The external and internal milieux are represented by the symbols o and i, respectively.

including the degree of coupling to the driver flow, the backflows through the carrier mechanism itself (reversibility), and nonspecific leaks through other pathways, as well as the rate of the unidirectional inflow. In polarized tissues, such as intestinal mucosa or kidney tubules, the segregation of Na^+-coupled and facilitated transporters at different faces of the cell membrane (e.g., mucosal and serosal) may lead to substantial underrepresentation of the efficiency of the cells' capacity to transduce Na^+ gradient energy into accumulation. Kimmich et al. (1977) showed that in chicken intestinal cells, the peak of methyl glucose accumulation was increased severalfold when transport at the basolateral membrane was inhibited. Experimental data with biological systems have shown very high degrees of coupling (~90%) (Kimmich, 1981; Heinz, 1978; Stein, 1986).

Considerable experimental evidence exists to support the four basic tenets of the model outlined above. In their detailed theoretical analysis, Jauch and Lauger (1986); Jauch et al. (1986); and Parent et al. (1992a,b) rigorously showed that Na^+ and amino acids (or other solutes) must be simultaneously cotranslocated on a common carrier. Moreover, it can be demonstrated that Na^+ and the cotransported solute are simultaneously required to protect against transport inactivation of the carrier by sulfyhdryl agents in solubilized membranes (Dudeck et al., 1987; McCormick and Johnstone, 1990). These protection studies further provide direct evidence that Na^+ and cotransported solute bind to a common carrier.

That Na^+ influences either V_{max} or K_s of the cotransported solute has been thoroughly documented (Curran et al., 1967; Goldner et al., 1969; Kimmich, 1981; Stein, 1986; Birnir et al., 1991; Parent et al., 1992b). Further, there is a reciprocal capacity of either Na^+ or coupled solute to drive the cotransported substance against its own electrochemical gradient (Hajjar et al., 1970; Curran et al., 1970; see also Heinz, 1978 and Stein, 1986 for theoretical treatments).

Membrane Potentials and Secondary Active Transport

In retrospect, early difficulties with the general acceptance of the idea that the energy source from the Na^+-gradient was adequate to bring about accumulation of an electrically neutral solute against its concentration gradient stemmed from: (a) a failure to recognize that the cotransport of an electrically neutral substance was frequently electrogenic, and (b) from an underestimation of the contribution of the potential difference to the total driving force of the Na^+-gradient.

Two experimental approaches were instrumental in arriving at the near universal acceptance that $\Delta\tilde{\mu}_{Na^+}$ was necessary and sufficient to account for the observed accumulation of cotransported solutes. These are: (a) the introduction of potential sensing dyes (Hoffman and Laris, 1974; Waggoner, 1976) to monitor potential changes during coupled transport, and (b) the use of a variety of plasma membrane vesicles incapable of producing metabolic energy (Colombini and Johnstone, 1974; Murer and Hopfer, 1974; Sigrist-Nelson et al., 1975). The representative data in Figures 2 and 3 show examples of the two types of studies.

In Figure 2, the results show that the membrane potential changes during Na^+-coupled neutral amino acid uptake in Ehrlich cells (Laris et al., 1978), the cells depolarizing when amino acids are taken up and hyperpolarizing when amino acids are lost by Na^+-dependent routes. The fluorescent dye technique permitted parallel assessment of the potential difference and solute accumulation and established that $\Delta\tilde{\mu}_{Na^+}$ could provide adequate energy for 10–30-fold solute accumulation (Gibb and

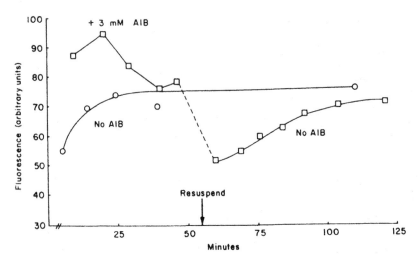

Figure 2. Change in membrane potential during uptake and efflux of a neutral amino acid. Ehrlich cells incubated with amino isobutyric acid (AIB) show a transient increase in fluorescence (depolarization) followed by a return to control levels. Upon resuspension (arrow) in AIB-free medium (which results in loss of cellular AIB) fluorescence is decreased transiently (hyperpolarization) followed by a gradual return to baseline values. (From Laris et al. (1978), with permission.)

Eddy, 1972; Kimmich et al., 1977, 1991; Laris et al., 1978). Earlier estimates of $\Delta\tilde{\mu}_{Na^+}$ had failed to show sufficient driving force for high (>10-fold) accumulation.

The results in Figure 3, obtained with membrane vesicles, show that imposition of a membrane potential difference (inside negative) greatly enhanced the peak uptake of a solute taken up by a Na^+-dependent route. The potential difference could be increased by using lipid soluble anions or valinomycin in vesicles with $K_i^+ \gg K_o^+$ (Colombini and Johnstone, 1974; Murer and Hopfer, 1974; Sigrist-Nelson et al., 1975; Lever, 1977; Hammerman and Sacktor, 1978; Hopfer, 1978; Wright et al., 1983; Kimmich et al., 1991).

Despite the fact that the fluorescent dyes had many unwanted characteristics, such as interference with mitochondrial ATP production, they were highly useful for the determination of the membrane potential, $\Delta\psi$, in cells capable of maintaining ATP levels by glycolytic means. Lipophilic cations have also been used to assess potentials in small cells and permitted concomitant measurement of a potential difference and solute accumulation (Skulachev, 1971; Schuldiner and Kaback, 1975; Lever, 1977).

Currently, patch clamping (Jauch and Lauger, 1986; Birnir et al., 1991; Parent et al., 1992a,b) has all but replaced lipophilic dyes and lipophilic cations as the method of choice in measuring membrane potential. The latter studies also address the site of action of the membrane potential in coupled transport and show that the

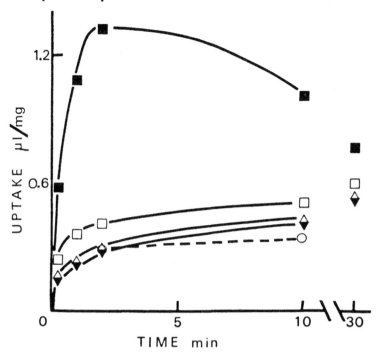

Figure 3. Na⁺ gradient stimulated AIB uptake. Vesicles preincubated in 100 mM KCl or NaCl were transferred to media containing 0.4 mM ^{14}C-labeled AIB in 100 mM Na⁺ or 100 ml KCl medium. $Na_o^+ = Na_i^+$ (□); $K_o^+ = K_i^+$ (△); $Na_i^+ = K_o^+$ (▼); $Na_o^+ = K_i^+$ (■); 3 methyl glucose uptake (○), equilibrium distribution. The units (ul/mg) indicate the volume of medium cleared of substrate per mg of vesicle protein. (From Colombini and Johnstone (1974), with permission.)

potential may have multiple sites of action including the relocation of the empty carrier and Na⁺ binding.

Na⁺-Coupling Ratios Greater Than 1.0

In the majority of cases reported to date, the coupling between Na⁺ and solute is 1:1 and sufficient free energy is available in the $\Delta\tilde{\mu}_{Na^+}$ to account for the observed accumulation (assuming near 100% coupling). However, in diverse tissues, examples have been cited where two Na⁺ ions are involved with the translocation of amino acids, succinate, and sugars (Vidaver, 1964a,b; Turner and Moran, 1982a,b; Wright et al., 1983; Weigensberg and Blostein, 1985; Kimmich et al., 1991). In Na⁺-dependent glucose transport in kidney tubules, the stoichiometry for transport is different along the length of the tubule (Turner and Moran, 1982a,b).

It has been proposed that the significance of the higher coupling ratio in the distal tubule of the kidney may be to ensure complete translocation of traces of remaining solute from the lumen (Turner and Moran, 1982a) by a low K_s, low V_{max} system after the bulk has been reabsorbed in the proximal tubule. Other possibilities have not yet been explored. The majority of systems which require 2 Na^+/solute are also Cl^--coupled.

NaCl-Coupled Amino Acid Uptake

To date, coupling to Na^+ and Cl^- (with the exception of Na^+/Cl^- cotransport) (Duffey et al., 1978; O'Grady et al., 1987), seems to be confined to a restricted number of amino acids and amines. Of the 20 or so amino acids which are incorporated into proteins and transported by Na^+ dependent routes, only glycine and glutamine uptake have been associated with a Cl^- dependence (Vidaver, 1964a; Weigensberg and Blostein, 1985; Mathew et al., 1993).

Cl^- dependence has been chiefly associated with amino acids which are not "typical" amino acids. These include taurine, β-alanine, γ-amino butyric acid (GABA), and betaine (Schon and Kelly, 1975; Kanner, 1978; Chesney, 1985; Turner, 1986; Kanner and Bendahan, 1990; Tiruppathi et al., 1992; Yamauchi et al., 1992). To a significant extent, some of these atypical amino acids (as well as glycine) are also neurotransmitters, although NaCl-coupled cotransport of these atypical amino acids also occurs outside the nervous system. The specificity for chloride is high. Of the anions tested (Cl^-, Br^-, I^-, NO_3^-, and glucuronate) as replacements, only sulphate has proved an effective substitute in a nominal way (25–50% of the activity in chloride; see Turner, 1986).

Since the majority of the amino acid substrates are near electrical neutrality at physiological pH, association with NaCl would presumably result in the formation of an electroneutral complex (depending on the net charge of the transporter itself), whose translocation (or relocation) might be insensitive to the transmembrane potential. Generally, the Cl^--dependent systems respond to the membrane potential and require coupling to 2 Na^+ (Kanner, 1978; Chesney, 1985; Turner, 1986). Thus, a five membered complex E-S:Na^+:Na^+:Cl^- is presumably involved in the translocation event. Whether ordered or random ligand binding occurs in complex formation has not been addressed to date.

The significance of the Cl^- dependence, unlike Na^+ dependence, is not self-evident. The biological significance and energy economy of Na^+ (or H^+) coupled solute cotransport is readily meaningful in terms of efficient transduction of metabolic energy. Influx of Cl^- has frequently been cited as an additional driving force. However, net Cl^- flow into the cell may be uphill as shown for the gallbladder (Duffey et al., 1978) and, therefore, would not contribute to the energization of the NaCl coupled solute flux. The not infrequent presence of these Cl^--coupled systems in the central nervous system, along with the fact that many of the amino acids are neurotransmitters, leads one to speculate that Cl^--dependent mechanisms may be important in restoring Cl^- to the intracellular milieu. In many cells, chloride concentration is under 10 mM. Given the fact that much Cl⁻ movement is electrically, as well as osmotically, silent, frequently involving exchange with HCO_3^-, these Cl⁻-coupled fluxes may provide a mechanism to maintain the cellular chloride concentration. Since chloride, as well as the cotransported organic solute, may both be moved uphill in the NaCl-coupled systems, the requirement for two sodium ions may reflect the need for an adequate driving force.

Na⁺+K⁺-Coupled Transport

From the earliest proposal for Na⁺-coupled transport, it was considered likely that K⁺ participated in the overall reaction (Crane, 1965). With the use of membrane vesicles where the internal and external K⁺ levels could be manipulated and replaced by other ions (Colombini and Johnstone, 1974; Murer and Hopfer, 1974; Sigrist-Nelson et al., 1975; Hopfer, 1978), it became clear that the apparent K⁺ requirement in intact cell systems was likely indirect and due to its requirement by the Na⁺/K⁺ ATPase to maintain the electrochemical potential difference for Na⁺.

A direct K⁺ requirement for translocation has, however, been reported for glutamic acid transport in brain (Kanner and Schuldiner, 1987; Carlson et al., 1989). The dicarboxylic amino acids appear to be transported largely by specific transporters which do not participate in neutral amino acid transport. Recent studies, both in reconstituted systems and the expression of the cloned transporter, have confirmed the K⁺ requirement (see below).

The physiological significance for this distinct K⁺ requirement remains enigmatic. It should be pointed out, however, that K⁺ in the above reports is not cotransported with Na⁺, but is countertransported. Glutamate transport coupled only to Na⁺ has also been reported (Lerner, 1987).

Summary

Coupling of solute flow to a Na⁺ (or H⁺) gradient remains the major means to energize the uptake of a large variety of solutes both organic and inorganic. The amino acids form the largest class of compounds transported by mammalian cells, the majority of which are cotransported (at least partly) with Na⁺. It has been conventional in discussing amino acid transport and cotransport to subdivide the amino acid transporters into groups as defined originally by Christensen, 1990 (e.g., A ASC, N system, etc.). These classifications have always been less than satisfactory because a single amino acid may be transported by several systems and these systems may differ in behavior between tissues and between species. Thus the A system in hepatocytes and hepatoma cells show differences with respect to sensitivity to different SH binding agents (Dudeck et al., 1987). Furthermore, except for tolerating α methyl α amino isobutyric acid (meAIB) as a substrate (which is used to define the A system), the classifications are largely made on the basis of selective inhibition, e.g., the ASC system is defined as Na^+-dependent but meAIB insensitive. Since the cells may be depolarized with meAIB (Laris et al., 1978) and transport through ASC is potential dependent (although electrically neutral (Bussolati et al., 1992)), the residual transport system(s) may not be measured under its optimal conditions. Thus, attributions such as different sensitivities to pH or tolerance for Li^+ (Christensen et al., 1967) may partly reflect the differences in the state of the system.

Except for meAIB no other truly specific substrate for an amino acid transport system has been defined. We deliberately chose not to discuss the properties of the putative different systems defined on the basis of selective inhibition. Instead the discussion has focussed on the nature of the ions coupled to amino acid uptake (e.g., Na^+, NaCl, etc.), since it now appears from cloning data that the nature of the cotransported ions identifies different families of transporters (see below).

TRANSPORT OF NEUROTRANSMITTER AMINES

A variety of amino acids or amines (directly derived from amino acids) are potent neurotransmitters. These include glycine, glutamate, γ-amino butyrate, and serotonin, and this list is by no means exhaustive (see Table 1). Neurotransmitters released into the postsynaptic cleft bind to the receptors at the postsynaptic membrane, triggering ion permeability changes which convey the electrical response. In many cases, termination of the signal occurs by reuptake of these compounds into the presynaptic terminals or into glial cells via a Na^+-coupled cotransporter. Although the significance is not clear, it has become apparent that the ion requirements for neurotransmitter transport are more complex than those associated with the majority of conventional α amino α carboxy amino acids, involving additional coupling to other ions such as K^+ or Cl^-. Given that so many of the neurotransmitters are recognized at concentrations in the μM range (or less) in contrast to nonneural tissues where K_s values in the mM range are not uncommon, it would not be unreasonable to speculate that the additional ions are involved in increasing carrier affinity for the solute in question.

Serotonin accumulation, as well as other biogenic amines in neuronal tissues, adds an additional order of complexity to ion coupled transport and accumulation. Three different monovalent ions appear to be involved in the formation of a functional serotonin transport complex. External Na^+ and Cl^- as well as internal K^+ are involved in serotonin uptake (Kanner, 1983; Graham and Langer, 1992). In this system the rapid relocation of the empty carrier is presumed to be the role for K^+. No experimental verification of this exists.

To date, neither the stoichiometry nor the electrogenicity of serotonin uptake has been resolved, the evidence gleaned from intact cells being different from that for isolated vesicle systems. A further complication to the study of biogenic amine transport is the fact that there are cellular granules which act as storage sites in neural cells (and other cells as well, e.g., platelets) so that a second level of transport occurs inside the cell. These intracellular granules have an ATP-ase which creates a pH gradient (inside acidic and positive) relative to the cytoplasm (Kanner, 1983; Graham and Langer, 1992). The amine appears to exchange with one or more protons in the granules leading to amine accumulation in the granules. The presence of carriers for the same substance at cell surfaces and in intracellular organelles adds to the complexity of dissociating the requirements of one system from the other in intact cells, giving rise to complications akin to those arising from the use of metabolizable substrates. The availability of relatively specific inhibitors for each of the systems, does permit assessment of one system in presence of the other (Keyes and Rudnick, 1982; Kanner, 1983; Graham and Langer, 1992). Nonetheless, this added complexity may make direct comparisons difficult between experiments using intact cell systems and isolated membrane fractions.

Evidence now exists which is consistent with the proposal that drugs like cocaine may exert their biological effects by preventing the reuptake of biogenic amines, thereby prolonging the response to these neurotransmitters and reducing the rate of ablation of the response. Cocaine inhibits dopamine transport in rat striatum (Ritz et al., 1987; Giros and Caron, 1993) and in nonneuronal cells expressing the cloned dopamine transporter (Kilty et al., 1991; Shimada et al., 1991). The cloned nor-epinephrine transporter (human) is also inhibited by cocaine (Pacholczyk et al., 1991), but with a lower affinity than the human dopamine transporter (Giros and Caron, 1993). In the absence of a functional dopamine transporter, agents like MPTP are less toxic implicating the transporter in MPTP uptake (Uhl and Kitayana, 1993). Although it is beyond the scope of this chapter to discuss the significance of these observations in any detail, it is self-evident that this new focus gives the Na^+-dependent reuptake mechanisms new physiological importance, in addition to providing a better understanding of the mode of action of some addictive drugs.

Given the substantial difference in structure between cocaine and the biogenic amines it will be a challenge to see whether derivatives can be synthesized which could be used to prevent cocaine binding without blocking the transporter mechanism itself.

H⁺-COUPLED FLUXES IN MAMMALIAN SYSTEMS

Although there is a great divide between animal cells and nonanimal cells (bacteria, plants, and fungi) with respect to the major cation (Na^+ or H^+) used for coupling in secondary active transport systems, there are growing numbers of examples of crossing over. In particular, in the small intestine, peptide transport has generally been shown to be accompanied by protons rather than by Na^+ (Ganapathy and Leibach, 1986; Calonge et al., 1989). That the intact peptide and not its constituent amino acids is translocated has been directly established in some cases (Calonge et al., 1989). In addition, the transport of a number of amino acids has also been shown to be linked to H^+ (Jessen et al., 1989). The physiological significance of this variation in cation-coupled flow in mammals is obscure. Although an appreciable $\Delta\tilde{\mu}_{H^+}$ can be established with sufficient free energy to drive the accumulation of the cosolute, the $[H^+]$ may be orders of magnitude less than $[Na^+]$ and only a fraction of the concentration of cosubstrate (except at cosubstrate concentrations $\leq 10^{-6}$M). The observed K_s for Na^+ is between 10^{-2}M and 3×10^{-2}M in many mammalian systems (Potashner and Johnstone, 1971; Kaunitz and Wright, 1984; Jauch and Lauger, 1986) far removed from the expected $[H^+]$ in the extracellular milieu of mammalian cells. Given a stoichiometric requirement for cation and cosolute, the bulk flow of H^+-driven transport will likely be small, even if K_s for H^+ is $\ll$ than K_s for Na^+. It may be that H^+-coupled flows are restricted to cosubstrates present at nanomolar levels.

Some transporters may function with either cation (Jessen et al., 1989). Cation coupled ATP-ases have also been shown to be capable of substituting H^+ for Na^+, depending on pH (Polvani and Blostein, 1988; Polvani et al., 1989). Such observations are consistent with the idea that the selectivity for H^+ or Na^+ reflects adaptation of a common, basic mechanism which has evolved subtle differences to exploit the variations in the ambient conditions.

PLASMA MEMBRANE VESICLES

Many ambiguities in the study of Na^+ (or H^+) coupled solute transport have been clarified by the introduction of isolated membrane vesicles to study transport. The inherent property of membranes to seal up and hence form a closed system for measuring translocation has made these preparations popular with investigators. A large variety of cell types and even greater number of solutes have been used in these systems.

The first reported instances using isolated plasma membrane vesicles to study Na^+-coupled transport were derived from brush borders of the small intestine (Murer and Hopfer, 1974; Sigrist-Nelson et al., 1975) and Ehrlich cells (Colombini and Johnstone, 1974). In rapid succession a number of other systems were established to study translocation of many solutes in many animal cell systems (Schuldiner and Kaback, 1975; Lever, 1977; Hammerman and Sacktor, 1978; Wright et al., 1983; Saier et al., 1988; see also Table 1).

By and large, studies with vesicles have confirmed the observations made with intact cells with respect to K_s for solute and sodium, effects of competing substrates and other inhibitors. Vesicle systems allow the examination of the translocation process in relative isolation from other cellular events.

With intact cells, the question of an additional metabolic energy source to drive transport was never completely resolved. Studies with transport competent vesicles showed that the systems behaved normally in the absence of any input of metabolic activity. Vesicle studies also bypassed the need to consider sequestration in organelles, possible binding to intracellular components, or conversion to other metabolites.

Perhaps more important than all these considerations was the fact that preparation of functional vesicular systems led to purification studies and reconstitution of transporters using artificial liposomes and imposed ion gradients to drive translocation. Vesicle systems also bypassed the need to measure binding as the one means to follow a transport system during purification. The ultimate assay for identification of a transporter protein is its ability, upon reconstitution, to show its anticipated behavior, i.e., rapid translocation of solute.

REGULATION OF Na$^+$-COUPLED AMINO ACID TRANSPORT

It has now been generally accepted that a major mechanism for the regulation of Na$^+$-independent glucose transport activity in insulin responsive adipose tissue and muscle is by the translocation of transporters from a cytoplasmic pool (Cushman and Wardzala, 1980; Suzuki and Kono, 1980).

No definitive evidence for such a phenomenon exists for amino acid transporters. Although there are many reports on the effects of a large number of hormones and other factors on amino acid transport, some going back over 30 years (e.g., Kipnis and Noall, 1958), no data have definitively shown a direct effect on the transporter itself. Many reports are in the category of metabolic responses, often associated with increased synthesis of the transporter in question.

The consensus that new protein synthesis is involved in regulating amino acid transport is based on the observations that inhibitors of protein and mRNA synthesis ablate the hormonal responses (Touabi and Jeanrenaud, 1969; Lever et al., 1976; Le Cam and Freychet, 1976; Guidotti et al., 1978). The primary target appears to be the A (meAIB sensitive) transport system. Even in hepatocytes, only the A transport system is elevated in response to liver regeneration (Fowler et al., 1992). There is also documented evidence that insulin enhances Na$^+$-dependent amino acid transport in muscle (Wool et al., 1965; Gumà et al., 1988) and Ehrlich ascites cells (Lin et al., 1993) without *de novo* protein synthesis. Moreover, in many instances the response to insulin is very rapid (Gumà et al., 1988) making it highly unlikely that protein synthesis is involved in the response. Such studies have given rise to the suggestion that insulin stimulates the relocation of an internal pool of transporters to the plasma membrane (Gumà et al., 1988; Lin et al., 1993), analogous to insulin's effect on some forms of the Na$^+$-independent glucose transporters.

In addition to hormones such as insulin and glucagon, the change in amino acid transport in response to depletion of cellular amino acids (generally increasing uptake) and to elevated amino acid levels (generally decreasing uptake) has been examined. Unfortunately, the experiments are not yet definitive with respect to showing a direct effect of the putative regulatory mechanism on the carrier itself (for reviews see Shotwell et al., 1983; Saier et al., 1988; Kilberg et al., 1993). Once more specific probes are available, the direct or indirect effects of many hormones, starvation, feeding, etc., will no doubt be reexamined to assess the nature of the regulation involved in amino acid transport.

STUDIES WITH ISOLATED CARRIER SYSTEMS

Solubilization and Reconstitution of Ion-Coupled Carriers

Studies on the purified carriers themselves and their reconstitution into artificial liposomes will likely lead to insights into the mechanism of action of ion-coupled carriers. The solubilization and reconstitution of membrane proteins is a multifac-

Table 2. Solubilization and Reconstitution Procedures

Protein	Tissue	Solubilization	Reconstitution Method	Reference
NaCl-coupled GABA transporter	Rat brain (synaptic plasma membranes)	2% cholic acid, 10% ammonium sulfate	Brain lipids/ asolectin & Sephadex G50 chroma-tography	Radian & Kanner (1985). J. Biol. Chem. 260, 11859–11865.
Na$^+$-K$^+$-coupled glutamate transporter	Rat brain (synaptic plasma membranes)	33mM CHAPS, 10% ammonium sulfate	Brain lipids/asolectin & Sephadex G50 chromatography	Danbolt et al. (1990). Biochemistry 29, 6734–6740.
Na$^+$-coupled phosphate transporter	Bovine kidney	1.5% CHAPS + brush border membrane lipids	Brush-border membrane lipids & Sephadex G50 column chromatography (Freeze/Thaw)	Vachon et al. (1991). Biochem. J. 278, 543–548.
Na$^+$-coupled glucose transporter	Intestinal brush-border membranes	1% CHAPS	Phosphatidyl choline/chole-sterol (80%: 20%) & detergent dilution/dialysis	Peerce & Clarke (1990). J. Biol. Chem. 265, 1731–1736.
Na$^+$-coupled amino acid transporter	Rat liver plasma membranes (A system)	2.5% cholic acid/4M urea	Asolectin: freeze-thaw/dilution/ sonication	Bracy et al. (1987). Biochim. Biophys. Acta 899, 51–58.
	Bovine kidney brush-border membranes	0.5% MEGA-10 (decanoyl-N-methyl-glucamide)	Egg yolk phospholipids, Sephadex G50 chromatography & freeze-thaw	Lynch & McGivan (1987). Biochem. J. 244, 503–508.
	Ehrlich ascites cell plasma membrane	2.5% cholic acid/4M urea	Asolectin, Sephadex G50 chromatography & freeze-thaw	McCormick et al. (1984). Arch. Biochem. Biophys. 231, 355–365.
Lactose carrier	E. coli membranes	1.25% octylglucoside	E. coli phospholipids & dilution	Newman & Wilson (1980). J. Biol. Chem. 255, 10583–10586.

eted procedure and previous reviews have described the theoretical and practical aspects of the many methods that can be employed (Helenius and Simons, 1975; Racker, 1979; Silvius, 1992). Table 2 describes a few of these procedures. Cholic acid or the zwitterionic derivative 3-[(3-Cholamidopropyl)-dimethylammonio]-1-propane sulfonate (CHAPS) are ideal agents for membrane solubilization (Koepsell, 1986; Silvius, 1992) given their high critical micellar concentrations and ease of removal by Sephadex G50 chromatography, dialysis, or dilution (Racker, 1979). Transporters may be stabilized during solubilization by the presence of substrates (Koepsell, 1986), phospholipids (Newman and Wilson, 1980; Vachon et al., 1991; In't Veld et al., 1992), or osmolytes (Vachon et al., 1991), and surprisingly by urea, a compound normally associated with protein denaturation. The functional recovery of the NaCl-coupled 5-hydroxytryptamine transporter from cholic acid solubilized placental membranes is irreversibly inactivated in the absence of urea (Ramamoorthy et al., 1992). Identical solubilization conditions (2.5% cholic and 4M urea) allowed the recovery of Na^+-coupled amino acid transport from rat liver (Bracy et al., 1987) and Ehrlich cells (McCormick et al., 1984), and a taurine transporter from placental membranes (Ramamoorthy et al., 1993) suggesting that urea may stabilize a number of ion-coupled carriers during solubilization (Ramamoorthy et al., 1992).

Removal of the solubilizing detergent and the presence of liposomes of a suitable lipid composition leads to proteoliposome formation. A low passive permeability and an adequate vesicle size are essential for measurable transport activity. A number of ion-coupled carriers work best upon reconstitution with some endogenous lipids, e.g., kidney renal brush-border Na^+-coupled phosphate transporter (Vachon et al., 1991), and the neural NaCl-coupled transporters for GABA and glycine (Radian and Kanner, 1985; López-Corcuera and Aragon, 1989). Since the driving forces and accumulation of solutes are transient, it is essential to optimize the internal vesicle volume to diminish the rate of collapse of the gradients. Vesicle size can be increased by a freeze–thaw cycle, a procedure that causes lipid vesicle fusion. The fusion events are influenced by a number of factors including cations and phospholipids (McCormick et al., 1985; McCormick and Johnstone, 1988b; Silvius, 1992). In reconstituted vesicles from Ehrlich cells, transport activity is directly related to vesicle size (McCormick et al., 1985). In addition, trace levels of phospholipids may have critical volume independent effects. For example, although removal of phosphatidic acid (PA) results in a loss of Na^+-dependent α amino isobutyric acid (α-AIB) transport and a reduction in vesicle size (McCormick and Johnstone, 1988b), careful reconstitution studies have shown that acidic phospholipids, PA, and cardiolipin (not phosphatidyl serine) directly activate the amino acid transporter in a manner unrelated to intravesicular volume changes (Lin et al., 1990). Restoration of transport activity is maximal when PA contains one β unsaturated fatty acid chain, suggesting that the β chain interacts with the carrier in the bilayer (Lin et al., 1990). There is also evidence from bacterial systems that acidic phospholipids stabilize the active conformation of a proton-coupled leucine transporter from *Lactococcus lactis* (In't Veld et al., 1992). To our knowledge the possibility that specific lipids may play important roles in the control of transporters has yet to be examined in detail.

A number of factors influence formation of proteoliposomes with a low passive ion permeability. Lin et al. (1990) have reported that polyethylene glycol (PEG) treatment of proteoliposomes lowers passive permeability to small molecules, enabling the ion-gradient to be maintained for longer periods. In several ion-coupled carrier reconstitution procedures (Fafournoux et al., 1989; Tamarappoo and Kilberg, 1991; Ramamoorthy et al., 1992; Ramamoorthy et al., 1993), solubilized proteins have been precipitated with PEG before incorporation into lipid vesicles. Trace amounts of PEG associated with the proteoliposomes in these procedures may reduce passive permeability and lead to high levels of ion-coupled transport (Ramamoorthy et al., 1993).

Permeability may also be reduced by introducing a freeze–thaw cycle into the reconstitution procedure (Pick, 1981; Anholt et al., 1982) leading to both larger and better sealed vesicles. Finally, Doyle and McGivan (1992) have reported that the incorporation of small quantities (5% mol/mol of phospholipid) of stearylamine into reconstitution procedures decreased cation permeability by increasing vesicle surface charge and improved Na^+-coupled alanine uptake by threefold.

Detection and Purification of Ion-Coupled Sugar Transporters

The most successful attempts to purify ion-coupled carriers have made use of high affinity ligands to follow the protein during purification prior to its incorporation into liposomes. The phlorizin-sensitive Na^+/glucose symporter from renal or intestinal cells has been followed by phlorizin binding to assay the transporter during purification (Peerce and Clarke, 1990). A photoactive derivative has been used to label components of the carrier in intact brush-border membranes (Semenza et al., 1984). However, since epithelial cells may contain more than one Na^+/glucose symporter, differing in their affinities for phlorizin (Turner and Moran, 1982a,b), purification or labeling strategies utilizing phlorizin may only target the carrier with the higher binding affinity.

Na^+-coupled glucose transporters have also been tentatively identified using substrate protection strategies with protein modifying reagents (Peerce and Wright, 1984) and by monoclonal antibodies raised against brush-border membrane extracts (Wu and Lever, 1987).

These indirect approaches suggest that a 75 kDa polypeptide is involved in Na^+/glucose symport and a carrier of this size is consistent with cloning data (see below). Procedures involving transporter purification, using reconstitution to follow Na^+/glucose symport activity, have not been very successful (Koepsell et al., 1983; Malathi and Takahashi, 1990). Moreover, the sequence identity (or similarity) of the partly purified 75 kDa peptide to the cloned glucose transporter has yet to be established (as of 1995).

More success has been obtained with the H^+/lactose carrier from *E. coli* which has been identified, purified, and extensively characterized. This transporter was identified as a product of the lac y gene, the gene cloned and its nucleotide sequence determined before the carrier was actually purified and reconstituted (Büchel et al., 1980). Purification of the protein to homogeneity was achieved after octylglucoside solubilization (Newman and Wilson, 1980). Upon reconstitution, the proteoliposomes exhibited membrane-potential-driven lactose transport and lactose counterflow. The amino acid composition of the purified protein was virtually identical to the sequence predicted from the lac y gene DNA sequence. More recently, using a novel procedure that runs counter to traditionally held ideas of membrane protein properties, Roepe and Kaback (1989) have shown that when lac permease is overexpressed in *E. coli*, a large portion of the newly synthesized transporter is not inserted into the lipid bilayer but remains associated with the membrane. After extraction from the membranes in 5M urea and purification, the protein can be inserted into proteoliposomes and shows the normal properties of H^+/lactose symport. The melibiose permease has been purified from *E. coli* in an almost identical manner (Roepe and Kaback, 1990), illustrating the general applications of this technique for the purification of transport proteins. The purification of the melibiose permease is of particular importance because this carrier can catalyze symport with Na^+, H^+, or Li^+, depending on the substrate (Pourcher et al., 1992). Clearly structure/function studies on this protein will be particularly informative as to the mechanism of ion-coupled transport.

Purification of Mammalian Ion-Coupled Amino Acid Transporters

The absence of high affinity ligands (substrates or inhibitors) has precluded to date even tentative identification of candidate amino acid carrier proteins. However, arginine and tyrosine specific reagents have been used to identify an intestinal Na^+-phosphate transporter (Peerce, 1989) and a protein with an identical molecular size (130 kDa) has been purified to homogeneity and shown to support Na^+-phosphate cotransport upon reconstitution (Peerce et al., 1993). Ion-coupled amino acid carriers have been solubilized, reconstituted, and purified from several mammalian cell types including Ehrlich ascites cells (McCormick and Johnstone, 1988a), rat liver (Fafournoux et al., 1989), and bovine renal brush-border membranes (Doyle and McGivan, 1992). Although the amino acid transporters from Ehrlich cells and brush-border membranes have distinctly different properties (Stevens et al., 1984), candidate carrier proteins with similar molecular sizes (120–130 kDa) have been purified. Procedures involving immunological depletion supported the conclusion for the involvement of a 120–130 kDa peptide (McCormick and Johnstone, 1988a; Doyle and McGivan, 1992). Moreover, when Ehrlich cell plasma membrane vesicles were irradiated, the loss of transport activity correlated with destruction of the 120–130 kDa peptide (McCormick et al., 1991).

Surprisingly, recent work has shown that in both the Ehrlich cell and bovine brush-border membranes, the polypeptides initially identified as the putative amino acid carriers contain proteins already identified, namely the integrin VLA-3 (McCormick and Johnstone, 1995) and aminopeptidase N, respectively (Plakidou-Dymock et al., 1993). Whether VLA-3 and aminopeptidase N function as components of a transport complex, regulators of Na^+-coupled amino acid carrier proteins, or whether these are merely fortuitous associations has not been clarified. Since aminopeptidase N binds amino acids, Plakidou-Dymock et al. (1993) have suggested that this protein may channel amino acids to the carrier binding site thus facilitating transport. The role of VLA-3 has not been fully characterized. However, K562 cells stably transfected with the cDNA for the α3 integrin show increased A system transport (McCormick and Johnstone, 1995).

Purification of Ion-Coupled Neurotransmitter Transporters

Although the purification to homogeneity of ion-coupled amino acid transporters from peripheral tissues has not been achieved, amino acid carriers have been successfully purified from brain tissue where these amino acids may function (see above) as neurotransmitters (Radian and Kanner, 1985; Radian et al., 1986; López-Corcuera and Aragon, 1989; Danbolt et al., 1990; López-Corcuera et al., 1991). In each case purification was based on a functional reconstitution assay and the purified proteins in proteoliposomes exhibited the expected ion requirements and kinetic constants for uptake determined in the cells of origin. The physical properties of the isolated carriers for glycine (López-Corcuera et al., 1991), GABA (Radian et al., 1986), and glutamate (Danbolt et al., 1990) are remarkably similar.

Differences in affinity of the carriers for their substrates in peripheral and neural tissues may account for the differences in ease of carrier purification. Na^+-linked amino acid carriers in peripheral tissues have K_m values in the mM range, while the Na^+-K^+ glutamate and NaCl-glycine and GABA transporters in neuronal tissues have K_m values less than 10 μM. This higher affinity may have favored the development of the elegant and sensitive reconstitution assays used during purification of these carriers.

CLONING AND EXPRESSION OF ION COUPLED TRANSPORTERS

Cloning of Ion-Coupled Transporters

The cloning and sequencing of ion-coupled carriers has produced evidence thus far of three major Na^+-dependent carrier families. The three families are defined by their sequence homology to: (a) the Na^+-glucose (Na^+-GLU) cotransporter from rabbit intestine (Hediger et al., 1987), (b) the Na^+-Cl^--coupled GABA transporter from rat brain (Guastella et al., 1990), and (c) a family of Na^+-K^+-coupled glutamate transporters expressed in brain and peripheral tissues (Kanai and Hediger, 1992). As noted above, assignments to the different families on the basis of sequence homology also correspond to assignments based on common ion requirements for transport, e.g., Na^+/solute; $2Na^+ + Cl^-$/solute, $Na^+ + K^+$/solute. There is limited sequence homology between transporters from the three different families.

The Na^+-GLU cotransporter from rabbit intestine was cloned by oocyte expression and was the first eukaryotic transporter cloned in this manner (Hediger et al., 1987). The isolated cDNA codes for a polypeptide with 662 amino acids (MW 73,080), consistent with the size for the Na^+-glucose cotransporter of 75 kDa, derived from purification and labeling experiments. From hydrophobicity analysis, the original secondary structural model predicted up to 13 postulated membrane spanning domains, the most popular model having 12 transmembrane domains (Wright et al., 1992). A 12 transmembrane domain structure is a common feature of a wide range of transport proteins in prokaryotic and eukaryotic cells (Griffith et al., 1992).

The cloning of the Na^+-glucose transporter, or any ion-coupled carrier, offers the potential for extensive structural/functional studies either through site-directed mutagenesis or over expression. Smith et al. (1992a) have reported the expression of the Na^+-glucose carrier in Sf9 cells (an insect cell line), which is functional despite reduced glycosylation. Other studies have also shown (Hediger et al., 1991; Hirayama and Wright, 1992) that glycosylation of the carrier, at Asn 248, is not necessary for transport activity. The expression of the cloned glucose transporter in *Xenopus* oocytes has allowed extensive kinetic and mechanistic studies (Parent et al., 1992a,b) and the electrophysiological properties of this system have been defined.

Na^+-dependent glucose transporters have now been cloned and sequenced from rabbit kidney (Coady et al., 1990), human intestine (Hediger et al., 1989), and LLC-PK$_1$ cells (Ohta et al., 1990) and these transporters show 100%, 85%, and 84% identity with the rabbit intestinal carrier (Wright et al., 1992). In addition, Na^+-coupled nucleoside (Pajor and Wright, 1992) and amino acid cotransporters (Kong et al., 1993) have been cloned from rabbit kidney and LLC-PK$_1$ cells. The amino acid sequence of the cloned nucleoside transporter shows 61% identity and 80% similarity to the Na^+-glucose cotransporter sequence.

A putative Na^+-coupled amino acid carrier from LLC-PK$_1$ cells (A system transporter) (Kong et al., 1993) showed a surprising 76% identity and 89% similarity with the amino acid sequence of the Na^+-glucose cotransporter from the same cell line (LLC-PK$_1$ cells). Transcripts for this transporter were detected in liver, skeletal muscle, and spleen, tissues that do not express high affinity Na^+/glucose cotransporter. This putative amino acid transporter has now been identified as a low affinity glucose transporter (Mackenzie et al., 1994).

The Na^+-myoinositol cotransporter is 46% identical with the glucose carrier (Kwon et al., 1992). The Na^+-proline (Nakao et al., 1987) and Na^+-pantothenate (Jackowski and Alix, 1990) carriers from *E. coli* show 28% and 25% identity, respectively (Pajor et al., 1992), to the Na^+ glucose transporter from rabbit intestine.

Other Na$^+$-coupled transporters show less relatedness. These include Na$^+$-coupled transporters for phosphate (Werner et al., 1991), bile acid (Hagenbuch et al., 1991), glutamate (Deguchi et al., 1990), and alanine (Kamata et al., 1992) cloned from rabbit kidney cortex, liver parenchymal cells, *E. coli* B, and the thermophilic bacterium PS3, respectively. These transporters are smaller (between 360 and 460 amino acid residues) and with the exception of the glutamate transporter, are proposed to have less than 12 transmembrane domain segments based on hydropathy plots. The Na$^+$-coupled glutamate, proline, glucose, and phosphate carriers do, however, contain a consensus sequence Gly--Ala-X-X-X-X-Leu-X-X-X-Gly-Arg proposed by Deguchi et al. (1990) to play a role in Na$^+$-binding (Werner et al., 1991). It is likely that additional families of carriers are yet to be defined. The observation that several Na$^+$-coupled carriers for diverse substrates are clearly structurally related offers an excellent experimental system for identifying regions of the proteins involved in Na$^+$-coupling and substrate binding.

Cloning of the NaCl-Coupled Transporters

Using a partial amino acid sequence of the purified transporter to construct an oligonucleotide probe (Guastella et al., 1990), the GABA transporter was cloned. The cDNA isolated was consistent with a protein of 67 kDa, the size of the deglycosylated transporter (Kanner et al., 1989). Neither the cytoplasmic amino nor the carboxy terminus appear to be necessary for transport activity (Mabjeesh and Kanner, 1992). Preliminary studies using site-directed mutagenesis in the transmembrane domain (implicated in ion-coupled transport) have found that Arg 69, located in the first transmembrane domain, is essential for NaCl-coupled transport (Pantanowitz et al., 1993). Interestingly, this residue is conserved in several other NaCl-coupled carriers related to the GABA transporter.

Consistent with the putative presence of multiple GABA transporters, a detailed screening of brain cDNA libraries with the GABA transporter cDNA or related sequences (López-Corcuera et al., 1992; Liu et al., 1993) has resulted in the isolation of several clones that express Na$^+$-dependent GABA transport in oocytes. The transporters differ in K_s values for GABA and in their sensitivity to inhibitors, but have a high degree of sequence homology. A series of carriers have recently been cloned with amino acid sequences homologous to the GABA carrier but which transport different substrates. These include transporters for taurine/β alanine (Liu et al., 1992a), glycine (Liu et al., 1992b; Smith et al., 1992b), dopamine (Shimada et al., 1991), betaine (Yamauchi et al., 1992), serotonin (Hoffman et al., 1991), and creatine (Guimbal and Kilimann, 1993). All the cloned transporters were shown to express NaCl-coupled transport of their respective substrates in oocytes (Liu et al., 1992a,b; Yamauchi et al., 1992) or after transfection of COS cells (Shimada et al., 1991; Smith et al., 1992b; Guimbal and Kilimann, 1993). The degree of homology of these proteins with the GABA transporter is ~40–50%.

Two additional carriers with extensive homology to the GABA transporter are the Na^+-coupled nor -epinephrine (Pacholczyk et al., 1991) and proline transporters (Fremeau et al., 1992) from human and rat brain, respectively. However, Cl^- dependence has not yet been definitively demonstrated for the cloned carriers.

In addition to the high sequence homology of these carriers, secondary structure models based on hydropathy analysis (Amara and Kuhar, 1993) are virtually superimposable. All have been assigned 12 transmembrane domains and each transporter is considered to have a large extracellular glycosylated loop between transmembrane domains three and four.

Cloning of $Na^+ + K^+$-Coupled Glutamate Transporters

The rat brain glutamate transporter has been purified to homogeneity (Danbolt et al., 1990) and cloned (Pines et al., 1992). The cloned and expressed transporter showed an absolute dependence on external Na^+ and internal K^+ consistent with the behavior of the native carrier.

The clone codes for a protein with 573 amino acids and a molecular weight of 64,000, identical to the value for the deglycosylated brain glutamate transporter (Pines et al., 1992). Hydropathy plots suggest at least eight transmembrane domains. A second high affinity glutamate transporter which shows K^+ counter-transport has been cloned from rabbit intestine but the mRNA is also present in specific neuronal structures in the central nervous system as well as in kidney, liver, and heart (Kanai and Hediger, 1992).

A third Na^+-dependent glutamate transporter has been cloned from rat brain using an oligonucleotide probe derived from a partial amino acid sequence of a protein that copurified with UDP galactose ceramide galactosyltransferase (Storck et al., 1992). The three cloned transporters show significant homology to one another (~50% identity) and to a proton-coupled glutamate transporter from *E. coli* (Tolner et al., 1992; Kanai et al., 1993). In addition, two Na^+-dependent neutral amino acid transporters recently cloned from human brain (Shafqat et al., 1993; Arriza et al., 1993) exhibit ~40% homology to the ion-coupled glutamate transporters. The proteins coded by these cDNAs display some, but not all, of the properties characteristic of the ASC system. The Na^+-coupled glutamate transporters show no significant homology either with the Na^+-glucose cotransporter and related proteins or with the NaCl-coupled transporters, in keeping with the conclusion that the $Na^+ + K^+$-coupled glutamate carriers belong to a separate family of proteins (Kanai et al., 1993).

BACTERIAL ION-COUPLED TRANSPORTERS

Several recent reviews discuss in detail the cloning of ion-coupled carriers from bacteria and the predicted structures and sequence homology of the encoded proteins (Henderson et al., 1992; Griffith et al., 1992; Haney and Oxender, 1992;

Kaback, 1992). The H^+-coupled lactose carrier from *E. coli* has been extensively studied and the structural and functional information generated may provide insight into the general mechanisms of ion-coupled transport. Extensive site-directed mutagenesis studies on this protein have identified residues that are important in substrate recognition and H^+-coupling (Kaback, 1988; Kaback, 1992) and it has been proposed (Kaback, 1992) that a His residue, located in a transmembrane domain, plays a central role in an H^+-relay system somewhat similar to the catalytic triad in serine proteases. In contrast, recent site directed mutagenesis of the Na^+-, H^+- or Li^+-coupled melibiose carrier from *E. coli* suggests that His is not involved in the transport process (Pourcher et al., 1992) and implicates acidic residues (Asp) in ion-binding and coupling (Pourcher et al., 1993; Zani et al., 1993). Clearly, further detailed studies are needed before a clear picture emerges of the mechanisms included in ion selectivity and coupling for different ion-coupled carriers.

CONCLUSIONS

After 40 years of descriptive work on the behavior of ion-coupled transport systems and the establishment of the near universal mechanism of energy transduction by coupling to ion flows, there is a sense that the next decade will see an information explosion on the mechanism of action of the transporters and their structures at the molecular level.

Molecular relatedness of these transporters seems to be associated with the nature of the ion requirements of the system, the differing ionic requirements representing different classes of transporters.

It is remarkable that a bacterial Na^+-coupled proline transporter is more homologous to a mammalian amino acid (or glucose transporter) than are the Na^+-dependent amino acid transport systems to NaCl-coupled amino acid transporters. The diversity of the ion requirement from a single Na^+ to three different ion requirements for transport of some neurotransmitter amines argues for an adaption of a biologically successful mechanism to achieve specialized functions with enhanced sensitivity and selectivity.

In the coming decade, progress in the understanding of ion-coupled transport systems is likely to evolve in three still poorly understood areas: (a) the mechanisms involved in acute and long-term regulation of these transporters; (b) the mechanism of the ion-coupled flow and energy transduction at the molecular level, and (c) the physiological significance of coupling to a single cation (Na^+ or H^+) versus multiple ions (e.g., $Na^+ + Cl^+$).

ACKNOWLEDGMENTS

The work attributed to the authors (R.M.J. and J.I.M.) was supported by grants from the Medical Research Council of Canada (MT 1984) to whom we express our thanks. To Marlene Gilhooly we express our appreciation for the highly skilled manuscript preparation.

REFERENCES

Amara, S.G. & Kuhar, M.J. (1993). Neurotransmitter transporters: Recent progress. Ann. Rev. Neurosci. 16, 73–93.

Anholt, R., Fredkin, D.R., Deerinck, T., Ellisman, M., Montal, M., & Lindstrom, J. (1982). Incorporation of acetylcholine receptors into liposomes. Vesicle structure and acetylcholine receptor function. J. Biol. Chem. 257, 7122–7134.

Arriza, J.L., Kavanaugh, M.P., Fairman, W.A., Wu, Y.-N., Murdoch, G.H., North, R.A., & Amara, S.G. (1993). Cloning and expression of a human neutral amino acid transporter with structural similarity to the glutamate transporter gene family. J. Biol. Chem. 268, 15329–15332.

Berlinguet, L., Begin, N., Babineau, L.M., Martel, F., Vallée, R., & Laferte, R.O. (1962). Biochemical studies of an unnatural and antitumor amino acid: 1-aminocyclopentanecarboxylic acid. Can. J. Biochem. and Physiol. 40, 425–432.

Birnir, B., Loo, D.D.F., & Wright, E.M. (1991). Voltage clamp studies of the Na^+/glucose cotransporter cloned from rabbit small intestine. Eur. J. Physiol. 418, 79–85.

Bracy, D.S., Schenerman, M.A., & Kilberg, M.S. (1987). Solubilization and reconstitution of hepatic system A-mediated amino acid transport. Preparation of proteoliposomes containing glucagon-stimulated transport activity. Biochim. Biophys. Acta 899, 51–58.

Büchel, D.E., Groneborn, B., & Müller-Hill, B. (1980). Sequence of the lactose permease gene. Nature (London) 283, 541–545.

Bussolati, O., Laris, P.C., Rotoli, B.M., Dall'Asta, V., Gazzola, G.C. (1992). Transport system ASC for neutral amino acids. An electroneutral sodium/amino acid cotransport sensitive to the membrane potential. J. Biol. Chem. 267, 8330–8335.

Calonge, M.L., Iludian. A., & Bolufer, J. (1989). Ionic dependence of glycylsarcosine uptake by isolated chicken enterocytes. J. Cell. Physiol. 138, 579–585.

Carlson, M.D., Kish, P.E., & Ueda, T. (1989). Characterization of the solubilized and reconstituted ATP-dependent vesicular glutamate uptake system. J. Biol. Chem. 264, 7369–7376.

Chesney, R.W. (1985). Factors affecting the transport of β-amino acids on rat renal brush-border membrane vesicles. The role of external chloride. Biochim. Biophys. Acta 812, 702–712.

Christensen, H.N. (1990). Role of amino acid transport and countertransport in nutrition and metabolism. Ann. Rev. Physiol. 70, 43–77.

Christensen, H.N., Liang, M., & Archer, E.G. (1967). A distinct Na^+ requiring transport system for alanine, serine, cysteine, and similar amino acids. J. Biol. Chem. 242, 5237–5246.

Christensen, H.N., Oxender, D.L., Liang, M., & Vatz, K. (1965). The use of N methylation to direct the route of mediated transport of amino acids. J. Biol. Chem. 240, 3609–3619.

Coady, M.J., Pajor, A.M., & Wright, E.M. (1990). Sequence homologies among intestinal and renal Na^+/glucose cotransporters. Am. J. Physiol. 259, C605–C610.

Colombini, M. & Johnstone, R.M. (1974). Na^+ gradient stimulated AIB transport in membrane vesicles from Ehrlich ascites cells. J. Membr. Biol. 18, 315–334.

Crane, R.K. (1960). Intestinal absorption of sugar. Physiol. Rev. 40, 789–825.

Crane, R.K. (1965). Na-dependent transport in the intestine and other animal tissues. Fed. Proc. 24, 1000–1006.

Crane, R.K. & Krane, S.M. (1956). On the mechanism of the intestinal absorption of sugars. Biochim. Biophys. Acta 20, 568–569.

Curran, P.F., Hajjar, J.J., & Glynn, I.M. (1970). Sodium alanine interactions in rabbit ileum. J. Gen. Physiol. 55, 297–308.

Curran, P.F., Schultz, S.G., Chez, R.A., & Fuisz, R.E. (1967). Kinetic relations of sodium amino acid interactions at the mucosal border of the intestine. J. Gen. Physiol. 50, 1261–1286.

Cushman, S.W. & Wardzala, L.J. (1980). Potential mechanism of insulin action on glucose transport in the isolated rat adipose cell. Apparent translocation of intracellular transport systems to the plasma membrane. J. Biol. Chem. 255, 4758–4762.

Danbolt, N.C., Pines, G., & Kanner, B.I. (1990). Purification and reconstitution of the sodium- and potassium-coupled glutamate transport glycoprotein from rat brain. Biochemistry 29, 6734–6740.

Deguchi, Y., Yamato, I., & Anraku, Y. (1990). Nucleotide sequence of gltS, the Na^+/glutamate symport carrier gene of *Escherichia coli* B. J. Biol. Chem. 265, 21704–21708.

Dibrov, P.A. (1991). The role of Na ion transport in *E. coli* energetics. Biochim. Biophys. Acta 1056, 209–224.

Doyle, F.A. & McGivan, J.D. (1992). Reconstitution and identification of the major Na^+-dependent neutral amino acid-transport protein from bovine renal brush-border membrane vesicles. Biochem. J. 281, 95–102.

Dudeck, K.L., Dudenhausen, E.E., Chiles, T.C., Fafournoux, P., & Kilberg, M.S. (1987). Evidence for inherent differences in system A carrier from normal and transformed liver tissue. J. Biol. Chem. 262, 12565–12569.

Duffey, M.E., Turnheim, K., Frizzell, R.A., & Schultz, S.G. (1978). Intracellular chloride activities in rabbit gallbladder: Direct evidence for the role of the sodium-gradient in energizing "uphill" chloride transport. J. Membr. Biol. 42, 229–245.

Eisenman, G. (1962). Cation selective glass electrodes and their mode of operation. Biophys. J. 2, 259–323.

Fafournoux, P., Dudenhausen, E.E., & Kilberg, M.S. (1989). Solubilization and reconstitution characteristics of hepatic system A-mediated amino acid transport. J. Biol. Chem. 264, 4805–4811.

Fowler, F.C., Banks, R.K., & Mailliard, M.E. (1992). Characterization of sodium-dependent amino acid transport activity during liver regeneration. Hepatology 16, 1187–1194.

Fremeau, R.T., Caron, M.G., & Blakely, R.D. (1992). Molecular cloning and expression of a high affinity L-proline transporter expressed in putative glutamatergic pathways of rat brain. Neuron 8, 915–926.

Ganapathy, V. & Leibach, F.H. (1986). Carrier-mediated reabsorption of small peptides in renal proximal tuble. Am. J. Physiol. 251, F945–F953.

Gibb, L.E. & Eddy, A.A. (1972). An electrogenic sodium pump as a possible factor leading to the concentration of amino acids by mouse ascites-tumour cells with reversed sodium ion concentration gradients. Biochem. J. 129, 979–981.

Giros, B. & Caron, M.G. (1993). Molecular characterization of the dopamine transporter. Trends in Pharmacol. Sci. 14, 43–49.

Goldner, A.M., Schultz, S.G., & Curran, P.F. (1969). Sodium and sugar fluxes across the mucosal border of rabbit ileum. J. Gen. Physiol. 53, 362–383.

Graham, D. & Langer, S. (1992). Advances in sodium-ion coupled biogenic amine transporters. Life Sci. 51, 631–645.

Griffith, J.K., Baker, M.E., Rouch, D.A., Page, M.G.P., Skurray, R.A., Paulsen, I.T., Chater, K.F., Baldwin, S.A., & Henderson, P.J.F. (1992). Membrane transport proteins: Implications of sequence comparisons. Curr. Opin. Cell Biol. 4, 684–695.

Guastella, J., Nelson, N., Nelson, H., Czyzyk, L., Keynan, S., Miedel, M.C., Davidson, N., Lester, H.A., & Kanner, B.I. (1990). Cloning and expression of a rat brain GABA transporter. Science 249, 1303–1306.

Guidotti, G.G., Borghetti, A.S., & Gazzola, G.C. (1978). The regulation of amino acid transport in animal cells. Biochim. Biophys. Acta 515, 329–366.

Guimbal, C. & Kilimann, M.W. (1993). A Na^+-dependent creatine transport in rabbit brain, muscle, heart and kidney. cDNA cloning and functional expression. J. Biol. Chem. 268, 8418–8421.

Gumà, A., Testar, X., Palacin, M., & Zorzano, A. (1988). Insulin-stimulated α-(methyl) aminoisobutyric acid uptake in skeletal muscle. Evidence for a short-term activation of uptake independent of Na^+-electrochemical gradient and protein synthesis. Biochem. J. 253, 625–629.

Hagenbuch, B., Stieger, B., Foguet, M., Lübbert, H., & Meier, P.J. (1991). Functional expression cloning and characterization of the hepatocyte Na^+/bile acid cotransport system. Proc. Natl. Acad. Sci. USA 88, 10629–10633.

Hajjar, J.J., Lamont, A.S., & Curran, P.F. (1970). Sodium alanine interaction in rabbit ileum. J. Gen. Physiol. 55, 277–296.

Hammerman, M. & Sacktor, B. (1978). Transport of β alanine in renal brush-border membrane vesicles. Biochim. Biophys. Acta 509, 338–347.

Haney, S.A. & Oxender, D.L. (1992). Amino acid transport in bacteria. Int. Rev. Cytol. 137A, 37–95.

Hediger, M.A., Coady, M.J., Ikeda, T.S., & Wright, E.M. (1987). Expression cloning and cDNA sequencing of the Na+/glucose co-transporter. Nature 330, 379–381.

Hediger, M.A., Mendlein, J., Lee, H-S., & Wright, E.M. (1991). Biosynthesis of the cloned intestinal Na+/glucose cotransporter. Biochim. Biophys. Acta 1064, 360–364.

Hediger, M.A., Turk, E., & Wright, E.M. (1989). Homology of the human intestinal Na+/glucose and *Escherichia coli* Na+/proline cotransporters. Proc. Natl. Acad. Sci. USA 86, 5748–5752.

Heinz, E. (1978). Mechanisms and Energetics of Biological Transport. In: Molecular Biochemistry and Biophysics, (Kleinzeller, A., Springer, G.F., and Wittman, H.G., eds.) Vol. 29, pp. 120–140. Springer Verlag, Berlin.

Helenius, A. & Simons, K. (1975). Solubilization of membranes by detergents. Biochim. Biophys. Acta 415, 29–79.

Henderson, P.J.F., Baldwin, S.A., Cairns, M.T., Charalambous, B.M., Dent, H.C., Gunn, F., Liang, W.-J., Lucas, V.A., Martin, G.E., McDonald, T.P., McKeown, B.J., Muiry, J.A.R., Petro, K.R., Roberts, P.E., Shatwell, K.P., Smith, G., & Tate, C.G. (1992). Sugar-cation symport systems in bacteria. Int. Rev. Cytol. 137A, 149–208.

Higgins, C.F. (1992). ABC transporters from microorganisms to man. Ann. Rev. Cell. Biol. 8, 67–113.

Hirata, H., Altendorf, K., & Harold, F.M. (1973). Role of an electrical potential in the coupling of metabolic energy to active transport by membrane vesicles of *E. coli*. Proc. Natl. Acad. Sci. USA 70, 1804–1808.

Hirayama, B.A. & Wright, E.M. (1992). Glycosylation of the rabbit intestinal brush-border Na+/glucose cotransporter. Biochim. Biophys. Acta 1103, 37–44.

Hoffman, B.J., Mezey, E., & Brownstein, M.J. (1991). Cloning of a serotonin transporter affected by antidepressants. Science 254, 579–580.

Hoffman, J.F. & Laris, P.C. (1974). Determination of membrane potentials in human and Amphiuma red blood cells by means of a fluorescent probe. J. Physiol. (Lond) 239, 519–522.

Hopfer, U. (1978). Transport in isolated plasma membranes. Amer. J. Physiol. 234, F89–F96.

In't Veld, G., De Vrije, T., Driessen, A.J.M., & Konings, W.N. (1992). Acidic phospholipids are required during solubilization of amino acid transport systems of Lactococcus lactis. Biochim. Biophys. Acta 1104, 250–256.

Jackowski, S. & Alix, J.-H. (1990). Cloning, sequence, and expression of the pantothenate permease (pan F) gene of *Escherichia coli*. J. Bacteriol. 172, 3842–3848.

Jauch, P. & Lauger, P. (1986). Electrogenic properties of the sodium-alanine cotransporter in pancreatic acinar cells. II. Comparison with transport models. J. Membr. Biol. 94, 117–127.

Jauch, P., Peterson, O.H., & Lauger, P. (1986). Electrogenic properties of the sodium-alanine cotransporter in pancreatic acinar cells. I. Tight-seal whole-cell recordings. J. Membr. Biol. 94, 99–115.

Jessen, H., Jorgensen, K.E., Roigaard-Petersen, H., & Sheikh, M.I. (1989). Demonstration of H+ and Na+ coupled cotransport of β alanine by luminal membrane vesicles of rabbit proximal tubule. J. Physiol. (Lond) 411, 517–528.

Johnstone, R.M. (1972). Shortfall of the potential energy from ion gradients for glycine accumulation. Biochim. Biophys. Acta 282, 366–373.

Kaback, H.R. (1988). Site-directed mutagenesis and ion-gradient driven active transport: On the path of the proton. Ann. Rev. Physiol. 50, 243–256.

Kaback, H.R. (1992). In and out and up and down with lac permease. Int. Rev. Cytol. 137A, 97–125.

Kamata, H., Akiyama, S., Morosawa, H., Ohta, T., Hamamoto, T., Kambe, T., Kagawa, Y., & Hirata, H. (1992). Primary structure of the alanine carrier of thermophilic bacterium PS3. J. Biol. Chem. 267, 21650–21655.

Kanai, Y. & Hediger, M.A. (1992). Primary structure and functional characterization of a high-affinity glutamate transporter. Nature 360, 467–471.

Kanai, Y., Smith, C.P., & Hediger, M.A. (1993). The elusive transporters with a high affinity for glutamate. Trends Neurosci. 16, 365–370.

Kanner, B.I. (1978). Active transport of γ-aminobutyric acid by membrane vesicles isolated from rat brain. Biochemistry 17, 1207–1211.

Kanner, B.I. (1983). Bioenergetics of neurotransmitter transport. Biochim. Biophys. Acta 726, 293–316.

Kanner, B.I. & Bendahan, A. (1990). Two pharmacologically distinct sodium- and chloride-coupled high affinity γ-aminobutyric acid transporters are present in plasma membrane vesicles and reconstituted preparations from rat brain. Proc. Natl. Acad. Sci. USA 87, 2550–2554.

Kanner, B.I. & Schuldiner, S. (1987). Mechanism of transport and storage of neurotransmitters. CRC Crit. Revs. Biochem. 22, 1–38.

Kanner, B.I., Keynan, S., & Radian, R. (1989). Structural and functional studies on the sodium-and chloride-coupled γ-aminobutyric acid transporter: Deglycosylation and limited proteolysis. Biochemistry 28, 3722–3728.

Kaunitz, J.D. & Wright, E.M. (1984). Kinetics of sodium D-glucose cotransport in bovine intestinal brush-border vesicles. J. Membr. Biol. 79, 41–51.

Keyes, S.R. & Rudnick, G. (1992). Coupling of transmembrane proton gradients to platelet serotonin transport. J. Biol. Chem. 257, 1172–1176.

Kilberg, M.S., Stevens, B.R., & Novak, D.A. (1993). Recent advances in mammalian amino acid transport. Ann. Rev. Nutrition 3, 137–165.

Kilty, J.F., Lorang, D., & Amara, S.G. (1991). Cloning and expression of a cocaine-sensitive rat dopamine transporter. Science 254, 578–579.

Kimmich, G.A. (1981). Intestinal absorption of sugar. In: Physiology of the Gastrointestinal Tract. (Johnston, L.R., ed.), pp. 1035–1061, Raven Press, New York.

Kimmich, G.A., Carter-Su, C., & Randles, J. (1977). Energetics of Na$^+$-dependent sugar transport by isolated intestinal cells: Evidence for a major role for membrane potentials. Amer. J. Physiol. 233, C357–C362.

Kimmich, G.A., Randles, J., & Bennett, E. (1991). Sodium dependent succinate transport by isolated chick intestinal cells. Am. J. Physiol. 260, C1151–C1157.

Kipnis, D.M. & Noall, M.W. (1958). Stimulation of amino acid transport by insulin in the isolated rat diaphragm. Biochim. Biophys. Acta 28, 226–227.

Koepsell, H. (1986). Methodological aspects of purification and reconstitution of transport proteins from mammalian plasma membranes. Rev. Physiol. Pharmacol. 104, 65–137.

Koepsell, H., Menuhr, H., Ducis, I., & Wissmuller, T.F. (1983). Partial purification and reconstitution of the Na$^+$- D-glucose cotransport protein from pig renal proximal tubules. J. Biol. Chem. 258, 1888–1894.

Kong, C.-T., Yet, S.-F., & Lever, J.E. (1993). Cloning and expression of a mammalian Na$^+$/amino acid cotransporter with sequence similarity to Na$^+$/glucose cotransporters. J. Biol. Chem. 268, 1509–1512.

Kwon, H.M., Yamauchi, A., Uchida, S., Preston, A.S., Garcia-Perez, A., Burg, M.B., & Handler, J.S. (1992). Cloning of the cDNA for a Na$^+$/myo-inositol cotransporter, a hypertonicity stress protein. J. Biol. Chem. 267, 6297–6301.

Laris, P.C., Bootman, M., Pershadsingh, H.A., & Johnstone, R.M. (1978). Influence of cellular amino acids and the Na$^+$:K pump on the membrane potential of the Ehrlich ascites tumor cells. Biochim. Biophys. Acta 512, 397–414.

Le Cam, A. & Freychet, P. (1976). Glucagon stimulates the A system for neutral amino acid transport in isolated hepatocytes of adult rat. Biochem. Biophys. Res. Commun. 72, 893–901.

Lerner, J. (1987). Acidic amino acid transport in animal cells and tissues. Comp. Biochem. Physiol. 87B, 443–457.

Lever, J.E. (1977). Membrane potential and neutral amino acid transport in plasma membrane vesicles from simian virus transformed mouse fibroblasts. Biochemistry 16, 4328–4334.

Lever, J.E., Clingan, D., & Jimenez de Asua, L. (1976). Prostaglandin F2alpha and insulin stimulate phosphate uptake and (Na^+, K^+) ATPase activity in resting mouse fibroblast cultures. Biochem. Biophys. Res. Commun. 71, 136–142.

Lin, G., McCormick, J.I., & Johnstone, R.M. (1994). Differentiation of two classes of "A" system amino acid transporters. Arch. Biochem. Biophys. 312, 308–315.

Lin, G., McCormick, J.I., Dhe-Paganon, S., Silvius, J.R., & Johnstone, R.M. (1990). Role of specific acidic lipids on the reconstitution of Na^+-dependent amino acid transport in proteoliposomes derived from Ehrlich cell plasma membranes. Biochemistry 29, 4575–4581.

Liu, Q.-R., López-Corcuera, B., Mandiyan, S., Nelson. H., & Nelson, N. (1993). Molecular characterization of four pharmacologically distinct α-aminobutyric acid transporters in mouse brain. J. Biol. Chem. 268, 2106–2112.

Liu, Q.-R., López-Corcuera, B., Nelson, H., Mandiyan, S., & Nelson, N. (1992a). Cloning and expression of a cDNA encoding the transporter of taurine and β-alanine in mouse brain. Proc. Natl. Acad. Sci. USA 89, 12145–12149.

Liu, Q.-R., Nelson, H., Mandiyan, S., López-Corcuera, B., & Nelson, N. (1992b). Cloning and expression of a glycine transporter from mouse brain. FEBS Lett. 305, 110–114.

López-Corcuera, B. & Aragón, C. (1989). Solubilization and reconstitution of the sodium-and-chloride-coupled glycine transporter from rat spinal cord. Eur. J. Biochem. 181, 519–524.

López-Corcuera, B., Liu, Q.-R., Mandiyan, S., Nelson, H., & Nelson, N. (1992). Expression of a mouse brain cDNA encoding novel gamma-aminobutyric acid transporter. J. Biol. Chem. 267, 17491–17493.

López-Corcuera, B., Vázquez, J., & Aragón, C. (1991). Purification of the sodium- and chloride-coupled glycine transporter from central nervous system. J. Biol. Chem. 266, 24809–24814.

Mabjeesh, N.J. & Kanner, B.I. (1992). Neither amino nor carboxyl termini are required for function of the sodium- and chloride-coupled γ-aminobutyric acid transporter from rat brain. J. Biol. Chem. 267, 2563–2568.

Mackenzie, B., Panayotova-Heiermann, M., Loo, D.D.F., Lever, J.E., & Wright, E.M. (1994) SAAT1 is a low affinity Na^+/glucose cotransporter and not an amino acid transporter. J. Biol. Chem. 269, 22488–22491.

Malathi, P. & Takahashi, M. (1990). Isolation and reconstitution of the sodium-dependent glucose transporter. Meth. Enzymol. 192, 438–447.

Mathew, A., Grdisa, M., & Johnstone, R.M. (1993). Nucleosides and glutamine but not glucose are energy sources for ATP production in chicken red cells. Biochem. & Cell. Biol. 71, 288–295.

McCormick, J.I. & Johnstone, R.M. (1995) Identification of the integrin $\alpha_3\beta_1$- as a component of a partially purified amino acid transporter from Ehrlich cell membrane. Biochem. J. 311, 743–751.

McCormick, J. & Johnstone, R.M. (1990). Evidence for an essential sulphydryl group at the substrate binding site of the A system transporter of Ehrlich cell plasma membranes. Biochem. Cell. Biol. 68, 512–519.

McCormick, J.I. & Johnstone, R.M. (1988a). Simple and effective purification of a Na^+-dependent amino acid transport system from Ehrlich ascites cell plasma membrane. Proc. Natl. Acad. Sci. USA 85, 7877–7881.

McCormick, J.I. & Johnstone, R.M. (1988b). Volume enlargement and recovery of Na^+-dependent amino acid transport in proteoliposomes derived from Ehrlich ascites cell membranes. J. Biol. Chem. 263, 8111–8119.

McCormick, J.I., Jetté, M., Potier, M., Beliveau, R., & Johnstone, R.M. (1991). Molecular size of a Na^+-dependent amino acid transporter in Ehrlich ascites cell plasma membranes estimated by radiation inactivation. Biochemistry 30, 3704–3709.

McCormick, J.I., Silvius, J.R., & Johnstone, R.M. (1985). Effect of alkali cations on freeze-thaw-dependent reconstitution of amino acid transport from Ehrlich ascites cell plasma membrane. J. Biol. Chem. 260, 5706–5714.

McCormick, J.I., Tsang, D., & Johnstone, R.M. (1984). A simple and efficient method for reconstitution of amino acid and glucose transport systems from Ehrlich ascites cells. Arch. Biochem. Biophys. 231, 355–365.

Mullins, L.J. (1975). Ion selectivity of carriers and channels. Biophys. J. 15, 921–931.

Murer, H. & Hopfer, U. (1974). Demonstration of electrogenic Na-dependent glucose transport in intestinal brush-border membranes. Proc. Natl. Acad. Sci. USA 71, 484–488.

Nakao, T., Yamato, I., & Anraku, Y. (1987). Nucleotide sequence of PutP, the proline carrier gene of E. coli K12. Mol. and Gen. Genet. 208, 70–75.

Newman, M.J. & Wilson, T.H. (1980). Solubilization and reconstitution of the lactose transport system from Escherichia coli. J. Biol. Chem. 255, 10583–10586.

Noall, M.W., Riggs, T.R., Walker, L.M., & Christensen, H.N. (1957). Endocrine control of amino acid transfer. Science 126, 1002–1005.

O'Grady, S.M., Palfrey, H.C., & Field, M. (1987). Characteristics and functions of Na-K-Cl cotransport in epithelial tissues. Am. J. Physiol. 253, C177–C192.

Ohta, T., Isselbacher, K.J., & Rhoads, D.B. (1990). Regulation of glucose transporters in LLC-PK1 cells: Effects of D-glucose and monosaccharides. Mol. Cell. Biol. 10, 6491–6499.

Pacholczyk, T., Blakely, R.D., & Amara, S.G. (1991). Expression cloning of a cocaine and antidepressant-sensitive human noradrenaline transporter. Nature 350, 350–354.

Pajor, A.M. & Wright, E.M. (1992). Cloning and functional expression of a mammalian Na$^+$/nucleoside cotransporter. A member of the SGLT family. J. Biol. Chem. 267, 3557–3560.

Pajor, A.M., Hirayama, B.A., & Wright, E.M. (1992). Molecular biology approaches to comparative study of Na$^+$-glucose cotransport. Am. J. Physiol. 263, R489–R495.

Pantanowitz, S., Bendahan, A., & Kanner, B.I. (1993). Only one of the charged amino acids located in the transmembrane α-helices of the γ-aminobutyric acid transporter (Subtype A) is essential for its activity. J. Biol. Chem. 268, 3222–3225.

Parent, L., Supplisson, S., Loo, D.F., & Wright, E.M. (1992a). Electrogenic properties of the cloned Na$^+$/glucose cotransporter. J. Membr. Biol. 125, 49–62.

Parent, L., Supplisson, S., Loo, D.F., & Wright, E.M. (1992b). Electrogenic properties of the cloned Na$^+$/glucose cotransporter: II. J. Membr. Biol. 125, 63–79.

Peerce, B.E. (1989). Identification of the intestinal Na-phosphate cotransporter. Am. J. Physiol. 256, G645–G652.

Peerce, B.E. & Clarke, R.D. (1990). Isolation and reconstitution of the intestinal Na$^+$/glucose cotransporter. J. Biol. Chem. 265, 1731–1736.

Peerce, B.E. & Wright, E.M. (1984). Sodium-induced conformational changes in the glucose transporter of intestinal brush borders. J. Biol. Chem. 259, 14105–14112.

Peerce, B.E., Cedilote, M., Seifert, S., Levine, R., Kiesling, C., & Clarke, R.D. (1993). Reconstitution of intestinal Na$^+$-phosphate cotransporter. Am. J. Physiol. 264, G609–G616.

Pick, U. (1981). Liposomes with a large trapping capacity prepared by freezing and thawing of sonicated phospholipid mixtures. Arch. Biochem. Biophys. 212, 186–194.

Pines, G., Danbolt, N.C., Bjørås, M., Zhang, Y., Bendahan, A., Eide, L., Koepsell, H., Storm-Mathisen, J., Seeberg, E., & Kanner, B.I. (1992). Cloning and expression of a rat brain L-glutamate transporter. Nature 360, 464–467.

Plakidou-Dymock, S., Tanner, M.J., & McGivan, J.D. (1993). A role for aminopeptidase N in Na$^+$-dependent amino acid transport in bovine renal brush-border membranes. Biochem. J. 290. 59–65.

Polvani, C. & Blostein, R. (1988). Protons as substitutes for sodium and potassium in the sodium pump reaction. J. Biol. Chem. 263, 16757–16763.

Polvani, C., Sachs, G., & Blostein, R. (1989). Sodium ions as substitutes for protons in the gastric H+,K+-ATPase. J. Biol. Chem. 264, 17854–17859.

Potashner, S.J. & Johnstone, R.M. (1971). Cation, gradients, ATP and amino acid accumulation in Ehrlich cells. Biochim. Biophys. Acta 233, 91–103.

Pourcher, T., Bassilana, M., Sarkar, H.K., Kaback, H.R., & Leblanc, G. (1992). Melibiose permease of *Escherichia coli*: Mutation of histidine-94 alters expression and stability rather than catalytic activity. Biochem. 31, 5225–5231.

Pourcher, T., Zani, M.-L., & Leblanc, G. (1993). Mutagenesis of acidic residues in putative membrane-spanning segments of the melibiose permease of *Esherichia coli*. J. Biol. Chem. 268, 3209–3215.

Racker, E. (1979). Reconstitution of membrane processes. Meth. Enzymol. 55, 699–711.

Radian, R. & Kanner, B.I. (1985). Reconstitution and purification of the sodium- and chloride-coupled γ-aminobutyric acid transporter from rat brain. J. Biol. Chem. 260, 11859–11865.

Radian, R., Bendahan, A., & Kanner, B.I. (1986). Purification and identification of the functional sodium- and chloride-coupled γ-aminobutyric acid transport glycoprotein from rat brain. J. Biol. Chem. 261, 15437–15441.

Ramamoorthy, S., Cool, D.R., Leibach, F.H., Mahesh, V.B., & Ganapathy, V. (1992). Reconstitution of the human placental 5-hydroxytryptamine transporter in a catalytically active form after detergent solubilization. Biochem. J. 286, 89–95.

Ramamoorthy, S., Kulanthaivel, P., Leibach, F.H., Mahesh, V.B., & Ganapathy, V. (1993). Solubilization and functional reconstitution of the human placental taurine transporter. Biochim. Biophys. Acta 1145, 250–256.

Reid, M. & Eddy, A.A. (1971). Apparent metabolic regulation of the coupling between the potassium ion gradient and methionine transport in mouse ascites-tumour cells. Biochem. J. 124, 951–952.

Ricklis, E. & Quastel, J.H. (1958). Effects of cations on sugar absorption by isolated surviving guinea pig intestine. Can. J. Biochem. Physiol. 36, 347–362.

Riggs, T.R., Walker, L.M., & Christensen, H.N. (1958). Potassium migration and amino acid transport. J. Biol. Chem. 233, 1479–1484.

Ritz, M.C., Lamb, R.J., Goldberg, S.R., & Kumar, M.J. (1987). Cocaine receptors on dopamine transporters are related to self-administration of cocaine. Science 237, 1219–1223.

Roepe, P.D. & Kaback, H.R. (1989). Characterization and functional reconstitution of a soluble form of the hydrophobic membrane protein lac permease from *Escherichia coli*. Proc. Natl. Acad. Sci. USA 86, 6087–6091.

Roepe, P.D. & Kaback, H.R. (1990). Isolation and functional reconstitution of soluble melibiose permease from *Escherichia coli*. Biochemistry 29, 2572–2577.

Saier, M.H. Jr., Daniels, G.A., Boerner, P., & Lin, J. (1988). Neutral amino acid transport systems in animal cells. Potential targets of oncogene action and regulators of cellular growth. J. Membr. Biol. 104, 1–20.

Schafer, J.A. & Heinz, E. (1971). The effect of reversal of Na+- and K+-electrochemical potential gradients on the active transport of amino acids in Ehrlich ascites tumor cells. Biochim. Biophys. Acta 249, 15–33.

Schon, F.E. & Kelly, J.S. (1975). Selective uptake of [3H]β-alanine by glia: Association with the glial uptake system for GABA. Brain Res. 86, 243–257.

Schuldiner, S. & Kaback, H.R. (1975). Membrane potential and active transport in membrane vesicles from *Escherichia coli*. Biochemistry 14, 5451–5460.

Seaston, A., Inkson, C., & Eddy, A.A. (1973). Absorption of protons with specific amino acids and carbohydrates by yeast. Biochem. J. 134, 1031–1043.

Semenza, G., Kessler, M., Hosang, M., Weber, J., & Schmidt, U. (1984). Biochemistry of the Na+, D-glucose cotransporter of the small intestinal brush-border membrane. The state of the art in 1984. Biochim. Biophys. Acta 779, 343–379.

Shafqat, S., Tamarappoo, B.K., Kilberg, M.S., Puranam, R.S., McNamara, J.O., Guadaño-Ferraz, A., & Fremeau, R.T. (1993). Cloning and expression of a novel Na+-dependent neutral amino acid transporter structurally related to mammalian Na$^+$/glutamate cotransporters. J. Biol. Chem. 268, 15351–15355.

Shimada, S., Kitayama, S., Lin, C.-L., Patel, A., Nanthakumar, E., Gregor, P., Kuhar, M., & Uhl, G. (1991). Cloning and expression of a cocaine-sensitive dopamine transporter complementary DNA. Science 254, 576–578.

Shotwell, M.A., Kilberg, M.S., & Oxender, D.L. (1983). The regulation of neutral amino acid transport in mammalian cells. Biochem. Biophys. Acta 737, 267–284.

Sigrist-Nelson, K., Murer, H., & Hopfer, U. (1975). Active alanine transport in isolated brush-border membranes. J. Biol. Chem. 250, 5674–5680.

Silvius, J.R. (1992). Solubilization and functional reconstitution of biomembrane components. Ann. Rev. Biophys. Biomol. Struct. 21, 323–348.

Skulachev, V.P. (1971). Energy transformations in the respiratory chain. Curr. Top. Bioenerg. 4, 127–190.

Smith, C.D., Hirayama, B.A., & Wright, E.M. (1992a). Baculovirus-mediated expression of the Na$^+$/glucose cotransporter in Sf9 cells. Biochim. Biophys. Acta 1104, 151–159.

Smith, K.E., Borden, L.A., Hartig, P.R., Branchek, T., & Weinshank, R.L. (1992b). Cloning and expression of a glycine transporter reveal colocalization with NMDA receptors. Neuron 8, 927–935.

Stein, W. (1986). Transport and Diffusion Across Cell Membranes, Chapter 5, pp. 374–439. Academic Press, New York.

Stevens, B.R., Kaunitz, J.D., & Wright, E.M. (1984). Intestinal transport of amino acids and sugars: Advances using membrane vesicles. Ann. Rev. Physiol. 46, 417–433.

Storck, T., Schulte, S., Hofmann, K., & Stoffel, W. (1992). Structure, expression, and functional analysis of a Na$^+$-dependent glutamate/aspartate transporter from rat brain. Proc. Natl. Acad. Sci. USA 89, 10955–10959.

Suzuki, K. & Kono, T. (1980). Evidence that insulin causes translocation of glucose transport activity to the plasma membrane from an intracellular storage site. Proc. Natl. Acad. Sci. USA 77, 2542–2545.

Tamarappoo, B.K. & Kilberg, M.S. (1991). Functional reconstitution of the hepatic system N amino acid transport activity. Biochem. J. 274, 97–101.

Tiruppathi, C., Brandsch, M., Miyamoto, Y., Ganapathy, V., & Leibach, F.H. (1992). Constitutive expression of the taurine transporter in a human colon carcinoma cell line. Amer. J. Physiol. 263, G626–G631.

Tolner, B., Poolman, B., Wallace, B., & Konings, W.N. (1992). Revised nucleotide sequence of the gltP gene, which encodes the proton-glutamate-aspartate transport protein of *Escherichia coli* K-12. J. Bact. 174, 2391–2393.

Touabi, M. & Jeanrenaud, B. (1969). α-aminoisobutyric acid uptake in isolated mouse fat cells. Biochim. Biophys. Acta 173, 128–140.

Turner, R.J. (1986). β amino acid transport across the renal brush-border membrane is coupled to both Na and Cl. J. Biol. Chem. 261, 16060–16066.

Turner, R.J. & Moran, A. (1982a). Heterogeneity of Na dependent D-glucose transport sites along the proximal tubule. Amer. J. Physiol. 242, F406–F414.

Turner, R.J. & Moran, A. (1982b). Further studies of proximal tubular brush border membrane D-glucose transport heterogeneity. J. Membr. Biol. 70, 37–45.

Uhl, G.R. & Kitayama, S. (1993). A cloned dopamine transporter. Potential insights into Parkinson's disease pathogenesis. Adv. Neurol. 60, 321–324.

Vachon, V., Delisle, M.-C., Laprade, R., & Beliveau, R. (1991). Reconstitution of the renal brush-border membrane sodium/phosphate co-transporter. Biochem. J. 278, 543–548.

Van Slyke, D.D. & Meyer, G.M. (1913). The fate of protein digestion products in the body III. The absorption of amino acids from the blood by the tissues. J. Biol. Chem. 16, 197–212.

Verzár, F. & Sullmann, H. (1937). Die bildung von phosphorsäureestern in der darmschleimhaut bei der resorption. Biochem. Z. 289, 323–340.

Vidaver, G.A. (1964a). Mucate inhibition of glycine entry into pigeon red cells. Biochemistry 3, 799–803.

Vidaver, G.A. (1964b). Some tests for the hypothesis that the sodium ion gradient furnishes the energy for glycine- active transport by pigeon red cells. Biochemistry 3, 803–808.

Waggoner, A.S. (1976). Optical probes of membrane potential. J. Membr. Biol. 27, 317–334.

Weigensberg, A. & Blostein, R. (1985). Na^+-coupled glycine transport in reticulocyte vesicles of distinct sidedness: Stoichiometry and symmetry. J. Membr. Biol. 86, 37–44.

Werner, A., Moore, M.L., Mantei, N., Biber, J., Semenza, G., & Murer, H. (1991). Cloning and expression of cDNA for a Na/Pi cotransport system of kidney cortex. Proc. Natl. Acad. Sci. USA 88, 9608–9612.

Wilbrandt, W. & Laszt, L. (1933). Untersuchungen über die Ursachen der selekiven Resorption der Zucker aus dem Darm. Biochem. Z. 259, 398–417.

Wool, I.G., Castles, J.J., & Moyer, A.N. (1965). Regulation of amino acid accumulation in isolated rat diaphragm: Effect of puromycin and insulin. Biochim. Biophys. Acta 107, 333–345.

Wright, E.M., Hager, K.M., & Turk, E. (1992). Sodium cotransport proteins. Curr. Opin. Cell. Biol. 4, 696–702.

Wright, S.H., Hirayama, B., Kaunitz, J.D., Kippen, I., & Wright, E.M. (1983). Kinetics of sodium succinate cotransport across renal brush-border membranes. J. Biol. Chem. 258, 5456–5462.

Wu, J-S.R. & Lever, J.E. (1987). Monoclonal antibodies that bind the renal Na^+/glucose symport system. Biochemistry 26, 5783–5790.

Yamauchi, A., Uchida, S., Kwon, H.M., Preston, A.S., Robey, R.B., Garcia-Perez, A., Burg, M.B., & Handler, J.S. (1992). Cloning of a Na^+- and Cl^--dependent betaine transporter that is regulated by hypertonicity. J. Biol. Chem. 267, 649–652.

Zani, M-L., Pourcher, T., & Leblanc, G. (1993). Mutagenesis of acidic residues in putative membrane-spanning segments of the melibiose permease of *Escherichia coli* II. Effect on cationic selectivity and coupling properties. J. Biol. Chem. 268, 3216–3221.

Chapter 4

Sodium-Calcium Exchangers and Calcium Pumps

EMANUEL E. STREHLER

INTRODUCTION: CALCIUM AS AN INTRACELLULAR SIGNAL

A bewildering multitude of biochemical processes is regulated by one and the same trigger: a change, usually an increase, in the concentration of intracellular free Ca^{2+}.

Principles of Medical Biology, Volume 4
Cell Chemistry and Physiology: Part III, pages 125–150.
Copyright © 1996 by JAI Press Inc.
All rights of reproduction in any form reserved.
ISBN: 1-55938-807-2

Ca^{2+} is thus considered one of the foremost "second messengers" of eukaryotic cells. Indeed, it is difficult, if not impossible, to name an important biological event where Ca^{2+} does not play an essential regulatory role. Processes as diverse as motility, secretion, gene expression, and cell division are controlled by the Ca^{2+} signal. The key to understanding the preeminent role of Ca^{2+} as an intracellular trigger lies in the hugely imbalanced distribution of this cation across cellular membranes: cytosolic free Ca^{2+} concentrations in resting cells are usually in the nanomolar range (typically around 100 nM), whereas extracellular Ca^{2+} (e.g., in the blood plasma) is about 3 millimolar. The low background concentration and the large inward-directed concentration gradient of Ca^{2+} insure that even a very limited, short-lived opening of the membrane will lead to a significant local increase in cytosolic free Ca^{2+}, and thus to an optimal signal-to-noise ratio. The Ca^{2+} signal must be recognized inside the cell and must then be accurately transmitted, processed, and terminated. Signal recognition must occur with very high specificity, effectively "feeling" comparatively small increases in free Ca^{2+} in the continuous presence of much (up to several orders of magnitude) higher concentrations of other cations (such as Mg^{2+}, Na^+, and K^+). This is accomplished through high-affinity Ca^{2+} binding proteins such as calmodulin which will bind Ca^{2+} ions tightly, yet reversibly, in the concentration range encountered inside a normal cell. Termination of the Ca^{2+} signal requires a resetting of the cytosolic free Ca^{2+} level to the resting state. Since the signaling agent (Ca^{2+}) itself can not be chemically destroyed (as is possible for other second messengers, e.g., cAMP or IP_3), terminating the signal either requires buffering of the Ca^{2+} and/or its permanent removal from the cytosol. The latter inevitably means that Ca^{2+} must be transported across a membrane, either into a Ca^{2+} sequestering intracellular organelle or out of the cell into the extracellular fluid.

THE INVOLVEMENT OF MEMBRANE-BOUND CALCIUM TRANSPORTERS IN INTRACELLULAR CALCIUM HOMEOSTASIS

Evidently, cells must be virtually impermeable to Ca^{2+}, or else the steep concentration gradient which normally exists for this cation across the plasma membrane (and most organellar membranes) would be impossible to maintain. Both Ca^{2+} influx into the cytosol along the concentration gradient, as well as Ca^{2+} extrusion against the gradient must therefore be carefully regulated and requires the presence of specially designed Ca^{2+}-specific membrane carriers. These carriers are integral membrane proteins that can be divided into three main categories according to their major functional characteristics: channels, exchangers, and pumps. All of these carriers can effectively participate in the control of intracellular Ca^{2+} concentrations because they facilitate transport of the cation from one side of the membrane to the other without being used up; i.e., a single transporter can repeatedly perform its task of allowing Ca^{2+} into, or of transporting it out of, the cell. Not surprisingly, therefore, the total number of Ca^{2+} transporter molecules in a given membrane is often not very high, a fact that has been at least partially responsible for the protracted difficulties in the biochemical and molecular characterization of these proteins.

Among the three categories of Ca^{2+} transporters, a (functional) distinction can be conveniently made between channels on the one hand and exchangers and pumps on the other. Ca^{2+} channels, as indicated by their name, form a pore (channel) in the membrane through which Ca^{2+} can flow down its concentration gradient. The opening and closing of the pore must be tightly controlled, e.g., by the membrane potential (voltage-gated Ca^{2+} channels) or by specific chemical substances (ligand-gated Ca^{2+} channels). In contrast, exchangers and pumps, although reversible, are mainly used by the cell to transport Ca^{2+} against its large concentration gradient. Exchangers trade Na^+ for Ca^{2+}, moving one ion along its concentration and the electrical gradient in exchange for the uphill movement of the other ion. Pumps are distinguished by their direct use of chemical energy (in the form of ATP) for the uphill transmembrane movement of the specifically transported cation.

DISTRIBUTION OF DIFFERENT CALCIUM TRANSPORTERS WITHIN A CELL

In accordance with their essential role in controlling intracellular Ca^{2+} homeostasis, a variety of different Ca^{2+} transporters are present in the plasma membrane surrounding a cell, as well as in the membranes of all intracellular organelles that participate in Ca^{2+} regulation. Figure 1 shows a schematic view of the major membrane locations of the diverse Ca^{2+} transporters in a typical animal cell. Channels are found mainly in the cell plasma membrane and in the endo-/sarco-plasmic reticulum (ER/SR) membrane. The outer nuclear envelope, which may be considered a specialized extension of the ER membrane, as well as other ER/SR-derived specialized compartments, such as the proposed Ca^{2+} storage organelles of nonmuscle cells (calciosomes) and the terminal cisternae of the SR in striated muscle cells, also contain Ca^{2+} channels with diverse functional and regulatory properties. Ca^{2+} pumps, providing Ca^{2+} transport in the opposite direction of the channels, are generally found in the same membranes (plasma membrane, ER/SR, outer nuclear envelope), whereas Na^+/Ca^{2+} exchangers are mainly present in the plasma membrane, as well as in the inner mitochondrial membrane of some cell types. Ca^{2+} import into mitochondria—which can serve as emergency and long-term Ca^{2+} storage organelles and in which numerous Ca^{2+}-dependent processes occur—is performed by an as yet poorly defined Ca^{2+} uniporter. This transporter uses the strong electrical gradient (about 180 mV) across the inner mitochondrial membrane to move Ca^{2+} from the positively charged outer (intermembrane) to the negatively charged inner (matrix) side of the membrane.

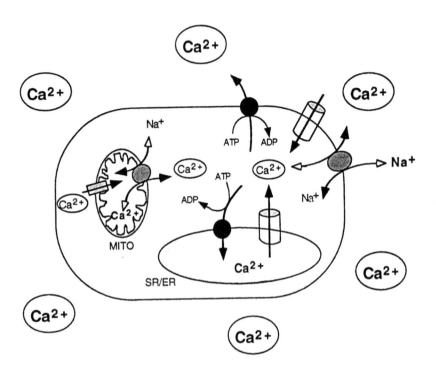

Figure 1. Cellular location of major membrane-intrinsic calcium transporters. A schematic view of an animal cell with two of its main Ca²⁺ regulating organelles. The extracellular free Ca²⁺ concentration is normally much higher than the cytosolic one (indicated by differently sized Ca²⁺ signs). Ca²⁺ fluxes across the plasma membrane are controlled by specific channels (cylinder with arrow showing direction of Ca²⁺ flux), Na⁺/Ca²⁺ exchangers (shaded oval with double arrows), and pumps (black ball with arrow indicating direction of Ca²⁺ pumping). Inner membranes of mitochondria (MITO) contain a Ca²⁺ uniporter (shaded rectangle with black arrow) for Ca²⁺ import and an exchanger mostly for Ca²⁺ efflux. Ca²⁺ efflux from, and reuptake into, the sarco-/endoplasmic reticulum (SR/ER) is performed by channels (cylinder) and pumps (ball), respectively. Channels and pumps are also found in the outer nuclear membrane (continuous with the ER) and in specialized ER/SR compartments (not shown). The presence of channels and pumps in the Golgi system and in lysosomes or other intracellular vesicles is also possible but not yet well documented. Open arrowheads on the Na⁺/Ca²⁺ exchangers indicate the possibility for reverse-mode operation; the solid arrowheads show the direction of ion fluxes in the forward mode.

In the following sections, we will focus on the two remaining types of Ca^{2+} transporters: exchangers and pumps.

SODIUM-CALCIUM EXCHANGERS: DISCOVERY, DISTRIBUTION, AND PROPERTIES

The existence of a Na^+ ion dependent transmembrane Ca^{2+} transport was first demonstrated over 25 years ago in studies using isolated cardiac muscles of guinea pigs (Reuter and Seitz, 1968) and giant axons of squids (Baker et al., 1969). This ion exchange system directly couples the transmembrane transport of Ca^{2+} to the transport of Na^+ in the opposite direction. The major driving force for this transport is provided by the electrochemical gradient for one cation across the membrane. In the forward mode, Ca^{2+} is transported out of the cell against its concentration gradient, whereas Na^+ enters the cell along its concentration gradient. Because the intracellular Na^+ concentration is only about 10 times lower than the extracellular one (about 10 mM vs. 140 mM), the Na^+ concentration gradient alone could hardly provide enough energy for the countertransport of Ca^{2+} against its more than 1,000-fold concentration gradient. Indeed, it was soon realized that the Na^+/Ca^{2+} exchange system is able to generate an electrical current, indicating that its operation entails a net charge movement across the membrane. It is now well established that one of the two major classes of Na^+/Ca^{2+} exchangers (exemplified by that of heart muscle) transports three Na^+ ions for every Ca^{2+} ion (Philipson and Nicoll, 1992). Because the resting potential of the heart sarcolemma is about −80 mV (negative on the intracellular side), the net inward transport of a positive charge (a Na^+ ion) during Na^+/Ca^{2+} exchange is energetically favorable. From the above it is clear that the Na^+/Ca^{2+} exchanger can also operate in the reverse mode, i.e., transport Ca^{2+} into and Na^+ out of the cell, depending on the membrane potential and the prevailing distribution of the two cations across the membrane. This unique feature of the exchanger is of great physiological importance since it allows a rapid reversal of the Ca^{2+} transport mode under changing conditions at the membrane. In addition, Na^+/Ca^{2+} exchangers can also operate in a self exchange mode, i.e., as Ca^{2+}/Ca^{2+} or Na^+/Na^+ exchangers.

Physiological, biochemical and, more recently, molecular studies have shown that there exist two major types of Na^+/Ca^{2+} exchangers (Philipson and Nicoll, 1992). In addition to the system studied mainly from cardiac muscles, a Na^+/Ca^{2+} exchanger with significantly different properties has been identified in the outer segment of rod photoreceptor cells of the vertebrate retina. The most salient difference between the cardiac- and the photoreceptor-type exchanger concerns the stoichiometry and identity of transported ions: the rod outer segment system normally transports four Na^+ ions in exchange for one Ca^{2+} and one K^+ ion (Schnetkamp, 1990). This transporter is thus more appropriately referred to as a $Na^+/Ca^{2+},K^+$ exchanger. A further characteristic of the $Na^+/Ca^{2+},K^+$ exchanger is its apparent flexibility of transport stoichiometry which can change from an electrogenic $4Na^+/1Ca^{2+},1K^+$ to an electroneutral $3Na^+/1Ca^{2+},1K^+$ exchange (Schnetkamp, 1990). This has been explained by the presence of multiple, at least partially independent, cation-binding sites whose state of occupancy is dependent on the distribution of the cations on either side of the membrane, leading to overall flexible $Ca^{2+}:K^+$ and $Ca^{2+}:Na^+$ coupling ratios. An electroneutral Na^+/Ca^{2+} exchange mechanism ($2Na^+/1Ca^{2+}$) has also been described in the inner mitochondrial membrane, but this exchanger has not yet been studied in biochemical or molecular detail.

The two types of plasma membrane Na^+/Ca^{2+} exchangers have been most extensively studied in the tissues where they are most abundant and where they clearly play a major role in transmembrane Ca^{2+} handling: in cardiac muscle and nerve cells as well as in the photoreceptor cells of the retina. Cardiac-type plasma membrane Na^+/Ca^{2+} exchangers are, however, not restricted to excitable cells but seem to be present in almost all tissues, albeit at highly variable levels. Judging from studies analyzing the messenger RNA encoding the human cardiac-type Na^+/Ca^{2+} exchanger, this protein is by far most abundant in the heart, present in significant amounts in the brain, kidney, lung, and pancreas, but also found at low levels in placenta, skeletal muscle, smooth muscle, and liver. Indeed, except in erythrocytes, this Ca^{2+} transporter has been found in all human tissues tested so far. In striking contrast, the rod outer segment $Na^+/Ca^{2+},K^+$ exchanger is highly specific for the retinal photoreceptor cells and messenger RNA coding for this transporter has only been detected in these cells. The only evidence for the existence of a K^+-cotransporting Na^+/Ca^{2+} exchange mechanism in a cell type other than the photoreceptors stems from ion transport studies on brain synaptic vesicle membranes and platelets.

MECHANISTIC ASPECTS OF SODIUM-CALCIUM EXCHANGERS

Although the exact mechanism by which the Na^+/Ca^{2+} exchanger accomplishes its task of ion transport is not yet known, a number of useful models have been proposed on the basis of theoretical considerations and experimental observations. A simple model that accommodates the results from various ion flux studies assumes that two separate ion binding sites exist in the cardiac-type Na^+/Ca^{2+} exchanger (Reeves, 1985): an "A-site" occupied by a single Ca^{2+} or by two Na^+ ions, and a "B-site" which is a nonselective alkali cation binding site when Ca^{2+} occupies the A-site, but becomes Na^+ selective when two Na^+ are bound to the A-site. Transport from this B-site will only occur when the A-site is occupied by Na^+ as well. An analogous model for the photoreceptor-type exchanger (Schnetkamp, 1990) postulates the existence of an A-site (occupied either by a Ca^{2+} or two Na^+ ions), an alkali ion-specific B-site, and a "K-site" specific for this type of exchanger. When Ca^{2+} is bound to the A-site, the K-site is selective for K^+ and K^+

will be cotransported with Ca^{2+} (under these conditions, the B-site is nonselective and Na^+ and K^+ may compete for this site). When the A-site contains two Na^+ ions, the B- and the K-sites both become selective for Na^+ and thus four Na^+ ions will be transported in exchange for a Ca^{2+} and a K^+. As in the two-site model for the cardiac-type exchanger, transport from the B-site can only occur when the A-site is occupied by two Na^+ ions (and the K-site by a fourth Na^+). The available data further indicate that the Ca^{2+}-occupied A-site can be translocated through the membrane independently of translocation of the K-site, a feature that helps explain the nonfixed $Ca^{2+}:K^+$ coupling ratio for the $Na^+/Ca^{2+},K^+$ exchanger.

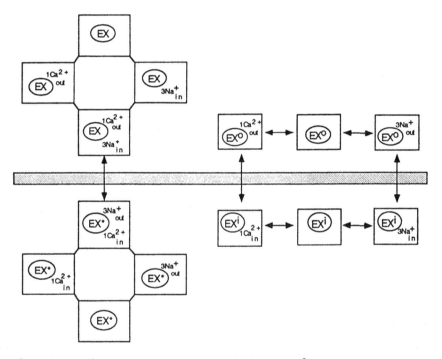

Figure 2. Simultaneous vs. consecutive model for Na^+/Ca^{2+} exchanger operation. A simultaneous exchanger (left) transports all Na^+ and Ca^{2+} ions in one membrane-crossing molecular rearrangement step (double arrow across shaded bar). In order for such an exchanger to operate in a Ca^{2+}/Ca^{2+} or Na^+/Na^+ self-exchange mode, the two sets of ion binding sites would have to be very similar. Ex and Ex* are two different exchanger configurations displaying preferential Ca^{2+} binding on the extra- and intracellular sides, respectively. In a consecutive exchanger (right), Ca^{2+} and Na^+ ions are translocated across the membrane in two independent membrane crossing steps. EX^o and EX^i are the two configurations of the exchanger displaying the transport ion binding sites on the external (extracellular) and internal (intracellular) side of the membrane, respectively. The models are drawn for a $3Na^+/1Ca^{2+}$ exchanger but can be similarly applied to a $4Na^+/1Ca^{2+},1K^+$ exchanger.

Two essentially different schemes have been developed for the kinetics of the exchange mechanism (Hilgemann, 1988): a simultaneous and a consecutive model (Figure 2). In the simultaneous scheme the Na^+/Ca^{2+} exchanger has two separate sets of sites, one accommodating a Ca^{2+} (and, in the case of the photoreceptor exchanger, a K^+) ion and a second set accommodating three Na^+ ions (four for the photoreceptor-type protein). During the exchange process these two sets of sites would trade places with respect to the plasma membrane surface. Such an exchanger would transport all Na^+ and Ca^{2+} (and K^+) ions simultaneously in one membrane-crossing step. Because the Na^+/Ca^{2+} exchangers can operate in a self-exchange mode (Ca^{2+}/Ca^{2+}, Na^+/Na^+, $Ca^{2+},K^+/Ca^{2+},K^+$) the two sets of sites would have to be very similar. In the consecutive scheme, only one set of sites is required which would alternately be exposed to the two different sides of the membrane and would accommodate either one Ca^{2+} ion (and one K^+ ion in the photoreceptor-type exchanger) or three Na^+ ions (four Na^+ in the photoreceptor exchanger). In this case, Ca^{2+} (and K^+ where applicable) moves across the membrane during one molecular rearrangement step and Na^+ during a separate, second step. The observed ability of the Na^+/Ca^{2+} exchangers to mediate Ca^{2+}/Ca^{2+} and Na^+/Na^+ self-exchange can be easily accommodated by the consecutive model. Numerous ion transport and electrophysiological studies currently favor a kinetic reaction scheme for the Na^+/Ca^{2+} exchanger resembling the consecutive model rather than the simultaneous one (Matsuoka and Hilgemann, 1992).

The single positive charge-carrying step during Na^+/Ca^{2+} exchange has been linked to the transport (and release) of the Na^+ ions. Thus, in the cardiac-type exchanger, the half-reaction carrying three Na^+ ions appears to be electrogenic, whereas the Ca^{2+} translocation step is electroneutral. This situation resembles that proposed for the Na^+/K^+ ATPase (which pumps three Na^+ out of the cell in exchange for two K^+) where the fully occupied ion-binding site (loaded with three Na^+ ions) carries a single positive charge and the empty, unliganded ion binding site may thus carry two negative charges.

MOLECULAR CHARACTERIZATION OF SODIUM-CALCIUM EXCHANGERS

For a more detailed understanding of the mechanism and the regulation of any protein, it is necessary to know more about its molecular structure, including as one of the first steps its primary amino acid sequence. Because the Na^+/Ca^{2+} exchangers are low abundance, labile membrane proteins that display no obvious and easily measurable enzymatic activity, their molecular characterization proved to be very difficult. Indeed, the first primary structure for a cardiac-type Na^+/Ca^{2+} exchanger (from a dog heart) was only reported in 1990 (Nicoll et al., 1990) and the first description of a $Na^+/Ca^{2+},K^+$ exchanger (from bovine rod photoreceptors) is an even more recent event (Reiländer et al., 1992). These new discoveries have added great excitement to the studies of the Na^+/Ca^{2+} exchangers.

The complete amino acid sequence for the dog (Nicoll et al., 1990) and human (Komuro et al., 1992) cardiac exchanger as deduced from cloned cDNAs comprises 970 and 973 residues, respectively, yielding calculated molecular weights of about 110 kDa for these proteins. Direct amino terminal amino acid sequencing of the purified protein showed, however, that a leader peptide is cleaved off during processing, leaving a mature exchanger of only 938 amino acids. The mature Na^+/Ca^{2+} exchangers of dog and human heart are more than 99% identical in their primary sequence, whereas their leader peptides are much more divergent, differing in length (32 and 35 residues, respectively) and sequence (only 66% identity over 32 residues). Hydropathy analysis of the primary sequence predicts 12 putative membrane-crossing hydrophobic stretches in the unprocessed protein. Because the first hydrophobic region corresponds to the leader sequence, the mature exchanger contains only 11 putative transmembrane segments and may thus have its N- and C-terminus on opposite sides of the membrane. The 11 hydrophobic segments are divided into two clusters separated by a large hydrophilic region comprising about 520 amino acid residues (more than half of the total mass of the protein) between putative transmembrane segments five and six (Figure 3). Remarkably, this arrangement of two clusters of hydrophobic stretches interrupted by a large hydrophilic segment is also found in other types of ion transporters, e.g., in the P-type cation pumps (see the sections below on the Ca^{2+} pumps).

It is reasonable to assume that multiple transmembrane segments form a three-dimensional pore structure through which the ions are translocated during the reaction cycle. In support of this proposal, several hydrophilic amino acid residues (particularly the hydroxyl-containing serine and threonine, as well as the acidic aspartate and glutamate residues) are interspersed at regular intervals in several of the putative transmembrane segments where they may participate in the formation of an ion conduction pathway. Hydrophilic and charged amino acids in transmembrane regions are also implicated in ion binding and transport in the Ca^{2+} pumps (see below). Experiments using antibodies raised against peptides corresponding to different areas of the Na^+/Ca^{2+} exchanger indicate that the large hydrophilic loop between the membrane-anchoring segments is located intracellularly. No precise function could as yet be assigned to this large part of the exchanger, although speculations linking this region to interactions with intracellular structures (e.g., the cytoskeleton) and various regulatory mechanisms seem plausible (Philipson and Nicoll, 1992). Rather intriguingly, a truncated cardiac Na^+/Ca^{2+} exchanger mutant lacking essentially all of the large hydrophilic segment has been shown to be functionally active; thus, the deleted region does not appear to play a major role in specific ion binding and transport (Matsuoka et al., 1993).

A synthetic peptide corresponding to a short stretch of about 20 amino acids in the hydrophilic loop just following the transmembrane segment five can act as an autoinhibitory peptide: its sequence, containing many basic and aromatic residues, is reminiscent of that of autoinhibitory, calmodulin-binding peptides in other proteins, notably in the plasma membrane Ca^{2+} pump (see below). The peptide does indeed bind calmodulin but only with moderate affinity, whereas it is a very potent inhibitor of both normal and reverse mode Na^+/Ca^{2+} exchange. Although calmodulin does not directly regulate the Na^+/Ca^{2+} exchanger, the presence of a potentially autoinhibitory amino acid region in the exchanger may indicate a common theme for autoregulation of the Ca^{2+} transporters. Cardiac-type Na^+/Ca^{2+} exchangers are positively regulated by Ca^{2+} and Mg^{2+}-ATP on the cytoplasmic side of the membrane. Accordingly, the existence of a regulatory Ca^{2+} binding site on the Na^+/Ca^{2+} exchangers has been proposed; such a site(s) may, in fact, be provided by some highly negatively charged (aspartate- and glutamate-rich) amino acid stretches within the large, cytoplasmic hydrophilic loop of these proteins (Nicoll et al., 1990). The mechanism of action of Mg^{2+}-ATP on the exchanger is not yet clear. Direct phosphorylation by any of several known protein kinases (protein kinase A, protein kinase C, Ca^{2+}-calmodulin dependent kinase) does not seem to be involved; proposed alternatives include indirect stimulation by phosphorylation of a regulatory protein or an effect on the asymmetric distribution of membrane phospholipids (Philipson and Nicoll, 1992). The exchanger—again, like the Ca^{2+} pump—is stimulated by anionic phospholipids whose preferred presence on the inner face of the membrane is maintained by an ATP-dependent aminophospholipid-translocase.

Analysis of the intact cardiac-type Na^+/Ca^{2+} exchanger by SDS gel electrophoresis reveals two major species of 120 and 160 kDa. Both molecular weights are considerably higher than the calculated mass for the mature exchanger of about 105 kDa. This discrepancy can at least in part be explained by the fact that the exchanger is glycosylated. Three potential N-linked glycosylation sites are present on putative extracellular domains of the protein, but judging from *in vitro* translation studies only the most N-terminal site (on the first extracellular segment) is actually glycosylated. Expression studies of the cloned cardiac-type exchanger suggest that the 120 kDa protein corresponds to a fully processed, mature, and functionally competent transporter (Philipson and Nicoll, 1992; Matsuoka et al., 1993).

In contrast to the cardiac-type exchanger, the purified (bovine) photoreceptor $Na^+/Ca^{2+},K^+$ exchanger migrates with a much higher apparent molecular mass of about 220 kDa. Its amino acid sequence as deduced from the cloned cDNA contains, however, only 1,199 residues, leading to a calculated mass of about 130 kDa (Reiländer et al., 1992). Moreover, the rod-type exchanger, like its cardiac-type counterpart, contains an N-terminal probable leader sequence of 65 residues, resulting in a processed protein of only 1,134 amino acids. The predicted transmembrane topology for this exchanger is very similar to the one for the cardiac-type transporter (Figure 3). The main difference between the two is a much larger first extracellular domain (382 vs. 39 residues) and a shorter (408 vs. 520 residues) intracellular hydrophilic loop domain in the rod exchanger. The extended N-terminal extracellular domain contains six possible N-glycosylation and several putative

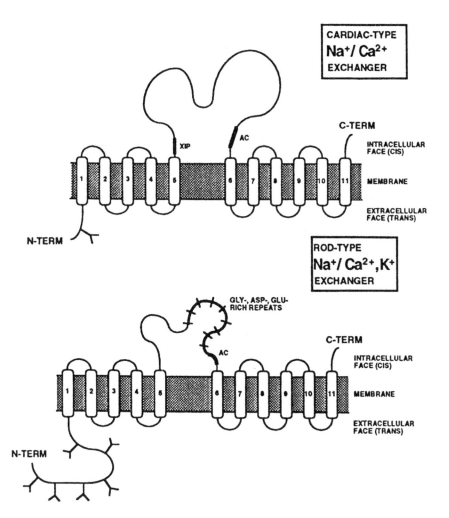

Figure 3. Putative transmembrane topology of the Na⁺/Ca²⁺ exchangers. Planar models for the cardiac-type (top) and rod photoreceptor-type (bottom) Na⁺/Ca²⁺ exchangers are shown. Note the two blocks of clustered transmembrane segments (1 to 5 and 6 to 11) and the overall similarity in the topology of the two exchanger types. Major differences reside in the N-terminal, extracellular domain and the major intracellular hydrophilic loop. The rod-type exchanger is much more heavily glycosylated (y-shaped structures) than the cardiac-type exchanger, and contains an extended region of glycine-, aspartate- and glutamate-rich repeats in its major intracellular loop. The intracellular hydrophilic loop is larger in size in the cardiac-type exchanger. A short segment of amino acids following transmembrane segment 5 can act as an exchanger inhibitory peptide (XIP). Both exchanger types contain a highly acidic stretch of amino acids (AC) that may be involved in regulatory Ca²⁺ binding.

O-glycosylation sites. The native photoreceptor-type Na^+, Ca^{2+},K^+ exchanger is indeed heavily glycosylated and this property may at least partially explain the significant difference between its calculated and observed molecular mass. A remarkable feature of the hydrophilic loop of the rod exchanger is an eightfold repeat of 24 amino acids mainly consisting of aspartate, glutamate, and glycine residues. In fact, most of the hydrophilic loop is enriched with these residues, and a remarkable segment of 26 consecutive acidic amino acids is found just at the N-terminal of the putative transmembrane domain six (Figure 3). There is surprisingly little primary sequence conservation between the cardiac and the photoreceptor Na^+/Ca^{2+} exchangers, the only exceptions being two relatively short stretches spanning the putative transmembrane segments 2/3 and 8. These conserved regions are likely to be important for Na^+/Ca^{2+} transport.

As suggested from their highly divergent sequences, the cardiac- and rod-type exchangers are encoded by different genes. Multiple variants of the cardiac-type exchanger are generated from a single gene (NCX1) through alternative splicing of its primary transcript (Kofuji et al., 1994; Lee et al., 1994). These alternatively spliced NCX1 isoforms differ from each other in a short segment of the cytoplasmic hydrophilic loop, and they are expressed in a tissue-specific manner (Reilly and Shugrue, 1992; Kofuji et al., 1994; Lee et al., 1994). Recently, a second gene (NCX2) encoding a "cardiac-type" exchanger has been identified (Li et al., 1994). Interestingly, this gene is predominantly expressed in brain and skeletal muscle, but not in cardiac muscle. By contrast, the rod-type exchanger appears to be the product of a single gene, and no alternative splice variants have so far been detected of this protein.

PHYSIOLOGICAL ROLE OF SODIUM-CALCIUM EXCHANGERS

The cardiac-type Na^+/Ca^{2+} exchanger plays an essential role in cardiac tissue where it is most abundantly expressed (Reeves, 1985). Depolarization of the heart sarcolemma leads to the influx of Ca^{2+} via the voltage gated Ca^{2+} channels. A transient rise in intracellular Ca^{2+} is necessary to trigger the so-called "Ca^{2+}-induced Ca^{2+} release" response from the sarcoplasmic reticulum (the major store of exchangeable free Ca^{2+}), leading to a rapid and dramatic rise in intracellular Ca^{2+} (from submicromolar to micromolar levels). This, in turn, activates muscle contraction. Under conditions of elevated intracellular Na^+ and membrane depolarization, the Na^+/Ca^{2+} exchanger may participate in Ca^{2+}-induced Ca^{2+} release by temporarily working in the reverse mode, thereby transporting Ca^{2+} into the cell. High Na^+ concentrations may occur under conditions (e.g., in the presence of digitalis glycosides) that inhibit the Na^+/K^+ pump (the major guardian of intracellular Na^+) or in cases of sustained opening of Na^+ channels. While the participation of the Na^+/Ca^{2+} exchanger in the early phases of excitation-contraction coupling remains controversial, this transporter certainly plays a major part in systole-diastole regulation by removing Ca^{2+}

from the cytosol upon membrane repolarization. For relaxation to occur at normal speed during beat-to-beat regulation, the bulk of the released cytosolic Ca^{2+} must be pumped back into the SR by the Ca^{2+} pump. However, under steady state conditions, all the extra Ca^{2+} that enters through the plasma membrane Ca^{2+} channels during stimulation must again be extruded from the cell. This task is essentially performed by the exchanger alone. Calculations have shown that the maximally active cardiac exchanger has a turnover number of more than $1,000\ Ca^{2+}$ per second per molecule and is capable of operation at a high rate of ion flux. With a half-time of a few hundred milliseconds for Ca^{2+} exchange during contraction-relaxation, the cardiac Na^+/Ca^{2+} exchanger constitutes a fast high capacity Ca^{2+} removal system. Its preferential localization in a defined subcellular Ca^{2+} compartment near the diadic region where the T-tubular sarcolemma and the SR terminal cisternae are in close juxtaposition (Philipson and Nicoll, 1992) further indicates its close linkage to the regulation of heart contractile function.

As mentioned earlier, the photoreceptor-type $Na^+/Ca^{2+},K^+$ exchanger is almost exclusively located in the outer segment membranes of rod photoreceptors. As in excitable heart muscle cells, highly dynamic Ca^{2+} fluxes also occur in the outer segment of photoreceptor cells (Schnetkamp, 1990). In the dark, Ca^{2+} permanently enters the cell through open cGMP-gated channels that allow both Na^+ and Ca^{2+} influx and maintain the membrane in a depolarized state. The Ca^{2+} influx is counteracted by the exchanger which is therefore highly active in these cells, whereas the Na^+ influx is opposed by the action of the Na^+/K^+ pump. The depolarized state of the membrane and the rather "shallow" Na^+ gradient (due to the constant Na^+ influx) would make a cardiac-type exchanger rather inefficient; the cotransport of K^+ with Ca^{2+} in the rod-type exchanger alleviates this problem by tapping the energy source stored in the outward-directed K^+ gradient. The rod exchanger has, in fact, an enormous capacity: it has been calculated that at maximum efficiency it can alter the total intracellular Ca^{2+} concentration by up to 0.5 mM per second. The need for a highly active Ca^{2+} extrusion system abruptly changes upon light stimulation of the photoreceptors. In the bright daylight, the retinal rods are saturated with light which leads to the closure of the cGMP-gated membrane channels via a cascade of molecular interactions beginning with the absorption of light (photons) by the retinal photopigment. The continued action of the Na^+/Ca^{2+} exchanger (as well as probably of the Ca^{2+} pumps) leads to a decrease of the intracellular free Ca^{2+} concentration below the steady state "dark" levels and to a hyperpolarization of the rod outer membrane. This in turn triggers a Ca^{2+}-sensitive light adaptation. The permanent operation of a highly efficient Na^+/Ca^{2+} extrusion system would, however, lead to undesirably low cytosolic Ca^{2+} levels. The uncoupling of K^+ cotransport in the rod Na^+/Ca^{2+} exchanger under certain conditions (as reflected in the flexible coupling ratios mentioned earlier) may be one of the measures to inactivate the exchanger. Thus, the photoreceptor-type

Na$^+$/Ca^{2+},K$^+$ exchanger may be uniquely qualified to handle the rapidly changing demands on Ca^{2+} fluxes across the outer segment membrane of these cells.

CALCIUM PUMPS: DISCOVERY, DISTRIBUTION, AND PROPERTIES

Ca^{2+} pumps (also called Mg^{2+}-dependent Ca^{2+}-ATPases or simply Ca^{2+}-ATPases) are responsible for the specific uphill transmembrane transport of Ca^{2+} against a large concentration gradient. They belong to the growing family of P-type ion-motive ATPases (Pedersen and Carafoli, 1987) which also includes eukaryotic Na$^+$/K$^+$, H$^+$/K$^+$, H$^+$, Cu$^+$ and prokaryotic K$^+$, Mg^{2+}, Cd^{2+} and other heavy metal ion pumps. P-type ATPases received their name from the fact that they form a phosphorylated intermediate during the reaction cycle in which the γ-phosphate of ATP is transferred to an invariant aspartate residue in the catalytic site. A further characteristic is their inhibition by orthovanadate [VO$_3$(OH)]$^{2-}$, a transition state analog of phosphate. Ca^{2+} pumps can be divided into two major classes: Ca^{2+} pumps of the surface (plasma) membrane and Ca^{2+} pumps of organellar membranes, in particular of the endoplasmic/sarcoplasmic (ER/SR) membrane. Accordingly, the terms PMCA and SERCA have been introduced for the plasma membrane calcium ATPases and sarcoplasmic/endoplasmic reticulum calcium ATPases, respectively.

The existence of an ATP-driven Ca^{2+} pumping mechanism in human erythrocyte membranes was first demonstrated some 30 years ago (Schatzmann, 1966), and the notion of a similar pump system in the sarcoplasmic reticulum membrane of striated muscle cells dates back to about the same time (Hasselbach and Makinose, 1961). Since then it has become clear that PMCAs and SERCAs are ubiquitous Ca^{2+} transporters of eukaryotic plasma and intracellular membranes, respectively. Their primary function is to remove Ca^{2+} from the cytosol and to pump it either out of the cell (PMCAs) or into Ca^{2+} storage organelles (SERCAs) against a 1,000- to 10,000-fold concentration gradient. While PMCAs are generally low abundance membrane proteins (0.01 to 0.1% of total membrane protein), SERCAs can account for up to 90% to 95% of the membrane protein in the sarcoplasmic reticulum of (fast-twitch skeletal) muscle cells. Most biochemical and biophysical studies on SERCAs have therefore been performed on the pump from muscle tissues; its extremely high concentration in the SR, where it can form almost perfect "crystals," has even allowed topological studies by x-ray and electron diffraction methods at a resolution of about 10 Å. In contrast, essentially all early biochemical and regulatory studies on the PMCA were done with human erythrocytes because of their lack of extensive contaminating intracellular membrane systems and because the PMCA appears to be the only active Ca^{2+} extrusion system in these cells.

In spite of their distinctly different membrane location and relative abundance, PMCAs and SERCAs share many structural and functional properties. On the other hand, they are distinguished by several regulatory and structural features unique to either one or the other type of Ca^{2+} pump (Carafoli and Chiesi, 1992; Penniston and Enyedi, 1996). Both transporters consist of a single polypeptide chain with a molecular mass of about 100 kDa for SERCAs and of about 130 kDa for PMCAs. The monomer has been shown to be fully functional in reconstituted systems but *in vivo* the predominant active pump may be a homodimer. In both types of Ca^{2+} pumps the hydrolysis of ATP is coupled to vectorial Ca^{2+} transport from the cis (cytosolic) to the trans (extracellular or intravesicular) side; however, the efficiency of the two pump types (at least in the purified state) is not the same viz. two Ca^{2+} ions are transported per ATP hydrolyzed by SERCAs, whereas only a single Ca^{2+} ion is pumped across the membrane for each ATP molecule hydrolyzed by PMCAs. In both the SERCA and PMCA pumps, countertransport of protons (H^+) in exchange for Ca^{2+} has been demonstrated. Because the stoichiometry appears to be $1H^+:1Ca^{2+}$, both pump systems are electrogenic (Hao et al., 1994; Inesi, 1994). Both pump types require the presence of Mg^{2+} on the cis side for full activity; however, this cation itself is not transported by the pumps. Evidently, the Ca^{2+} affinity of a Ca^{2+} pump must be very different on the cis and the trans side of the membrane because its primary function is to recognize and pick up Ca^{2+} in the cytosol at submicromolar concentrations and then to release the same ion on the opposite side of the membrane where Ca^{2+} concentrations may be millimolar. Indeed, one of the landmarks of the fully activated SERCA and PMCA pumps is their very high Ca^{2+} affinity on the cytosolic side (K_{Ca} 0.5 µM or less) and a much lower one on the trans side (K_{Ca} around 5 mM). Compared with the Na^+/Ca^{2+} exchangers, the Ca^{2+} pumps do not possess a very high transport capacity; their turnover number is on the order of 20 per second.

The continuous operation of the Ca^{2+} pumps would be very expensive for a cell (because of the high metabolic cost of generating the necessary ATP) and might also lead to undesirably low cytosolic free Ca^{2+} levels in the resting steady state. The activities of SERCAs and PMCAs are therefore carefully regulated *in vivo* and the pumps are kept in a relatively inactive, inhibited state until they are needed. Different ways of regulation of the Ca^{2+} pumps are discussed in more detail below in the section on their molecular characterization.

MECHANISTIC ASPECTS OF CALCIUM PUMPS

The fact that the SERCA pump is highly abundant in the SR and can therefore be relatively easily purified in large quantities has greatly facilitated detailed studies on the kinetics of its reaction (Inesi, 1985). Such studies are much more difficult with the PMCA because of its low abundance, but the functional (and to a large extent structural) similarity between these two Ca^{2+} pump types suggests that many mechanistic details worked out for the SERCA are valid for the PMCA pumps as well. A generalized reaction scheme for the Ca^{2+} pumps is depicted in Figure 4. The pumps can exist in essentially two major different states which are often referred to as E_1 and E_2 conformations (Inesi, 1985). In the E_1 state, high-affinity Ca^{2+} binding/transport sites are available on the cis side. Under optimal conditions (i.e., when the pumps are fully activated) the Ca^{2+} dissociation constants are in the

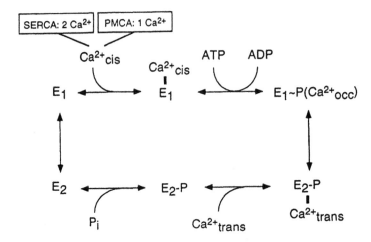

Figure 4. Simplified scheme for the reaction cycle in Ca^{2+} pumps. The pumps may adopt two major conformations E_1 and E_2. The E_1 conformation shows high affinity for two Ca^{2+} (SERCA pumps) or one Ca^{2+} (PMCA pumps) on the cis side. Ca^{2+} binding greatly enhances the pumps' ATPase activity, leading to the rapid formation of the high-energy phosphorylated intermediate $E_1{\sim}P$ and occlusion (occ) of the transported Ca^{2+} ion(s). Ca^{2+} translocation across the membrane presumably occurs concomitantly with the release of energy stored as conformational constraint during the transition from the $E_1{\sim}P$ to the low-energy E_2-P conformation. Ca^{2+} affinity on the trans side is low and Ca^{2+} is therefore released. This is followed by hydrolysis of the phosphoenzyme and a poorly understood rearrangement step(s) from the E_2 to the E_1 conformation.

0.1 to 1 micromolar range. The two Ca^{2+} ions of the SERCA pump bind sequentially and cooperatively to the E_1 conformer. Ca^{2+} loading of the E_1 form strongly activates its ATPase activity, leading to the rapid formation of the high-energy phosphorylated intermediate E_1-P. This is followed by a slower step in which the Ca^{2+} ion(s) become(s) occluded (i.e., inaccessible to the aqueous phase from either side of the membrane) and the pump molecule undergoes a further conformational change to the lower-energy E_2-P form. Although it is not yet clear at which step the bound Ca^{2+} ions are actually moved through the lipid bilayer, ion translocation is likely to be directly linked to the conformational change that accompanies the E_1-P to E_2-P transition. At the end of this step, the Ca^{2+} transport sites display only low affinity for the translocated cation, thereby allowing its release on the trans side. Finally, hydrolysis of the E_2 phosphoenzyme leads to the E_2 conformation of the pump which then recycles back to the resting E_1 state. The chemical energy of ATP is thus initially stored as conformational constraint in the E_1-P state of the pump, and its dissipation through a series of conformational changes results in a dislocation of the bound Ca^{2+} and a disruption of the high affinity binding site for this cation. Since the catalytic (ATP binding and phosphorylation) and the Ca^{2+} binding sites are spatially widely separated in the pump molecules (see below), long range interactions must occur to transmit the conformational changes (Inesi, 1994).

MOLECULAR CHARACTERIZATION OF THE CALCIUM PUMPS

Complete primary structures for SERCA pumps were first reported in 1985 (MacLennan et al., 1985) and for PMCAs in 1988 (Shull and Greeb, 1988; Verma et al., 1988), both from molecular cloning studies. Extensive cDNA and peptide sequence analyses revealed a multitude of isoforms for both Ca^{2+} pump types in mammals. SERCA pumps contain about 1,000 amino acid residues yielding a molecular mass of about 110 kDa. One obvious difference between SERCA and PMCA pumps is the significantly larger size of the PMCAs which contain about 1,200 residues with a calculated Mr of around 135 kDa. This size difference is essentially due to a C-terminal extension in the PMCAs when compared with the SERCAs. Amino acid sequence comparisons, computer-assisted secondary structure predictions, direct protein labeling, and biochemical and biophysical experiments with synthetic peptides have been used to design a rather detailed model for the topological arrangement of the Ca^{2+} pumps in the membrane. These approaches have permitted the assignment of structural, functional and regulatory domains in these proteins (Green and MacLennan, 1989; Strehler, 1991; Carafoli and Chiesi, 1992). The Ca^{2+} pumps are built according to a general principle common to all P-type cation pumps (Green and MacLennan, 1989). The models shown in Figure 5 for a SERCA and a PMCA pump illustrate the high overall structural similarity between these two pump types, but also highlight some of their structural and regulatory differences. The overall two- (and presumably three-) dimensional similarity of SERCAs and PMCAs is not immediately evident from a comparison of their primary amino acid sequences; in fact, the two pump types show a surprisingly low degree (less than 30%) of sequence identity. There are, however, a number of regions with a very high degree of primary sequence conservation, not only in the two Ca^{2+} pump types, but also in the other known P-type ATPases (Green and MacLennan, 1989). These sequences clearly represent invariable important functional or structural domains, such as the nucleotide binding site and the site of phosphorylated intermediate formation (Figure 5). The Ca^{2+} pumps are predicted to traverse the membrane 10 times, with four transmembrane regions clustered in the N-terminal half and the remaining six located towards the C-terminus. In contrast to the mature Na^+/Ca^{2+} exchangers, which contain 11 putative transmembrane domains, the N- and the C-termini of the Ca^{2+} pumps are located on the same (cytosolic) side of the membrane. The clustering of transmembrane segments into two major hydrophobic domains is, however, a shared property of the different Ca^{2+} transporters. Indeed, this feature appears to be of a general nature

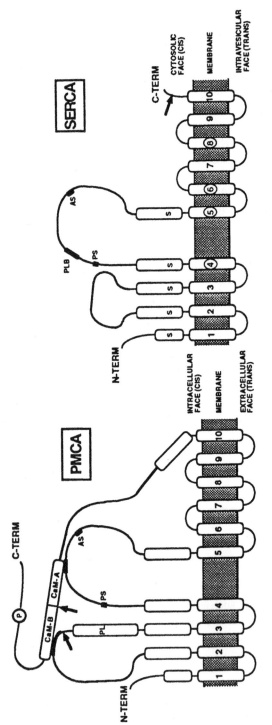

Figure 5. Suggested overall topology of the Ca²⁺ pumps. Models comparable to those shown in Figure 3 for the Na⁺/Ca²⁺ exchangers are shown for the PMCA (left) and SERCA (right) pumps. Note the high overall similarity between the two pump types, and the fact that the bulk of the protein mass is located on the cis-side of the membrane. As in the Na⁺/Ca²⁺ exchangers, the transmembrane segments are clustered in two blocks (segments 1 to 4 and 5 to 10). Major differences between the two pump types include a larger loop between transmembrane segments two and three (due to an acidic-phospholipid-sensitive domain PL), and an extended C-terminal regulatory domain in PMCAs when compared to SERCAs. The PMCA is shown in the autoinhibited state in which the calmodulin binding domain (subdomains CaM-A and CaM-B) is bound to two intramolecular sites on the first and second hydrophilic loop. In SERCA, the region interacting with the inhibitory protein phospholamban is labeled PLB. The regions adjoining transmembrane segments one to five form a stalk (S) in SERCAs, and probably in PMCAs too. The four transmembrane segments known to participate in Ca²⁺ binding and translocation in SERCAs are circled. The catalytic phosphorylation site (PS) and a stretch of the ATP binding site (AS) are indicated. The site of possible regulatory phosphorylation of the PMCAs by cAMP-dependent protein kinase and/or protein kinase C is also shown (P). Arrows point to the regions where isoform diversity is created by alternative RNA splicing.

142

since it is found in many other membrane carriers as well (e.g., in voltage-gated cation channels).

Why are SERCA pumps exclusively located in intracellular membranes and PMCAs in the plasma membrane? The answer to this question is not yet known: unlike the Na^+/Ca^{2+} exchangers, neither of the Ca^{2+} pump types contains a leader sequence, and no specific targeting sequences have been identified in any of the pumps. Studies with expressed chimeras consisting of recombined parts of SERCA and PMCA pumps may eventually provide an answer.

The vast majority of the mass of the Ca^{2+} pumps is located on the cytosolic face of the membrane and only short loops connecting pairs of transmembrane segments are predicted to be exposed to the extracellular (PMCAs) or intravesicular (SERCAs) face (Figure 5). Glycosylation, which is apparently prominent in the $Na^+/Ca^{2+}(K^+)$ exchangers, has not been detected in the Ca^{2+} pumps. The structural model for the pumps predicts little room for direct pump-modifying interactions from the trans side; rather, regulation of these transporters appears to occur from the cis side. The hydrophilic regions of the pumps are separated—at least in the primary sequence—into two (SERCAs) or three (PMCAs) major segments. Extensions of the first five transmembrane helices into the cytosol form a so-called stalk on which the bulk of the hydrophilic portion of the pump sits like the head of a mushroom. The first cytosolic loop between the membrane-spanning segments two and three may fold into an extended antiparallel β-structure domain and appears to be involved in transmitting the conformational change induced by the catalytic site phosphorylation to the Ca^{2+} translocating region close to and within the membrane. The large cytoplasmic region between transmembrane segments four and five contains two distinct but communicating domains, one encompassing the catalytic site with its phosphorylation-sensitive aspartate residue and one containing the nucleotide (ATP) binding site. An additional cytosolic domain is characteristic for the PMCAs; it consists of approximately 150 C-terminal residues involved in the regulation of this type of Ca^{2+} pump.

One of the most interesting questions concerns, of course, the location and structure of the high affinity Ca^{2+} binding sites. No unequivocal clues are given by the primary sequences of the pumps. However, elegant site-directed mutagenesis work on the SERCA pump combined with protein expression and functional analysis of the mutants has provided convincing evidence that several polar residues located within the otherwise highly hydrophobic transmembrane segments 4, 5, 6, and 8 play a crucial role in high affinity Ca^{2+} binding and ion translocation in the SERCA pumps (Green and MacLennan, 1989; MacLennan et al., 1992). Based on these studies, a model can be proposed in which these four transmembrane segments (as well as perhaps some of the others) are arranged in a three-dimensional cylindrical channel structure that forms the Ca^{2+} translocation domain (MacLennan et al., 1992). Significantly, polar residues are found at most corresponding (but not always identical) positions in the transmembrane segments of the PMCAs as well.

When the pumps are in the active E_1 conformation, Ca^{2+} appears to have access to these high affinity sites; phosphorylation at the (remote) catalytic site will then lead to long-range-transmitted conformational alterations in the geometry of the Ca^{2+} translocation domain, resulting in Ca^{2+} ion occlusion and finally, after the E_2 conformation has been attained, in exposure of the ion(s) to the trans side.

While general structural, functional, and mechanistic aspects are undoubtedly shared between the SERCA and the PMCA pumps, details of their regulation and mechanism of action are distinct. In addition, multiple isoforms exist for both the SERCA- and the PMCA-types of Ca^{2+} pumps (Missiaen et al., 1991; Strehler, 1991) and even the isoforms within a pump family show differences in structural and regulatory properties. One of the salient differences between SERCAs and PMCAs is the regulation of the PMCAs by direct interaction with the well-known intracellular Ca^{2+}-binding protein calmodulin (Strehler, 1991; Carafoli and Chiesi, 1992). The interaction with calmodulin occurs in a short region of the extra C-terminal domain present only in PMCAs (Figure 5). The effect of the binding of Ca^{2+}-calmodulin to the PMCA is the release of the autoinhibition of the pump and results in a dramatic increase of the pump's Ca^{2+} affinity (K_{Ca} shifts from about 20 μM to 0.5 μM or less). In the absence of bound calmodulin, the calmodulin-binding domain of the PMCAs serves as an autoinhibitory sequence which interacts intramolecularly with two areas of the molecule located in the first cytosolic loop and on the C-terminal side of the catalytic site (Carafoli, 1994) (Figures 5 and 6). The inhibitory domain thus acts like a lid covering the active site of the pump and preventing high affinity binding of Ca^{2+} to the transport sites. Although a corresponding autoinhibitory domain does not exist in SERCAs, a remarkably analogous type of regulation occurs in at least some of the SERCA isoforms (Missiaen et al., 1991): in heart, slow skeletal, and smooth muscle sarcoplasmic reticulum, a small pentameric membrane protein called phospholamban inhibits the SERCA by interacting with a sequence close to the catalytic site (Carafoli and Chiesi, 1992). The inhibitory interaction depends on the state of phosphorylation of phospholamban: in the unphosphorylated state, the protein acts as an inhibitor of the pump; upon phosphorylation (by the cAMP- and the Ca^{2+}-calmodulin-dependent protein kinases) the inhibitory effect is released and the SERCA is highly activated (Figure 6). The positive inotropic effect of β-adrenergic stimulation in heart muscles can thus at least be partially explained by an increased phosphorylation of phospholamban by the cAMP-dependent kinase, leading to higher SERCA activity, and thus to faster relaxation rates. Both SERCAs and PMCAs may be regulated by a number of additional factors (Missiaen et al., 1991; Carafoli and Chiesi, 1992). For the PMCAs these include interaction with acidic phospholipids, phosphorylation by cAMP dependent protein kinase, and protein kinase C, as well as limited proteolysis (removal of the autoinhibitory C-terminal domain) by a Ca^{2+} activated protease called calpain. No specific inhibitors are currently known for the PMCAs (vanadate

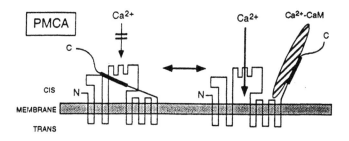

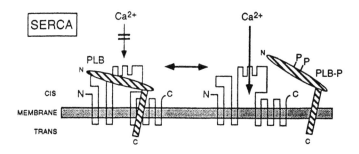

Figure 6. Regulation of PMCAs and SERCAs by calmodulin and phospholamban, respectively. Top: In the absence of calmodulin (left), the calmodulin binding domain of the PMCA acts as an autoinhibitory sequence and prevents high affinity Ca^{2+} binding. Binding of Ca^{2+}-calmodulin (striped cigar) releases the inhibition, allowing high affinity Ca^{2+} binding and pump activity (right). Bottom: Interaction of unphosphorylated phospholamban (PLB, striped shapes) with the SERCA pump prevents high affinity Ca^{2+} binding, leading to a relatively inactive pump (left). Phosphorylation of phospholamban (PLB-P) by cAMP- or Ca^{2+}/calmodulin-dependent protein kinase releases the inhibitory interaction, thereby allowing increased Ca^{2+} binding and pump activity (right).

is a potent inhibitor but will inhibit all P-type ATPases), whereas a specific inhibitor for the SERCAs is the naturally occurring sesquiterpene lactone, thapsigargin.

In mammals, Ca^{2+} pumps are generated from multigene families consisting of at least three SERCA (Missiaen et al., 1991) and four PMCA (Strehler, 1991) members. The genes within a given family are structurally highly conserved and clearly derived from a single common ancestor, but even the different pump gene families still show evidence for common evolutionary roots. Within a pump family, the amino acid sequences encoded by different genes show 75% to 85% identity. Not unexpectedly, the areas of highest sequence divergence are clustered in a few regions (particularly at the N- and C-termini) presumably corresponding to specialized (e.g., regulatory) domains of the pump isoforms. A similar statement can be made concerning the differences of isoforms encoded by alternatively spliced mRNAs generated from the same primary transcript. Alternative splicing occurs at two major, conserved locations in the PMCA gene transcripts (Strehler, 1991): one affects a sequence rich in basic residues N-terminal of the proposed phospholipid-sensitive domain in the cytosolic loop between transmembrane segments two and three, and the other affects the calmodulin-binding regulatory region close to the C-terminus (Figure 5). Because of these variations, some of the alternatively spliced PMCA variants show distinctly different affinities for calmodulin, different pH sensitivities, and a different potential for regulation by phosphorylation via the cAMP dependent protein kinase. In the SERCA1 and 2 genes, alternative splicing of their primary transcripts affects the extreme C-terminal coding sequence (Missiaen et al., 1991), but the functional significance of these alterations is not yet known. One of the most striking differences among the SERCA isoforms is a severalfold lower apparent Ca^{2+} affinity in SERCA3 than in SERCA1 or SERCA2 (MacLennan et al., 1992). The reason for this may be a shift in the equilibrium between the two major conformational states E_1 and E_2 toward the (low Ca^{2+} affinity) E_2 conformation. A similar finding has been made for two PMCA4 isoforms, but in this case the effect is even more intriguing because the two pump isoforms are generated from alternatively spliced transcripts of the same gene and differ only in a very short region at their C-termini.

PHYSIOLOGICAL ROLE OF CALCIUM PUMPS

The importance of the SERCA-type Ca^{2+} pumps in Ca^{2+} uptake into the SR in striated muscles is obvious. The speed of relaxation after contraction is directly dependent on the efficiency of this pump. Accordingly, disturbances in its expression will have a profound effect on the contractile properties of a muscle. A decreased activity of SERCA in cardiac muscles has been found in several forms of heart insufficiency such as hypothyroid disease and pressure-load hypertrophy. This is due to a reduced level of SERCA and/or to a change in the expression pattern of its natural regulator phospholamban. The reduced velocity of pumping by the SERCA will lead to a delayed relaxation and may result in higher intracellular Ca^{2+} levels and increased end-diastolic pressure. In cells with a less extensive sarco/endoplasmic reticulum network, such as in smooth muscles, the relative involvement of the PMCAs in the removal of free cytosolic Ca^{2+} becomes more relevant. Smooth muscles show much slower rates of contraction and relaxation and an extensive high-capacity Ca^{2+} removal system is therefore not needed. Accordingly, the observed elevated cytoplasmic Ca^{2+} levels in hypertensive aortic smooth muscle cells are probably caused by a decreased level and/or a decreased efficiency of both the PMCAs and the SERCAs. In all cases, Ca^{2+} that entered the cell from the extracellular side (i.e., through Ca^{2+} channels) must be removed again across the plasma membrane in order to restore steady-state resting Ca^{2+} levels. In heart and other excitable cells with high transmembrane Ca^{2+} fluxes, Ca^{2+} transport across the plasma membrane is primarily performed by the high-capacity Na^+/Ca^{2+} exchanger, whereas the SERCA is responsible for Ca^{2+} reuptake into the SR. However, in other cell types, when the free Ca^{2+} concentrations reach levels below 1 μM, the high Ca^{2+}-affinity PMCAs may be the sole transporters operating in the Ca^{2+} extrusion mode. Clearly, the importance of Ca^{2+} extrusion across the plasma membrane varies and is most significant in those cells that rely mainly on extracellular Ca^{2+} for the generation of their Ca^{2+} signals. Besides playing a role in the recovery from Ca^{2+} activation, the PMCAs are also involved in setting the local subplasmalemmal Ca^{2+} concentration and, through long-term changes in their expression pattern, in the modulation of steady-state free Ca^{2+} levels. The fine-tuning of intracellular Ca^{2+} regulation to the specific physiological needs of each cell type may be accomplished by expressing different sets of SERCA and PMCA isoforms.

SUMMARY

Ca^{2+} acts as a universal second messenger in intracellular signaling in eukaryotes. The maintenance and regulation of the free intracellular Ca^{2+} concentration requires the presence of specialized Ca^{2+} transporters in the plasma membrane and in intracellular organellar membranes. Na^+/Ca^{2+} exchangers and ATP-driven Ca^{2+} pumps are the two major systems responsible for uphill Ca^{2+} transport against its large concentration gradient. Na^+/Ca^{2+} exchangers are high capacity transporters involved in the handling of large and dynamic Ca^{2+} fluxes across the plasma membrane. Cardiac-type exchangers are ubiquitous but are most abundant in excitable cells such as in heart and nervous tissue. They use the energy stored in the electrochemical Na^+ gradient to exchange one Ca^{2+} for three Na^+ per reaction cycle. In the forward mode they expel Ca^{2+} from the cell in exchange for Na^+. However, depending on the prevailing Na^+ and Ca^{2+} ion distribution and on the membrane potential they can also operate in the reverse mode. The photoreceptor exchanger is highly specific for the outer membrane segment of vertebrate retinal rods and normally operates with a $4Na^+/1Ca^{2+},1K^+$ stoichiometry. The cardiac- and photoreceptor-type exchangers are single polypeptides with native molecular masses of about 120 and 220 kDa, and cDNA-derived calculated masses of about 105 and 130 kDa, respectively. The differences are mainly due to glycosylation on their N-terminal extracellular segments. Both exchanger types contain a leader peptide which is cleaved during maturation, and they show extensive structural similarities in their predicted transmembrane topology: two clusters of five and six transmembrane segments close to the N- and C-termini, respectively, are separated by a large cytosolic hydrophilic domain.

Ca^{2+} pumps are divided into two distinct families of plasma membrane and sarco/endoplasmic reticulum membrane Ca^{2+} ATPases (PMCAs and SERCAs). In tissues with a low Na^+/Ca^{2+} exchanger activity and under conditions unfavorable for Na^+/Ca^{2+} exchange, PMCAs are the sole mechanism of specific Ca^{2+} export. They play an essential role in the long-term maintenance and the resetting of resting free Ca^{2+} levels. SERCAs are responsible for the rapid Ca^{2+} reuptake into the endoplasmic/sarcoplasmic reticulum and its specialized compartments after Ca^{2+} signaling. Ca^{2+} pumps belong to the class of P-type pumps which are characterized by the formation of a phosphorylated intermediate in which the energy derived from ATP is transiently stored as conformational constraint. Ca^{2+} pumps shuttle between two major conformational states E_1 and E_2. In the E_1 state they bind Ca^{2+} on the cytosolic (cis) side with very high affinity ($K_{Ca} \approx 0.5\ \mu M$). Upon Ca^{2+} binding their ATPase activity is stimulated, leading to the formation of the phosphorylated intermediate and followed by occlusion of the transported Ca^{2+}, transition to the low energy, low Ca^{2+} affinity E_2 state and release of Ca^{2+} on the trans side. Two Ca^{2+} ions are transported per ATP split in SERCAs but only one in the PMCAs. SERCAs and PMCAs are single polypeptides with molecular masses of about 100 and 130 kDa. Despite limited primary sequence identity their predicted transmembrane topologies are very similar and consist of two clusters of four and six transmembrane segments separated by a large cytosolic domain containing the catalytic and ATP binding sites. Polar amino acid side chains within at least four transmembrane segments participate in forming the high affinity Ca^{2+} binding/transport sites in SERCAs, and probably in PMCAs, too. Long-range conformational interactions must occur in the pumps to enable changes in the catalytic site region to affect Ca^{2+} binding and translocation in the transmembrane region. PMCAs, but not SERCAs, are directly stimulated by Ca^{2+}-calmodulin; indeed, most of the additional mass of PMCAs with respect to SERCAs is due to a C-terminal regulatory region containing the autoinhibitory, calmodulin-binding domain. Multiple isoforms of SERCAs and PMCAs are generated from multigene families and from alternatively spliced mRNAs. The tissue and developmental stage-specific expression of several SERCA and PMCA isoforms suggests that specific functional and/or regulatory properties of these pumps are adapted to the physiological needs of a given cell type.

REFERENCES

Baker, P.F., Blaustein, M.P., Hodgkin, A.L., & Steinhardt, R.A. (1969). The influence of calcium on sodium efflux in squid axons. J. Physiol. 200, 431–458.

Carafoli, E. (1994). Plasma membrane calcium ATPase: 15 years of work on the purified enzyme. FASEB J. 8, 993–1002.

Carafoli, E. & Chiesi, M. (1992). Calcium pumps in the plasma and intracellular membranes. Curr. Top. Cell. Regul. 32, 209–241.

Green, N.M. & MacLennan, D.H. (1989). ATP driven ion pumps: An evolutionary mosaic. Biochem. Soc. Trans. 17, 819–822.

Hao, L., Rigaud, J.-L., & Inesi, G. (1994). Ca^{2+}/H^+ countertransport and electrogenicity in proteoliposomes containing erythrocyte plasma membrane Ca-ATPase and exogenous lipids. J. Biol. Chem. 269, 14268–14275.

Hasselbach, W. & Makinose, M. (1961). Die Calciumpumpe der "Erschlaffungsgrana" des Muskels und ihre Abhängigkeit von der ATP-Spaltung. Biochem. Z. 333, 518–528.

Hilgemann, D.W. (1988). Numerical approximation of sodium-calcium exchange. Progr. Biophys. Mol. Biol. 51, 1–45.

Inesi, G. (1994). Teaching active transport at the turn of the twenty-first century: Recent discoveries and conceptual changes. Biophys. J. 66, 554–560.

Inesi, G. (1985). Mechanism of calcium transport. Ann. Rev. Physiol. 47, 573–601.

Kofuji, P., Lederer, W.J., & Schulze, D.H. (1994). Mutually exclusive and cassette exons underlie alternatively spliced isoforms of the Na/Ca exchanger. J. Biol. Chem. 269, 5145–5149.

Komuro, I., Wenninger, K.E., Philipson, K.D., & Izumo, S. (1992). Molecular cloning and characterization of the human cardiac Na^+/Ca^{2+} exchanger cDNA. Proc. Natl. Acad. Sci. USA 89, 4769–4773.

Lee, S.L., Yu, A.S., & Lytton, J. (1994). Tissue-specific expression of Na^+-Ca^{2+} exchanger isoforms. J. Biol. Chem. 269, 14849–14852.

Li, Z., Matsuoka, S., Hryshko, L.V., Nicoll, D.A., Bersohn, M.M., Burke, E.P., Lifton, R.P., & Philipson, K.D. (1994). Cloning of the NCX2 isoform of the plasma membrane Na^+-Ca^{2+} exchanger. J. Biol. Chem. 269, 17434–17439.

MacLennan, D.H., Brandl, C.J., Korczak, B., & Green, N.M. (1985). Amino-acid sequence of a Ca^{2+} + Mg^{2+}-dependent ATPase from rabbit muscle sarcoplasmic reticulum, deduced form its complementary DNA sequence. Nature 316, 696–700.

MacLennan, D.H., Toyofuku, T., & Lytton, J. (1992). Structure-function relationships in sarcoplasmic or endoplasmic reticulum type Ca^{2+} pumps. Ann. New York Acad. Sci. 671, 1–10.

Matsuoka, S. & Hilgemann, D.W. (1992). Steady-state and dynamic properties of cardiac sodium-calcium exchange. Ion and voltage dependencies of the transport cycle. J. Gen. Physiol. 100, 963–1001.

Matsuoka, S., Nicoll, D.A., Reilly, R.F., Hilgemann, D.W., & Philipson, K.D. (1993). Initial localization of regulatory regions of the cardiac sarcolemmal Na^+-Ca^{2+} exchanger. Proc. Natl. Acad. Sci. USA 90, 3870–3874.

Missiaen, L., Wuytack, F., Raeymaekers, L., DeSmedt, H., Droogmans, G., Declerck, I., & Casteels, R. (1991). Ca^{2+} extrusion across plasma membrane and Ca^{2+} uptake by intracellular stores. Pharmac. Ther. 50, 191–232.

Nicoll, D.A., Longoni, S., & Philipson, K.D. (1990). Molecular cloning and functional expression of the cardiac sarcolemmal Na^+-Ca^{2+} exchanger. Science 250, 562–565.

Pedersen, P.L. & Carafoli, E. (1987). Ion motive ATPases. I. Ubiquity, properties, and significance to cell function. TIBS 12, 146–150.

Penniston, J.T., & Enyedi, A. (1996). Comparison of ATP-powered Ca^{2+} pumps. In: Ion Pumps. Advances in Molecular and Cell Biology, JAI Press, Inc., Greenwich, CT, in press.

Philipson, K.D. & Nicoll, D.A. (1992). Sodium-calcium exchange. Curr. Op. Cell Biol. 4, 678–683.

Reeves, J.P. (1985). The sarcolemmal sodium-calcium exchange system. Curr. Top. Membr. Transport 25, 77–127.

Reiländer, H., Achilles, A., Friedel, U., Maul, G., Lottspeich, F., & Cook, N.J. (1992). Primary structure and functional expression of the Na/Ca,K-exchanger from bovine rod photoreceptors. EMBO J. 11, 1689–1695.

Reilly, R.F. & Shugrue, C.A. (1992). cDNA cloning and expression of a renal Na^+/Ca^{2+} exchanger. Am. J. Physiol. 262, F1105–1109.

Reuter, H. & Seitz, N. (1968). The dependence of calcium efflux from cardiac muscle on temperature and external ion composition. J. Physiol. 195, 451–470.

Schatzmann, H.J. (1966). ATP-dependent Ca^{2+} extrusion from human red cells. Experientia 22, 364–368.

Schnetkamp, P.P.M. (1990). Na-Ca or Na-Ca-K exchange in rod photoreceptors. Prog. Biophys. Mol. Biol. 54, 1–29.

Shull, G.E. & Greeb, J. (1988). Molecular cloning of two isoforms of the plasma membrane Ca^{2+} transporting ATPase from rat brain. Structural and functional domains exhibit similarity to Na^+, K^+- and other cation transport ATPases. J. Biol. Chem. 263, 8646–8657.

Strehler, E.E. (1991). Recent advances in the molecular characterization of plasma membrane Ca^{2+} pumps. J. Membr. Biol. 120, 1–15.

Verma, A.K., Filoteo, A.G., Stanford, D.R., Wieben, E.D., Penniston, J.T., Strehler, E.E., Fischer, R., Heim, R., Vogel, G., Mathews, S., Strehler-Page, M.-A., James, P., Vorherr, T., Krebs, J., & Carafoli, E. (1988). Complete primary structure of a human plasma membrane Ca^{2+} pump. J. Biol. Chem. 263, 14152–14159.

RECOMMENDED READINGS

Allen, T.J.A., Noble, D., & Reuter, H. (eds.) (1989). Sodium-Calcium Exchange. Oxford University Press, Oxford, UK.

Blaustein, M.P., DiPolo, R., & Reeves, J.P. (eds.) (1991). Sodium-Calcium Exchange. Ann. New York Acad. Sci., Vol. 639, The New York Academy of Sciences, New York.

Carafoli, E. (1987). Intracellular calcium homeostasis. Ann. Rev. Biochem. 56, 395–433.

Carafoli, E. & Penniston, J.T. (1985). The calcium signal. Scientific American 253, 70–78.

Pedersen, P.L. & Carafoli, E. (1987). Ion motive ATPases. I. Ubiquity, properties, and significance to cell function. TIBS 12, 146–150.

Pedersen, P.L. & Carafoli, E. (1987). Ion motive ATPases. II. Energy coupling and work output. TIBS 12, 186–189.

Scarpa, A., Carafoli, E., & Papa, S. (eds.) (1992). Ion-motive ATPases: Structure, function, and regulation. Ann. New York Acad. Sci., Vol. 671, The New York Academy of Sciences, New York.

Chapter 5

The Na$^+$/H$^+$ Exchanger

CHRISTIAN FRELIN and PAUL VIGNE

Principles of Medical Biology, Volume 4
Cell Chemistry and Physiology: Part III, pages 151–167.
Copyright © 1996 by JAI Press Inc.
All rights of reproduction in any form reserved.
ISBN: 1-55938-807-2

INTRODUCTION

The existence of an Na^+/H^+ exchanger in intestine and kidney brush-border membrane vesicles was first suggested in 1976 by Mürer and his coworkers. It was then realized that the system is present in virtually all cell types, and that it is involved in a large variety of cell functions. This chapter summarizes our present knowledge of the structure, properties, and functions of the Na^+/H^+ exchanger. (For reviews see Benos, 1982; Frelin et al., 1988; Kleyman and Cragoe, 1988; Grinstein et al., 1989, and Wakabayashi et al., 1992). Two books (Grinstein, 1988; Cragoe et al., 1992) on the Na^+/H^+ exchanger have been published and should be consulted for more detailed information.

EVIDENCE FOR THE PRESENCE OF H^+ PUMPS IN EUKARYOTIC CELLS

Maintenance of unequal concentrations of ions across membranes is a fundamental property of living cells. In most cells, the concentration of K^+ inside the cells is about 30 times that in the extracellular fluids, while sodium ions are present in much higher concentration outside the cells than inside. These concentration gradients are maintained by the Na^+-K^+-ATPase by means of the expenditure of cellular energy. Since the plasma membrane is more permeable to K^+ than to other ions, a K^+ diffusion potential maintains membrane potentials which are usually in the range of -30 to -90 mV. H^+ ions do not behave in a manner different from that of other ions. If passively distributed across the plasma membrane, then the equilibrium intracellular H^+ concentration can be calculated from the Nernst equation via

$$E_{H^+} = \frac{RT}{F} \ln \frac{[H^+]i}{[H^+]o} \qquad (1)$$

where E is the membrane potential, R the gas constant, T the absolute temperature, F the Faraday, and $[H^+]i$ and $[H^+]o$, the H^+ concentrations inside and outside, respectively. Equation (1) can be rearranged to yield (at physiological temperatures):

$$E \ (mV) = -60 \ (pHo-pHi)$$

This simple calculation indicates that if H^+ were passively distributed across the plasma membrane, and for a mean membrane potential of -60 mV, intracellular pH values (pHi) would be expected to be 1 pH unit lower than the external pH values (pHo). Numerous experiments performed on a large variety of cells indicate, however, that pHi values are usually close to pHo values, or slightly below (Roos and Boron, 1981). This implies that H^+ ions are not passively distributed across the plasma membrane and that H^+ pumps drive H^+ out of equilibrium. Different H^+ pumps have been identified in eukaryotic cells: an ubiquitous Na^+/H^+ exchanger, and pumps that have a more restricted cellular distribution. These are (H^+)ATPases, a Cl⁻/HCO₃⁻ exchanger, a Cl⁻/NaHCO₃ exchanger and a $Na^+/(HCO_3^-)n$ cotransporter where n > 1.

Physicochemical buffering is a property of weak acids and bases whereby these compounds minimize changes in pH by reacting with exogenous H^+. Cells are filled with proteins that are polyelectrolytes and that buffer against pH changes. Cytoplasmic buffering power, defined as the amount of strong acid or strong base that is necessary to change pHi by 1 pH unit, is usually in the range of 10 to 50 mmole/l per pH unit (Roos and Boron, 1981). These values mean that the H^+ ions are highly buffered and that a considerable amount of H^+ has to be produced or secreted by the cells to change pHi by 1 pH unit. Physicochemical buffering provides only a short term protection against cellular acidosis. Long-term protection is achieved by membrane transport mechanisms for H^+ equivalents such as the Na^+/H^+ exchanger and bicarbonate transport systems.

BIOCHEMICAL PROPERTIES OF THE Na⁺/H⁺ EXCHANGER

The Na^+/H^+ exchanger is usually considered to catalyze the exchange of Na^+ for H^+. It should be noted, however, that because the kinetics of H^+ binding are identical to the kinetics of OH^- de-binding (by virtue of the small dissociation constant of water), it cannot be determined whether the system acts as an Na^+/H^+ exchanger or as an Na^+/OH^- cotransporter.

Stoichiometry

In eukaryotic cells, the Na^+/H^+ exchanger catalyzes a 1/1, electroneutral and reversible, exchange of Na^+ for H^+. As a consequence, rates of ion transport are independent of the value of the membrane potential and are governed by the Na^+ and H^+ gradients across the membrane. Prokaryotic cells also have an Na^+/H^+ exchanger but the system differs from that found in eukaryotic cells in that it is electrogenic. For an electroneutral exchange mechanism, the driving force resides in the ionic concentrations (or activities) of the transported ions and, at equilibrium, the following relationship holds:

$$\frac{[H^+]i}{[H^+]o} = \frac{[Na^+]i}{[Na^+]o}$$

Assuming then a Na^+ concentration ratio of 10, the Na^+/H^+ exchanger could achieve a pHi value 1 unit higher than pHo. The fact that such a steady state pHi value is rarely observed indicates that the activity of the exchanger is counterbalanced by cell acidifying mechanisms such as electrogenic membrane leaks for acid equivalents (H^+, OH^-, HCO_3^-) and the metabolic production of H^+.

Kinetic Properties

The properties of interaction with external and internal Na$^+$ and H$^+$ have been extensively analyzed and reviewed (Aronson, 1985; Montrose and Mürer, 1988). They can be summarized as follows:

The External pH Dependence of the Na$^+$/H$^+$ Exchanger

An increase in pHo increases the rate of exchange of Na$^+$ for H$^+$. Hill coefficients characterizing the interaction of H$^+$ at the external face of the exchanger are close to one, indicating simple Michaelis-Menten kinetics. Half maximum activation of Na$^+$ transport is observed at pHo values in the range of 7.3 to 7.5. Although the pH of circulating fluids is usually maintained constant around these values, there are situations where changes in pHo alter the activity of the Na$^+$/H$^+$ exchanger. For instance, under ischemic conditions (hypoxic conditions associated with reduced blood flow), the acidification of extracellular spaces prevents H$^+$ efflux by the Na$^+$/H$^+$ exchanger and further amplifies the intracellular acidification generated by the anaerobic conditions of metabolism.

In excitable cells, the properties of interaction of external H$^+$ ions with the exchanger are independent of the presence of Na$^+$, thus suggesting the presence of distinct binding sites for Na$^+$ and H$^+$ at the external face of the exchanger protein. Conversely, in kidney brush-border membranes, changing the external Na$^+$ concentration modifies the pHo dependence of the exchanger, suggesting that external Na$^+$ and H$^+$ compete for the same binding site. These different properties may be accounted for by the existence of different isoforms of the Na$^+$/H$^+$ exchanger.

The Internal H$^+$ Dependence

Na$^+$ influx and H$^+$ efflux by the exchanger are increased at low pHi values. For this simple reason, the main function of the system is to protect cells against cellular acidosis. When pHi decreases, Na$^+$/H$^+$ exchange increases; that is, more H$^+$ ions are pumped out of the cells, thus leading to a compensatory cellular alkalinization. The pHi for half-maximum activation of the Na$^+$/H$^+$ exchanger differs greatly in various cell types. In some cells (e.g., resting fibroblasts), pHi values for half-maximum activation of the exchanger can be as low as 6.2. In such cells, the exchanger is nearly quiescent at normal pHi values (7.0–7.2), and it does not normally contribute much to Na$^+$ influx. Yet it can protect cells against an intracellular acidification by virtue of the simple fact that an intracellular acidification activates H$^+$ efflux. In other cell types (e.g., activated fibroblasts and excitable cells), the pHi for half-maximum activation of the exchanger is close to the resting pHi. Under these conditions, the system is operating at physiological pHi values and contributes to a significant Na$^+$ load which is compensated for by the activity of the Na$^+$-K$^+$-

ATPase. In these cells, the system can still be activated by an intracellular acidification, and hence contributes to the protection of cells against acute acid loads.

It was first demonstrated by Aronson et al. (1982) that internal H^+ stimulates the net influx of Na^+ with sigmoidal kinetics rather than with classical Michaelis-Menten kinetics. Such a cooperative interaction of internal H^+ with the Na^+/H^+ exchanger has been found in most cell types with apparent values of the Hill coefficient ranging from 1.2 to almost 3. A cooperative interaction of internal H^+ allows the system to respond rather markedly to a small variation in pHi. This property is ideal for a fine tuning of the pHi.

The External Na⁺ Dependence

Activation of Na^+/H^+ exchange by external Na^+ is usually well fitted by simple Michaelis-Menten kinetics. The concentration of Na^+ that produces half-maximum activation of the exchanger lies between 10 and 50 mM. Thus, at the Na^+ concentration prevailing in biological fluids (140 mM), a large fraction of the external Na^+ binding sites on the exchanger are saturated with Na^+. Both competitive and noncompetitive interactions of H^+ and Na^+ for their binding to the Na^+ binding sites have been observed in different cell types. The only monovalent cation that can substitute for Na^+ is Li^+. Although Li^+ presents a higher affinity than Na^+ for the Na^+/H^+ exchanger, it supports lower maximum rates of exchange than Na^+.

The Internal Na⁺ Dependence

The activity of the Na^+/H^+ exchanger is high at resting $[Na^+]i$ values (10–20 mM). However, high intracellular Na^+ concentrations inhibit H^+ extrusion. This results in cellular acidification. Similarly, inhibition of Na^+/H^+ exchange by amiloride and its analogs also produces cellular acidification. The properties of interaction of internal Na^+ with the exchanger are probably complex.

Pharmacological Properties

Reversible inhibition by the diuretic drug amiloride is a general characteristic of the Na^+/H^+ exchanger. Amiloride (Figure 1) is a pyrazine carbonyl-guanidine analogue discovered in 1965 at Merck, Sharp & Dohme during a search for nonsteroidal K^+ sparing diuretics. When administered to dogs, rats, or humans, amiloride produces a modest diuretic response and significantly reduces potassium excretion. Clinically, amiloride is used in combination with more potent diuretics such as hydrochlorothiazide for the treatment of hypertension and congestive heart failure. This pharmacological action of amiloride is due to the blockade of the Na^+ conductive ion channel that is located in the distal and collecting renal tubules. As a consequence, less Na^+ is reabsorbed (hence the diuretic effect) and less K^+ is

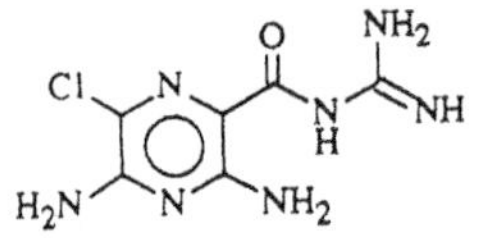

Non selective

Amiloride

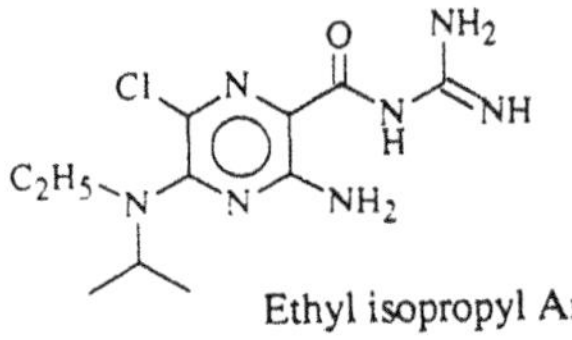

Na/H antiporter inhibitor

Ethyl isopropyl Amiloride

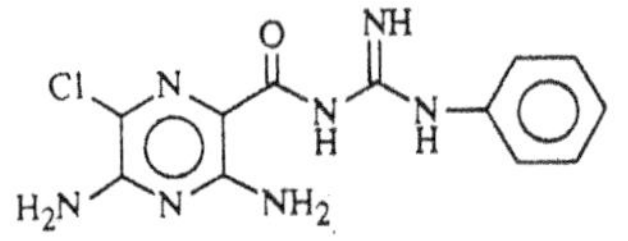

Na channel inhibitor

Phenamil

Figure 1. Structures of amiloride, ethyl isopropyl amiloride, and phenamil.

secreted (hence the antikaliuretic effect). Micromolar concentrations of amiloride are sufficient to block renal epithelial Na^+ channels.

The Na^+/H^+ exchanger is a second target of amiloride. Its sensitivity to amiloride is however less than that of the epithelium Na^+ channel. Considering, in addition, that the action of amiloride is competitively antagonized by Na^+, millimolar concentrations of amiloride are required to block Na^+/H^+ exchange activity under physiological conditions of external Na^+ concentration (140 mM). This explains why few of the pharmacological actions of amiloride observed *in vivo* can be attributed to the inhibition of Na^+/H^+ exchange activity.

A very large number of derivatives of amiloride synthesized by E.J. Cragoe, Jr. at Merck, Sharp & Dohme have made it possible to determine in detail the structure-activity relationships for the Na^+/H^+ exchanger and the epithelium Na^+ channel. The following trends have been observed (Vigne et al., 1984; Frelin et al., 1987):

1. The presence of a halogen atom at position six of the pyrazine ring is important for activity. Its replacement by -H drastically reduces the ability of the molecule to inhibit both the epithelium Na^+ channel and the Na^+/H^+ exchanger.
2. Alkyl group substitutions at ring position five result in a tremendous decrease in inhibitory activity of the epithelium Na^+ channel, but greatly

potentiate the ability of the parent compound to inhibit the Na^+/H^+ exchanger. Derivatives bearing two substituents on the 5-amino nitrogens are more potent inhibitors of the Na^+/H^+ exchanger than those having the same number of carbon atoms in a single substituent.

3. Modifications at the terminal guanidine nitrogen greatly enhance inhibitory activity of the epithelium Na^+ channel. They markedly decrease that of the Na^+/H^+ exchanger.

Figure 1 presents the structure of amiloride and that of amiloride analogues that are known to be the most potent inhibitors of epithelium Na^+ channels and the Na^+/H^+ exchanger. Other derivatives of guanidinium that are unrelated to amiloride and that have antihypertensive properties also block the Na^+/H^+ exchanger (Frelin et al., 1986).

Finally, it should be stressed that amiloride derivatives must be used with caution. They may affect a number of other ion transport mechanisms (e.g., the Na^+/Ca^{2+} exchanger), or they may affect cellular activities (Cragoe et al., 1992), because of their weak base properties and capacity to accumulate in acidic intracellular compartments such as lysosomes. Amiloride derivatives may also be toxic to cells.

MOLECULAR PROPERTIES OF THE Na⁺/H⁺ EXCHANGER

In an elegant study of reverse genetics, Pouyssegur and his colleagues cloned an ubiquitous form of Na^+/H^+ exchanger (Sardet et al., 1989). The strategy was first to select mutants defective in Na^+/H^+ transport activity using a proton suicide technique. Mutagenized cells were loaded with Li^+ or Na^+ and then exposed to a Na^+- or Li^+-free saline solution buffered at an acidic pH (5.5). These conditions favor reverse operation of the Na^+/H^+ exchanger, i.e., an efflux of Na^+ (or Li^+) and an influx of H^+. The conditions used induce a large acid load that kills cells that possess a Na^+/H^+ exchanger. Only cells that lack the Na^+/H^+ exchanger survive the selection procedure. A similar strategy has been used to isolate cells that express the Na^+/H^+ exchanger with altered kinetic properties or that overexpress Na^+/H^+ transport activity.

The mutant cells lacking Na^+/H^+ exchange activity were then complemented by transfection with human genomic DNA (Sardet et al., 1989). This approach led to the successful identification of a cDNA (NHE-1) that codes for a 85 kDa protein comprising 815 amino acids. The protein present in plasma membranes is glycosylated and has a molecular weight of 110 kDa. Two distinct domains are distinguished in the protein. The amino terminal part of the protein is amphipathic and shows 10 stretches of hydrophobic amino acid residues that likely represent transmembrane spanning domains. The carboxy terminus is highly hydrophilic and represents a large cytoplasmic domain with several potential phosphorylation sites. NHE-1, which is an ubiquitous form of the Na^+/H^+ exchanger, can be activated by growth factors and agonists of phospholipase C, and is expressed at the basolateral surface of intestinal epithelial cells. A different protein kinase A activatable, isoform of NHE-1 is found in trout erythrocytes (Borgese et al., 1992).

Three other forms of the Na^+/H^+ exchanger have subsequently been cloned. Screening of a rabbit kidney proximal tubule library with an NHE-1 probe (using low stringency hybridization) has led to the identification of two other forms of Na^+/H^+ exchanger (Tse et al., 1992) that are entirely (NHE-3) or predominantly (NHE-2) expressed in the intestine and kidney. A fourth form of Na^+/H^+ exchanger (NHE-4) has been identified in the gastrointestinal tract (Orlowski et al., 1992). These isoforms are less sensitive than NHE-1 to amiloride derivatives.

REGULATION OF Na^+/H^+ EXCHANGE ACTIVITY

The activity of the Na^+/H^+ exchanger can be modified by many distinct stimuli including growth factors, neurotransmitters, vasoconstrictors, and several physical factors. Effectors of the Na^+/H^+ exchanger differ widely in the time course of their action. Some act within seconds or minutes of their application to cells. Others require hours or even days to develop their action. Activation of the Na^+/H^+ exchanger activity results in an increase in pHi. Conversely, an inhibition of the Na^+/H^+ exchanger leads to a cellular acidification.

Short-term regulation of the Na^+/H^+ exchange activity is achieved by growth factors whose receptor have an intrinsic tyrosine kinase activity (Epidermal growth factor [EGF], Insulin, platelet derived growth factor [PDGF] and fibroblast growth factor [FGF]) and by agonists of G protein coupled receptors that activate phospholipase C. Activation is also obtained with drugs that interfere with signaling pathways such as phorbol esters, which are potent agonists of protein kinase C, and okadaic acid, which is a potent inhibitor of protein phosphatases 1 and 2A.

The activation of NHE-1 is associated with a phosphorylation of the exchanger protein on serine residues that are located in the C terminal cytoplasmic region of the protein (Sardet et al., 1990). The serine residue whose phosphorylation regulates Na^+/H^+ exchange and the kinase involved have not yet been identified. Phosphorylation of NHE-1 results in an increase in affinity of the exchanger protein for internal H^+. As a consequence the dose-response curve for the activation of Na^+ uptake by pHi shows a shift to higher pHi values.

The properties of other isoforms of the Na^+/H^+ exchanger have been less thoroughly studied but they may be different from those of NHE-1. For instance, the NHE-1 isoform of mammalian fibroblasts is insensitive to cAMP, whereas the β-NHE-1 isoform of trout erythrocytes is sensitive to the effector. Conversely, the isoform which is present in the apical membrane of epithelial cells of the small intestine and kidney (NHE-2 or NHE-3) is inhibited rather than activated by phorbol esters.

In some cells, short-term activation of Na$^+$/H$^+$ exchange involves an increase in maximum velocity of the system. The molecular mechanism underlying this regulation has not yet been elucidated.

Slow (adaptive) changes in Na$^+$/H$^+$ exchange activity have been observed in response to several hormonal effectors under pathological situations (Sacktor and Kinsella, 1988). These actions are mediated by changes either in the kinetic properties of the system or in the number of functional exchangers.

OTHER pHi REGULATING MECHANISMS

The Na$^+$/H$^+$ exchanger is only one mechanism for the membrane transport of H$^+$. Other mechanisms have been identified and contribute to the regulation of pHi.

H$^+$-ATPases

At least three types of ATP driven H$^+$ pumps may be potentially involved in the regulation of pHi. These are: (a) an electrogenic H$^+$ translocating ATPase found in the mammalian distal nephron; (b) an electroneutral K$^+$-H$^+$ exchange ATPase that is responsible for acid secretion by gastric cells; and (c) a vacuolar H$^+$-ATPase which is responsible for the low pH of endocytic vesicles. The vacuolar H$^+$-ATPase can be inhibited by the macrolide antibiotic bafilomycin A1. The gastric K$^+$-H$^+$-ATPase is inhibited by antiulcer drugs such as omeprazole.

Bicarbonate Transport Systems

Three types of transport mechanisms for bicarbonate have been identified: (a) an electroneutral Na$^+$ dependent Cl$^-$/HCO$_3^-$ cotransporter; (b) an electroneutral, Na$^+$ independent Cl$^-$/HCO$_3^-$ cotransporter; and (c) an electrogenic Na$^+$/HCO$_3^-$ n (n>1) cotransporter. Unlike the Na$^+$/H$^+$ exchanger which is ubiquitous, bicarbonate transport mechanisms are expressed in a cell specific manner. In most cell types, however, the Na$^+$/H$^+$ exchanger coexists with one or several bicarbonate transport systems. As a consequence, pHi regulation can be achieved by various means in different cell types.

Because the inward gradient for Cl$^-$ usually exceeds that for HCO$_3^-$, the Cl$^-$/HCO$_3^-$ exchanger catalyzes the net efflux of HCO$_3^-$ and therefore acts as a cell acidifying mechanism that counteracts the action of the Na$^+$/H$^+$ exchanger. Three forms of Cl$^-$/HCO$_3^-$ exchangers have been cloned, including erythrocyte band 3 (Alper, 1991). The Na$^+$-dependent Cl$^-$/HCO$_3^-$ exchanger mainly acts as a cell alkalinizing mechanism. This is because the inward Na$^+$ gradient favors HCO$_3^-$ influx rather than efflux. It thus operates in parallel with the Na$^+$/H$^+$ exchanger to protect cells against an intracellular acidosis. Finally, the electrogenic Na$^+$/(HCO$_3^-$) n (n > 1) cotransporter promotes an influx (e.g., in bovine corneal epithelial cells) or efflux (e.g., in the renal proximal convoluted tubule) of HCO$_3^-$, depending on the membrane potential and its stoichiometry. It therefore acts either as a cell acidifying or alkalinizing mechanism.

As in the case of the Na^+/H^+ exchanger, the activity of bicarbonate transport systems is regulated by both protein kinase A and protein kinase C (Vigne et al., 1988; Ganz et al., 1989).

H^+ Channels

The existence of H^+ conducting channels has been established in snail neurons and axolotl oocytes. In these tissues, the channels are inactive at the resting membrane potential but open when the membrane is depolarized. Voltage gated H^+ channels are inhibited by divalent cations, such as Cd^{2+} and Zn^{2+}. Their opening induces an influx or an efflux of H^+ which depends on both the membrane potential and transmembrane pH gradient. Palytoxin, one of the most potent natural toxins known so far, may act by opening a membrane H^+ channel (Frelin et al., 1990a).

PHYSIOLOGICAL FUNCTIONS OF THE Na^+/H^+ EXCHANGER

pHi Regulation

The primary function of the Na^+/H^+ exchanger is to maintain pHi relatively constant. Its kinetic properties and possible regulation by phosphorylation reactions are ideally suited to protect the cell against both acute and chronic cellular acidosis.

Resting cells usually face rather small acid loads due to their low metabolic activity and electrogenic H^+ influx. The production of intracellular H^+ is counterbalanced by a low activity of the Na^+/H^+ exchanger. Following an acute acid load, more H^+ binds to the exchanger protein and Na^+/H^+ exchange activity increases. H^+ is pumped out of the cells until pHi returns to its original steady state (often referred to as the set-point of the exchanger). In activated cells, a high and sustained activity of Na^+/H^+ exchanger may be necessary to compensate for an increased acid load due. Phosphorylation of the exchanger protein, which sensitizes the exchanger to internal H^+, provides a mechanism by which an increased acid load can be compensated for without modifying the steady state pHi value as illustrated in Figure 2. If phosphorylation of the exchanger protein shifts its pHi dependence to higher values, then cellular alkalinization can be observed.

A concerted cellular response involving several pHi regulating mechanisms has been observed in some instances. In mesangial cells, arginine vasopressin enhances pHi regulation by stimulating three pHi regulating mechanisms: the Na^+/H^+ exchanger, a Cl^-/HCO_3^- exchanger, and a Na^+-dependent Cl^-/HCO_3^- exchanger without much altering the steady state pHi value (Ganz et al., 1989). Another well-documented example is provided by activated neutrophils. Neutrophils are attracted to sites of infection where they destroy invading organisms by a variety of microbicidal ways, including phagocytosis, degranulation, and the production

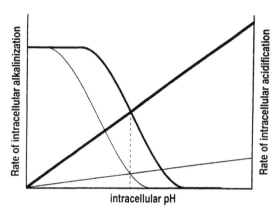

Figure 2. The role of the Na⁺/H⁺ exchanger in pHi regulation. A steady state pHi is attained when the rate of cellular acidification is equal to the rate of cellular alkalinization. In resting cells, the rate of intracellular acidification is low and low activity of the Na⁺/H⁺ exchanger is sufficient to maintain a stable pHi. An acute acid load activates H⁺ efflux, thus allowing cells to recover their basal pHi. When, however, cells face chronic intracellular acidification (for instance as a result of an increased metabolic activity), a lower steady state is attained. Sensitization of the Na⁺/H⁺ exchanger to H⁺ following its phosphorylation increases H⁺ efflux and provides a mechanism for maintaining a stable pHi in spite of the increased acid load.

of toxic oxygen metabolism. Activation is accompanied by a large acid load due to the stimulation of NADPH oxidase and the associated acceleration of the hexose monophosphate shunt. It is also accompanied by the activation of different cell alkalinizing mechanisms that maintain pHi at or above the resting pHi in spite of the massive generation of intracellular H⁺. These are the Na⁺/H⁺ exchanger, a H⁺ conductance and bafilomycin A1 sensitive vacuolar type H⁺-ATPase. These three processes are mediated by protein kinase C and are mimicked by phorbol esters.

High and sustained metabolic activity is usually associated with chronic acid loads. For instance, in cardiac cells, increasing the frequency of contractions leads to a rate-dependent acid load. Acid equivalents are generated metabolically via stimulation of glycolysis and the production of lactate. In such cells, a higher activity of the Na⁺/H⁺ exchanger is required to maintain stable pHi values than in resting, metabolically inactive cells. This is achieved by constitutive phosphorylation of the exchanger protein. Evidence for such a mechanism is provided by the observation that depletion of cellular ATP reversibly reduces the phosphorylation state of the exchanger, reduces its affinity for internal H⁺, and decreases its activity at physiological pHi values. Finally, it should be noted that regulation of pHi by the Na⁺/H⁺ exchanger is achieved at the expense of the membrane Na⁺ gradient and that a high activity of the Na⁺/H⁺ exchanger imposes an Na⁺ load on the cells which may have its own consequences (e.g., alteration of Ca^{2+} homeostasis via a modified activity of the Na^+/Ca^{2+} exchanger). Direct evidence that the Na^+/H^+ exchanger contributes significantly to the uptake of Na^+ by cardiac cells is provided by the observation that amiloride prevents the cardiotoxic effects of cardiac glycosides (Frelin et al., 1984).

Transepithelial Acid-Base Transport

In epithelia, a major function of the Na^+/H^+ exchanger is to promote transepithelial acid-base transport. A standard model of acid secretion is the renal proximal tubule (Figure 3). Acid secretion into the lumen is achieved by the Na^+/H^+ exchanger. It induces acidification of the luminal fluid whereupon H^+ titrates the available HCO_3^- to form H_2CO_3. Membrane-bound carbonic anhydrase, in turn, causes catalytic dehydration to give off CO_2 and H_2O. The CO_2 equilibrates very rapidly across the luminal membrane, and the CO_2 reaching the cell interior is bound by the OH^- formed as a result of the splitting of water, an action catalyzed by cytosolic carbonic anhydrase. The newly formed HCO_3^- crosses the contraluminal membrane, thereby reaching the capillaries. The H^+ formed as a result of the splitting of water is extruded by the luminal Na^+/H^+ exchanger.

Volume Regulation

When cells are swollen hypoosmotically and returned to an isoosmotic solution, they shrink to volumes less than those of the control and then gradually adjust their volume back to normal. Regulatory volume increase (RVI) proceeds by the

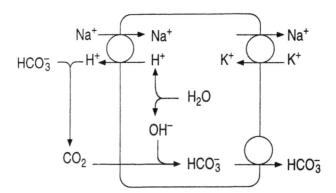

Figure 3. A schematic diagram of the renal proximal tubule showing how the Na^+/H^+ exchanger, by acidifying tubular fluids, promotes the regeneration of bicarbonate. Interconversions of CO_2, water, H^+, OH^- and HCO_3^- are catalyzed by carbonic anhydrase (membrane bound and cytosolic).

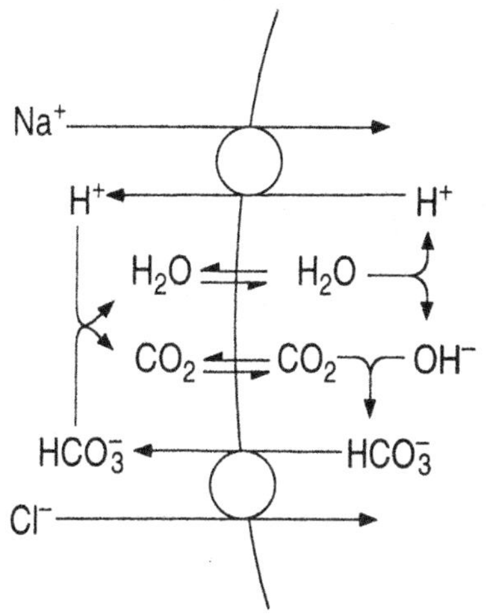

Figure 4. A schematic diagram showing how a Na^+/H^+ exchanger coupled to a Cl^-/HCO_3^- exchanger results in net movement of NaCl into cells during regulatory volume increase.

combined operation of the Na^+/H^+ and Cl^-/HCO_3^- exchangers, the net result being the gain of NaCl by the cells, with the obligatory uptake of water (Figure 4). Such a mechanism is most prominent in cell types that have a robust Cl^-/HCO_3^- exchanger and a high buffering power (Cala and Grinstein, 1988). In other cell types, RVI involves an activation of the $Na^+/K^+/2Cl^-$ cotransporter. During RVI, marked changes in the cellular ion content and volume occur in the absence of major changes in pHi (see Chapter 7).

Intracellular Signaling

An early rise in pHi, mediated by activation of the Na^+/H^+ exchanger, appears to be a fairly common response of metabolically quiescent cells to various agonists of phospholipase C and mitogens (Moolenaar, 1986). These observations have led to the hypothesis that changes in pHi resulting from the activation of the Na^+/H^+ exchanger mediate the intracellular actions of various external stimuli. However, further studies showed that changes in pHi are only observed when cells are incubated in the absence of bicarbonate (i.e., under conditions of reduced pHi buffering) and that pHi is likely to facilitate intracellular messenger actions. It could be that activation of the Na^+/H^+ exchanger only helps cells to regulate their pHi in the face of an increased acid load resulting from an increased metabolic demand. Yet, as is known, an alkaline pHi seems to have a permissive effect on cellular activation.

THE Na$^+$/H$^+$ EXCHANGER IN PATHOLOGY

Cardiac Ischemia

A striking example of a pathological situation involving the Na$^+$/H$^+$ exchanger is that of reperfusion induced arrhythmias. Under conditions of reduced coronary blood flow, cellular concentrations of high energy phosphate compounds decrease to very low levels, and cells acidify as a consequence of the accumulation of lactic acid produced under anaerobic conditions of metabolism. The coronary blood flow being reduced, weak acids accumulate in extracellular spaces since they cannot be removed from the circulation. The result is extracellular acidification. Under these conditions, the Na$^+$/H$^+$ exchanger is inactive and cannot prevent cells from acidifying. This is due to the fact that a reduced pHo limits the activity of the exchanger even under conditions of intracellular acidosis. If, however, the ischemia is not of long duration, then it does not result in extensive cell damage. On the other hand, reperfusion of ischemic zones causes arrhythmias and irreversible cell damage that can lead to cell death. The mechanism of reperfusion-induced cardiotoxicity (often called the oxygen paradox) is explained in the following manner. During reperfusion of ischemic zones with a solution at physiological pH, a large H$^+$ gradient (acid inside, alkaline outside) develops across the plasma membrane. This condition favors maximum activity of the Na$^+$/H$^+$ exchanger and a large influx of Na$^+$. If the period of ischemia is short (i.e., if some ATP is still available), then the Na$^+$-K$^+$-ATPase is able to cycle Na$^+$ back to the blood and the combined operation of the Na$^+$/H$^+$ exchanger and the Na$^+$-K$^+$-ATPase allows cells to recover their normal pHi without accumulating Na$^+$. After prolonged periods of ischemia, however, the activity of the Na$^+$-K$^+$-ATPase may be reduced by low concentrations of ATP. Thus, Na$^+$ accumulates inside the cells and this activates the Na$^+$/Ca^{2+} exchanger in the reverse mode. Ca^{2+} overloading then leads to arrhythmias, extensive cell lysis, and eventually cardiac arrest (Lazdunski et al., 1985). Reperfusion-induced arrhythmias can be prevented to a large extent by amiloride and its derivatives, and by low Na$^+$ solutions. They can also be prevented by slowly raising the pH of the perfusate solutions.

Essential Hypertension

Studies have led to evidence suggesting that two types of Na$^+$ transport mechanisms may be defective in vascular smooth muscle cells in some forms of hypertension. These are (a) an increased cell membrane permeability to Na$^+$, and (b) decreased active pumping of Na$^+$ at the prevailing intracellular Na$^+$ concentration, resulting from elevated blood levels of an endogenous inhibitor of the Na$^+$-K$^+$-ATPase, the so-called natriuretic hormone. The increase in intracellular Na$^+$ concentration, in turn, promotes Ca^{2+} influx by the operation of the Na$^+$/Ca^{2+} exchanger in the reverse mode, and hence, a higher contractile state of vascular smooth muscle cells. It has, furthermore, been hypothesized that in essential hypertension, the link between perturbed Na^+ homeostasis and abnormal Ca^{2+} regulation is a hyperactive Na^+/H^+ exchange system, and that this explains a number of abnormalities including vasoconstriction, low plasma renin activity, and increased proximal tubular Na^+ reabsorption. Although Na^+/H^+ exchange activity is elevated in platelets and leukocytes of patients with essential hypertension, in lymphocytes, and the aorta of spontaneously hypertensive rats, a defect in Na^+/H^+ exchange activity is unlikely to be the major cause leading to the development of hypertension. First, the activity of the Na^+/Ca^{2+} exchanger in vascular smooth muscle cells is probably too low to tightly link abnormalities in Na^+ transport to changes in Ca^{2+} homeostasis (Frelin et al., 1990b). And second, no evidence for polymorphism in the NHE-1 exchanger gene that might be associated with inherited hypertension has hitherto been obtained (Lifton et al., 1991).

SUMMARY AND CONCLUSIONS

The most important conclusions can be summarized as follows: H^+ ions are not passively distributed across the plasma membrane. Physicochemical buffering provides a mechanism for the short-term maintenance of a stable pHi. Long-term pHi regulation is provided by different membrane H^+ transport mechanisms, including an ubiquitous Na^+/H^+ exchanger. The existence of at least four different forms of Na^+/H^+ exchanger and of additional bicarbonate dependent pHi regulating mechanisms that are expressed in a cell specific manner provides a wide range by which cells regulate their pHi in different situations. The main role of the Na^+/H^+ exchanger is to protect cells against acute cellular acidosis. The phosphorylation of the exchanger protein by protein kinases and the resulting sensitization of the exchanger to internal H^+ provides a mechanism by which H^+ efflux is controlled by metabolic activity. In some cells, the Na^+/H^+ exchanger serves other functions such as transepithelial acid transport and volume regulation.

Finally, there are several instances in which the activity of the Na^+/H^+ exchanger contributes to pathological states. The most clearly defined state is reperfusion-induced cardiac injury.

REFERENCES

Alper, S. (1991). The band-3 related anion exchanger (AE) gene family. Ann. Rev. Physiol. 53, 549–564.
Aronson, P.S. (1985). Kinetic properties of the plasma membrane Na^+/H^+ exchanger. Ann. Rev. Physiol. 47, 545–560.
Aronson, P.S., Nee, J., & Suhm, M.A. (1982). Modifier role of internal H^+ in activating the Na^+/H^+ exchanger in renal microvillus membrane vesicles. Nature 299, 161–163.
Benos, D.J. (1982). Amiloride: a molecular probe for sodium transport in tissues and cell. Am. J. Physiol. 242, C131–C145.

Borgese, F., Sardet, C., Cappadoro, M., Pouyssegur, J., & Motais, R. (1992). Cloning and expression of a cAMP activatable Na$^+$/H$^+$ exchanger. Evidence that the cytoplasmic domain mediates hormonal regulation. Proc. Natl. Acad. Sci. USA 89, 6765–6769.

Cala, P.M. & Grinstein, S. (1988). Coupling between Na$^+$/H$^+$ and Cl$^-$/HCO$_3$ exchange in pHi and volume regulation. In: Na$^+$/H$^+$ Exchange (Grinstein, S., ed.), pp. 201–208. CRC Press, Inc., Boca Raton, Florida.

Cragoe, E.J., Jr., Kleyman, T.R., & Simchowitz, L. (eds.) (1992). Amiloride and its analogs. Unique cation transport inhibitors. VCH Publishers Inc., New York.

Frelin, C., Vigne, P., & Breittmayer, J.P. (1990a). Palytoxin acidifies chick cardiac cells and activates the Na$^+$/H$^+$ antiporter. FEBS Lett. 264, 63–66.

Frelin, C., Vigne, P., & Lazdunski, M. (1984). The role of the Na$^+$/H$^+$ exchange system in cardiac cells in relation to the control of the internal Na$^+$ concentration. A molecular basis for the antagonistic effect of ouabain and amiloride on the heart. J. Biol. Chem. 259, 8880–8885.

Frelin, C., Vigne, P., & Lazdunski, M. (1990b). The Na$^+$/H$^+$ exchange system in vascular smooth muscle cells. In: Advances in Nephrology (Grunfeld, J.P., Bach, J.F., Funck-Brentano, J.L., & Maxwell, M.H., eds.), pp. 17–29. Year Book Medical Publishers Inc., Chicago.

Frelin, C., Vigne, P., Barbry, P., & Lazdunski, M. (1986). Interaction of guanidinium and guanidinium derivatives with the Na$^+$/H$^+$ exchange system. Eur. J. Biochem. 154, 241–245.

Frelin, C., Vigne, P., Barbry, P., & Lazdunski, M. (1987). Molecular properties of amiloride action and of its Na$^+$ transporting targets. Kidney Int. 32, 785–793.

Frelin, C., Vigne, P., Ladoux, A., & Lazdunski, M. (1988). The regulation of the intracellular pH in cells from vertebrates. Eur. J. Biochem. 174, 3–14.

Ganz, M.B., Boyarsky, G., Sterzel, R.B., & Boron, W.F. (1989). Arginine vasopressin enhances pHi regulation in the presence of HCO$_3^-$ by stimulating three acid base transport systems. Nature 337, 648–651.

Grinstein, S. (ed.) (1988). Na$^+$/H$^+$ exchange. CRC Press, Inc., Boca Raton, Florida.

Grinstein, S., Rotin, D., & Mason, M.J. (1989). Na$^+$/H$^+$ exchange and growth factor induced cytosolic pH changes. Role in cellular proliferation. Biochim. Biophys. Acta 988, 73–97.

Kleyman, T.R. & Cragoe, E.J. (1988). Amiloride and its analogs as tools in the study of ion transport. J. Memb. Biol. 105, 1–21.

Lazdunski, M., Frelin, C., & Vigne, P. (1985). The Na$^+$/H$^+$ exchange system in cardiac cells. Its biochemical and pharmacological properties and its role in regulating internal concentrations of sodium and internal pH. J. Mol. Cell. Cardiol. 17, 1029–1042.

Lifton, R.P., Hunt, S.C., Williams, R.R., Poussegur, J., & Lalouel, J.M. (1991). Exclusion of the Na$^+$/H$^+$ antiporter as a candidate gene in human essential hypertension. Hypertension 17, 8–14.

Montrose, M.H. & Mürer, H. (1988). In: Na$^+$/H$^+$ Exchange (Grinstein, S., ed.), pp. 57–75. CRC Press, Inc., Boca Raton, Florida.

Moolenaar, W. (1986). Effects of growth factors on intracellular pH regulation. Ann. Rev. Physiol. 48, 363–376.

Mürer, H., Höpfer, U., & Kinne, R. (1976). Sodium/proton antiport in brush border membrane vesicles isolated from rat small intestine and kidney. Biochem. J. 154, 597–604.

Orlowski, J., Kandasamy, R.A., & Shull, G.E. (1992). Molecular cloning of putative members of the Na$^+$/H$^+$ exchanger family. J. Biol. Chem. 267, 9331–9339.

Roos, A. & Boron, W.F. (1981). Intracellular pH. Physiol. Rev. 61, 296–434.

Saktor, B. & Kinsella, J. (1988). Regulation of Na$^+$/H$^+$ exchange activity by adaptative mechanism. In: Na$^+$/H$^+$ exchange (Grinstein, S., ed.), pp. 307–324, CRC Press, Inc., Boca Raton, Florida.

Sardet, C., Counillon, L., Franchi, A., & Pouyssegur, J. (1990). Growth factors induce phosphorylation of the Na$^+$/H$^+$ antiporter, a glycoprotein of 110 kD. Science 247, 723–726.

Sardet, C., Franchi, A., & Pouyssegur, J. (1989). Molecular cloning, primary structure and expression of the human growth factor activatable Na$^+$/H$^+$ antiporter. Cell 56, 271–281.

Tse, C.M., Brant, S.R., Walker, S., Pouyssegur, J., & Donowitz, M. (1992). Cloning and sequencing of a rabbit cDNA encoding an intestinal and kidney specific Na⁺/H⁺ exchanger isoform. J. Biol. Chem. 267, 9340–9346.

Vigne, P., Breittmayer, J.P., Frelin, C., & Lazdunski, M. (1988). Dual control of the intracellular pH in aortic smooth muscle cells by a cAMP sensitive HCO₃/Cl antiporter and a protein kinase C sensitive Na⁺/H⁺ antiporter. J. Biol. Chem. 263, 18023–18029.

Vigne, P., Frelin, C., Cragoe, E.J., & Lazdunski, M. (1984). Structure-activity relationships of amiloride and certain of its analogues in relation to the blockade of the Na⁺/H⁺ exchange system. Mol. Pharmacol. 25, 131–136.

Wakabayashi, S., Sardet, C., Fafournoux, P., Counillon, L., Meloche, S., Pages, G., & Pouyssegur, J. (1992). Structure and function of the growth factor activatable Na⁺/H⁺ exchanger (NHE1). Rev. Physiol. Biochem. Pharmacol. 119, 157–186.

Chapter 6

Oxidase Control of Plasma Membrane Proton Transport

FREDERICK L. CRANE, IRIS L. SUN, RUTH A. CROWE, and HANS LÖW

Principles of Medical Biology, Volume 4
Cell Chemistry and Physiology: Part III, pages 169–186.
Copyright © 1996 by JAI Press Inc.
All rights of reproduction in any form reserved.
ISBN: 1-55938-807-2

INTRODUCTION

Plasma membranes contain electron transport systems. Some of these systems, such as the superoxide-generating NADPH oxidase of neutrophils, are well-known. Others, such as the NADH oxidase found in all cells, are not fully defined. Still others, such as glutathione oxidase, have barely been recognized. Electron transport across the plasma membrane results in proton release from cells. It is clear that some of this release in some cells is based on activation of the Na^+/H^+ exchanger by electron flow. Activation of other channels or the proton-exporting ATPase may also be involved. Direct proton transport through the electron transport components may also account for a small part of the proton movement in plasma membranes. The activation of the proton movement by plasma membrane electron transport shows interaction with hormones or growth factors. It can be controlled by oncogene products associated with the plasma membrane and is susceptible to inhibition by known antitumor agents.

PROTON TRANSPORT CHANNELS

The well-recognized exchange systems for proton movement across the plasma membrane are HCO_3^-/Cl^-, Na^+/H^+, H^+-ATPase, and K^+/H^+. These have been discussed in detail in the preceding chapter.

Proton transport through the exchange channels and H^+/ATPase is controlled by hormones and growth factors. This control is based on the status of membrane hormone and growth factor receptors in the plasma membrane, the structural status for interaction of the receptors and channels, the control of kinases or phosphatases and the status of cytosolic control systems such as G proteins or internal pH. The source of reductants for activation of proton movement by the oxidation-reduction enzymes is NADH- or NADPH-generated in the cytosol by glycolysis or the pentose shunt. Glutathione reduction can be based on NADPH-dependent glutathione reductase (Figure 1).

In this chapter the evidence for activation of some of these known channels by plasma membrane electron transport or the direct action of the oxidoreductase in proton movement is discussed.

REDOX-DRIVEN PROTON TRANSPORT

In other membranes responsible for energy transduction, oxidation-reduction reactions are the basis for proton movement across the membrane. The classical example is found in mitochondria where electron flow through the respiratory system drives proton movement from the inner matrix to the intermembrane space. The resultant protonmotive force is used to drive ATP synthesis or to maintain mitochondrial membrane potential (negative inside). A similar association between electron transport in a membrane oxidation reduction chain and proton movement

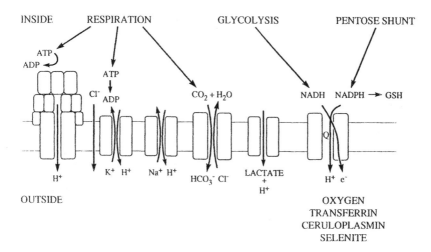

Figure 1. The channels which can be available for proton release in different cells. These may be activated by ligands attached to receptors or signals generated by the electron transport system. The electron transport across the membrane can also be accompanied by proton movement, depending on the orientation of electron transport, but this movement would be limited because of the slow rate of electron transport compared to the rapid rate which can be elicited through channels. Any possible relation of oxidase control to the H^+-ATPase or the H^+-K^+-ATPase has not been tested by inhibitors such as bafilomycin or omeprazole, respectively (Swallow et al., 1990).

through the membrane is found in chloroplasts where the electron transport drives proton movement from the exterior into the lumen of the thylakoid cisternae. The protonmotive force or concentration of protons inside the lumen is then released by export through an ATP synthase channel to drive ATP formation. There is now good evidence for similar types of proton movement associated with electron transport reactions in plasma membranes.

Mitochondria

In mitochondria there are two types of mechanisms for coupling the electron transport to the movement of protons across the membrane. The first is based on anisotropic reduction and oxidation of a lipid-soluble quinone inside the membrane. The quinone, coenzyme Q, becomes protonated upon reduction and diffuses to an oxidation site on the other side of the membrane where removal of electrons leads to proton release. This is essentially a proton carrier system with the hydroquinone acting as the proton carrier in the lipid phase of the membrane. A further refinement of this system in mitochondria provides for a coenzyme Q redox cycle where the movement of one electron through the chain allows for two protons to cross the

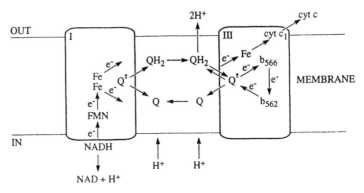

Figure 2. Scheme for proton transport across the mitochondrial inner membrane based on a Q-cycle between the NADH dehydrogenase complex (I) and the cytochrome bc_1 complex (III). Oxidation of the first hydroquinone QH_2 releases protons and transfers electrons to both the iron and cyt b_{566}. The second semiquinone formed by oxidation at the iron is reduced to hydroquinone by b_{562} and takes up protons which are also released across the membrane when it is reoxidized. The quinone is reduced in two one-electron steps by the NADH dehydrogenase to take up protons and reform the hydroquinone. The bound semiquinone at each site is advantageous because it will not be oxidized by oxygen to form superoxide as it would if free floating in the lipid phase. Based on proposals by Krabe and Wikström (1987), Mitchell and Moyle (1985) with modifications to keep the semiquinone in bound condition as known in chloroplasts. The net result is two protons transferred for each electron (Crane, 1990). It should be noted that the dehydrogenase has 12 subunits and the bc_1 complex has 11 subunits, so they are much more complex than the oxidases in the plasma membrane which may have two or three subunits.

membrane on the hydroquinone. In other words, with a simple two-electron reduction of quinone (with protonation) on one side of the membrane, and a two-electron oxidation on the other side, the stoichiometry would be $2H^+/2e^-$. By using the coenzyme Q cycle, with the semiquinone serving as the oxidant for the second electron from the hydroquinone (Figure 2), two protons can be transferred across the membrane for each electron to give a stoichiometry of $2H^+/e^-$.

The second proton transfer mechanism involves protonation of carboxyl or histidyl groups associated with electron carriers in the membrane and release of protons from these sites through proposed channels when the electron carrier is oxidized. This is essentially a proton channel system with movement through the channel gated by the oxidation-reduction state of the prosthetic group on the electron transport protein. The classical example of this is seen in cytochrome c oxidase (Figure 3).

Chloroplasts

In chloroplasts, there are two parts of the electron transport system that are responsible for oxidation-reduction based proton movement (Figure 4). There are

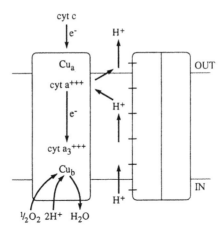

Figure 3. Scheme for proton pumping through cytochrome oxidase coupled to electron transport across the membranes through cytochromes a–a_3. As electrons reduce, cytochrome a protons are taken up from the inside. Transfer of the electron to cytochrome a_3 allows release of the proton through the channel to the outside. Transfer through the channel would be driven by redox reversible proton binding associated with cytochrome a. Simplified scheme based on Krab and Wikström (1987). Since one proton per electron is taken up by internal H_2O formation and one proton per electron is pumped out, the total charge is $2H^+/1e^-$. Cytochrome oxidase (complex IV) has 13 subunits, but 2 subunits alone may be sufficient for redox-driven proton transfer.

two photoreactions in photosynthetic electron transport which generate reducing potential for electron flow. In the first system, called photosystem II, electron flow is through a quinone-hydroquinone pool, and the quinone involves the lipophilic plastoquinone. The quinone is reduced by a sequential two electron movement to produce hydroquinone which is protonated on the outer side of the thylakoid membrane. The protonated plastohydroquinone is then oxidized near the interior surface of the thylakoid with release of protons into the lumen of the thylakoid. At this time it is not clear if the plastoquinone oxidation involves a cycle to increase proton/electron stoichiometry. In the second light reaction, photosystem I electrons are transferred through a less lipophilic quinone, vitamin K_1, but the orientation of the oxidation reduction and protonation of vitamin K_1 in the membrane has not yet been established. On the basis of quinone function in both photosystems, it appears that the proton movement in the thylakoid membranes is based on hydroquinone acting as a transmembrane proton carrier. Other protons are released by photooxidation of water (Figure 4).

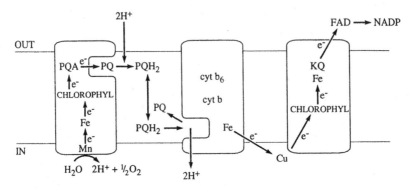

Figure 4. Scheme for proton transfer by plastoquinone as a mobile carrier in membrane lipid. Electrons are transferred one by one to a bound plastoquinone A (PQA) which in turn reduces external plastoquinone. When reduced, the anionic plastoquinone takes up protons to become a hydroquinone which is oxidized by the cytochrome $b_6 f$ complex on the inside of the membrane to release protons. A second quinone, vitamin K_1 (KQ) is also involved in chloroplast electron transport, but its role in proton movement is not known.

PLASMA MEMBRANE REDOX-LINKED
PROTON TRANSPORT

In addition to transport through channels in response to the energy derived from ion gradients or ATP hydrolysis, proton movement across membranes can also be mediated by electron flow in membrane proteins or cofactors. This type of proton movement is best seen in mitochondria or chloroplasts where the movement of electrons in the membrane provides for movement of protons against a concentration gradient. The proton gradient established by this process can then be discharged through membrane channels associated with ATP synthesis to drive ATP formation.

Neutrophil NADPH Oxidase

Since other membranes have an integral transmembrane electron transport system, the question arises whether these electron carriers can be involved in an oxidation-reduction driven proton movement. In neutrophil as well as macrophage plasma membranes, the answer is already yes. The superoxide-producing NADPH oxidase in these membranes is associated with a channel for proton movement to accompany the electron flow when internal NADPH is oxidized by external oxygen to produce superoxide (Nanda et al., 1993). This is a relatively simple electron transport system which contains a heterodimeric cytochrome b which also binds flavin. Thus, two proteins in a transmembrane electron transport system can transfer protons across the membrane.

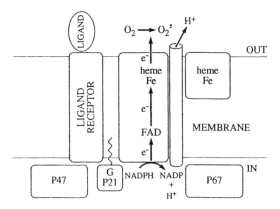

Figure 5. Scheme for proton transport associated with the superoxide-generating NADPH oxidase of the neutrophil plasma membrane. Attachment of a ligand such as FMLP to the receptor and mobilization of cytosolic proteins P47 and P67 at the membrane leads to activation of the oxidase. A GTP-binding protein, P21-rac, is required for message transmission. Activation of protein kinase C is also involved since phorbol myristate acetate activates directly without ligand attachment. Proton transport is through a channel near the cytochrome b_{557} (P91 and P22 heme proteins) (Nanda et al., 1993). The channel is specifically blocked by Zn^{++} and Ca^{++} and activated by arachidonate (Henderson and Chappell, 1992).

This electron transport and proton flow is controlled by at least three other proteins on the cytosolic side of the membrane, P47, P67, and P21rac (Figure 5). Note that this is a much simpler enzyme complex than the complex III (cyt bc_1) of mitochondria which drives proton export from the interior of mitochondria based on quinol oxidation. Despite its simplicity, the neutrophil enzyme may have similarities to the complex three, since it essentially carries out an oxidation of a protonated two electron flavin by a nonprotonated cytochrome b complex with two heme sites. This is the essence of the mitochondrial enzyme in that the two electron quinol is oxidized by a cytochrome b with two heme components.

Relation to Endomembranes

The association of proton movement with electron transport is not reflected in the fatty acyl desaturase system universal to endomembranes. In these enzymes the dehydrogenase, NADH cyt b_5 reductase and the cytochrome b_5 (a single heme cytochrome) are associated exclusively with the cytosolic side of the membrane by acyl groups and have no transmembrane segment. The cytochrome b_5 oxidase associated with the desaturation of fatty acyl CoA may be transmembranous but has not been associated with proton movement. It is an iron-containing protein. The other type of endomembrane cytochromes are the P-450 group of cytochrome bs

involved in oxidative detoxification reactions. It is not clear if any proton movement is associated with this system. All of the endomembrane enzymes are present in small amounts in plasma membrane.

Ligand-Activated NADH Oxidase

The plasma membranes of all cells contain a unique ligand-activated NADH oxidase, the components of which are largely undefined (Brightman et al., 1992). This oxidase has been shown to be associated with proton movement in all types of cells (plant and animal), so it is a basis either for proton transfer itself or for activation of proton transfer channels of other types with independent energization.

TRANSPLASMA MEMBRANE NADH OXIDASE

Components of the transplasma membrane oxidase have been isolated from purified rat liver cell plasma membrane. The components found have been an NADH oxidase fragment with two proteins (36 and 75 kDa) and an NADH ferricyanide reductase which carries out transmembrane electron transport from cytosolic NADH to external ferricyanide. The reductase is different from the NADH cytochrome b_5 reductase also found in plasma membranes. From study of the amino acid composition of a transplasma membrane NADH dehydrogenase in yeast, this transmembrane enzyme can have a molecular weight approaching 68 kDa with six transmembrane helices (Dancis et al., 1992). The logical arrangement of the electron transport system can be based on the requirement for coenzyme Q for both NADH ferricyanide and NADH oxygen oxidoreductase activity (Sun et al., 1992). The dehydrogenase would act as a coenzyme Q reductase, and the oxidase segment would act as a reduced coenzyme Q oxidase. With this scheme, the enzyme could use the coenzyme Q as a proton carrier analogous to the mitochondrial dehydrogenase cytochrome bc_1 system (Figure 6).

OXIDASE ACTIVATION OF PROTON CHANNELS

At least three types of proton channel systems are recognized in animal cells. These include the Na^+/H^+ exchanger, the H^+-ATPase, and the HCO_3^-/Cl^- exchanger. It is clear that a major part of proton release by some cells in response to transplasma membrane electron transport is by activation of the Na^+/H^+ exchanger. This is clear from the characteristics of the proton movement elicited and the magnitude of H^+ release in relation to electron flow when electron transport is activated. Activation of electron transport can be elicited by addition of di-ferric transferrin to activate the transmembrane NADH oxidase activity or by electron flow to external ferricyanide from internal NADH. Addition of di-ferric transferrin to certain cells, especially pineal cells, elicits a remarkable proton release and internal alkalinization. The stoichiometry of H^+ release to iron reduced is more than 100 to 1 (Sun et

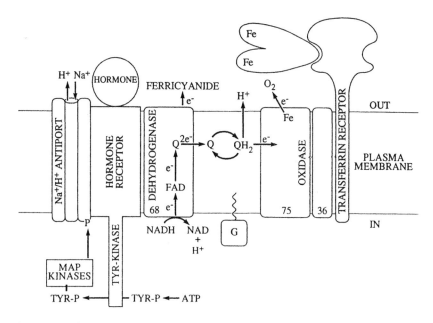

Figure 6. Scheme to represent known aspects of the plasma membrane NADH oxidase and its association with proton release. The oxidase is activated when hormones or ferric transferrin bind receptors. Oxidase may activate tyrosine kinase which can activate MAP kinases to result in phosphorylation of the exchanger leading to Na^+/H^+ exchange. Oxidation of quinol in the membrane can also release protons to the outside equal to the number of electrons transferred. External ferricyanide can activate electron flow by accepting electrons at the quinone. G proteins (GTP binding proteins) such as ras-activate electron transport and proton release in some way and may be a link to kinase activation (McCormick, 1993). Semiquinone formation in the membrane could lead to superoxide and peroxide formation by one electron reduction of oxygen.

Table 1. The Stoichiometry of Oxidant-Induced Proton Transport From HeLa Cells

Oxidant Used to Induce H^+ Release	Rate of Reduction of the Oxidant (nmole min^{-1} gww^{-1})	Rate of Proton Release in Response to Oxidant (nequiv H^+ min^{-1} gww^{-1})	Ratio of H^+ Release per Electron (H^+/e^-)
Ferricyanide	114	273	2.4
Diferric transferrin	15	764	48.4

Note: Reduction and proton release measured under the same conditions in 100 mM NaCl, 83.5 mM sucrose, 1.5 mM Tris HCl at starting pH 7.4 at 22 °C, 0.1 mM ferricyanide or 17 µM di-ferric transferrin added to start the reactions.

Table 2. Inhibition of Oxidant-Induced Proton Release by Amiloride

Cell	Activation	Amiloride Concentration	Inhibition (%)	Reference
HeLa	FeTf	0.2 nM	55	Sun et al., 1987
HeLa	Fe(CN)$_6$	0.2 mM	47	
Transfer Pineal SV40	FeTf	0.2	75	Sun et al., 1988
	Fe(CN)$_6$	0.2	88	
C3H10T1/2	Fe(CN)$_6$	0.3	0	
Ha ras	Fe(CN)$_6$	0.3	49	
HeLa	Fe(CN)$_6$	0.3	25	Toole-Simms, 1988
HeLa	FeTf	0.3	32	Toole-Simms, 1988
Xenopus Oocyte	Na$^+$	0.1	100	Sasaki et al., 1992
Guinea Pig Erythrocyte	PMA	0.3	13	Zhao & Willis, 1993

Note: A problem in the interpretation of the amiloride inhibition of proton release is the amiloride inhibition of di-ferric transferrin and ferricyanide reduction (Sun et al., 1987). The question arises: does inhibition of the exchanger inhibit electron transport or does inhibition of electron transport inhibit a H$^+$ release not dependent on the exchanger? Exchanger inhibition is most likely because of the Na$^+$ dependence for part of the H$^+$ release (Sun et al., 1988). As an alternative, proton transport may be necessary for electron transport (Stahl and Anst, 1993).

al., 1988; Table 1). Hormones such as insulin and EGF can also control the electron transport (Crane et al., 1991).

Activation of the Na$^+$/H$^+$ exchanger is revealed by the characteristic inhibition by amiloride and N-substituted amilorides, lack of effect by benzamil, and specific requirement for external Na$^+$ or Li$^+$ with low activity with Cs$^+$. With pineal and HeLa cells, the inhibitions indicate that a major part of the proton transport is based on activation of the exchanger (Table 2).

INTERNAL ALKALINIZATION

In HeLa cells, the di-ferric transferrin-stimulated proton release is accompanied by an increase in the pH inside the cell as measured by change in fluorescence of the internal indicator BCECF incorporated into the cells (Figure 7). This alkalinization is consistent with exchanger activation to export protons. Most of the internal pH increase is inhibited by amiloride or dimethylamiloride (Mrkic et al., 1992; Yun et al., 1993). The internal change in pH is prevented if Cs$^+$ is substituted for Na$^+$ or Li$^+$ in the external media (Toole-Simms, 1988). Ferricyanide-stimulated proton release by HeLa cells is not accompanied by an internal pH increase. The basis for this difference from di-ferric transferrin has not been investigated. It is possible that the rapid electron transport in response to ferricyanide leads to oxidation of sufficient NADH, which produces protons inside the cell, to counter

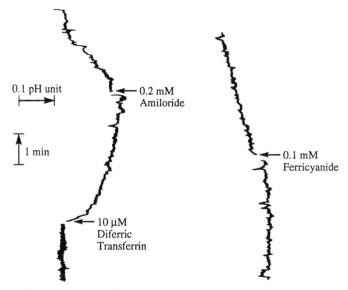

Figure 7. Change in intracellular pH induced by oxidant addition to HeLa cells. Internal pH measured by change in fluorescence of BCECF incorporated into cells. Modified from Toole-Simms (1988).

any activation of the Na^+/H^+ exchanger. Since the proton release in the presence of ferricyanide is partly amiloride-sensitive, there is evidence that the exchanger is activated. The internal acidification must derive from an internal source such as NADH (Figure 7).

MECHANISM OF EXCHANGER ACTIVATION

The mechanism by which the plasma membrane electron transport activates the exchanger is not established. Since the electron transport system in erythrocytes activates tyrosine phosphorylation, specifically on band three protein (Harrison et al., 1991), and the exchanger can be activated by phosphorylation through tyrosine kinase coupled to MAP kinases, which ultimately phosphorylate serine on the exchanger, it is likely that the activation of the exchanger by the electron transport system is based on activation of protein tyrosine kinase (Sardet et al., 1991).

A simple approach to oxidant control of the Na^+/H^+ exchanger would be monomer-dimer transition based on SH to S-S transition. This linkage has been observed in isolated membranes without change in activity. On the other hand, the exchanger activity in membrane vesicles from eye lens is activated by H_2O_2 which could be based on oxidation of SH groups (Ye and Zadunaisky, 1992).

An alternative control would be direct interaction between the electron carrier protein and the protein exchanger, but it is not clear that there is sufficient oxidase available to interact by direct conformational change with the exchanger molecules.

A few oxidase molecules could activate many exchanger molecules by kinase activation. Since most evidence at this time favors oxidant activation of a protein tyrosine kinase, the activation of the exchanger would be through a series of MAP type kinases which are activated by tyrosine phosphorylation to activate serine/threonine kinase. This, in turn, would phosphorylate a serine on the exchanger to change its pH set point for H^+ exchange (Sardet et al., 1991). Unfortunately, the effect of genistein, which inhibits tyrosine kinase, on di-ferric transferrin activation of the exchanger has not been tested. Activation by EGF which activates the exchanger through the tyrosine kinase pathway is inhibited by genistein.

AMILORIDE-INSENSITIVE PROTON RELEASE

In pineal and HeLa cells there is a slow H^+ release which is not inhibited by amiloride or amiloride analogs (Table 2). Part of the differential response to amiloride may express the presence of different types of Na^+/H^+ exchanger molecules since three different isoforms have been identified (Gris-Yun et al., 1993). Differential response to amiloride and amiloride analogs has been demonstrated in other cell lines (Viniegra et al., 1992). The basis for the amiloride-insensitive activity is apparently not HCO_3^-/Cl^- exchange since the stilbene derivatives SITS and DIDS, which inhibit the bicarbonate-chloride exchanger have no effect on the oxidant-activated proton movement. The H^+-ATPase or stoichiometric H^+ release based on anisotropic oxidoreduction remain as alternative mechanisms for the residual activity in HeLa or pineal cells. If the H^+-ATPase is responsible for some of the oxidant-induced proton release from these cells, the activity should be inhibited by thiol reagents such as N-ethylmaleimide or by balfilomycin which is a specific inhibitor of the ATPase. Unfortunately, these agents have not been tested (Swallow et al., 1990).

In C3H10T1/2 cells, the proton release induced by ferricyanide is insensitive to amiloride analogs. The rate of H^+ release is also very low to the extent that it is equal to or less than the electron transport (Table 3). Therefore, the proton movement from these cells may be based on anisotropic oxidation-reduction of

Table 3. Comparison of H^+ Release Rates and Ferricyanide Reduction Rates of C3H10T1/2 and C3H10T1/2 Ha-ras Transfected Cells

	Ferricyanide Reduction Rate (nmole/min/gww)	Proton Release Rate with $Fe(CN)_6$ (nmoles min^{-1} gww^{-1})	Ratio (H^+/e^-)
C3H10T1/2	72	100	1.4
C3H10T1/2 Ha-ras	208	170	0.8

Note: Ferricyanide reduction rate calculation based on 400×10^6 cells per gram.

protonated electron transport carriers. As an alternative, the activity may represent activation of a H^+-ATPase. The stoichiometric process would be consistent with lack of pH change observed by fluorescent dye inside the cells.

ONCOGENE EFFECTS

If C3H10T1/2 cells are transfected with the Ha-ras oncogene, proton release in response to ferricyanide is greatly increased and the proton release becomes quite sensitive to amiloride (Table 2). N-ras and other control transfectants do not show this increased amiloride-sensitive proton release, so the presence of the mutant p21 ras protein appears to induce a coupling between the electron transport and the exchanger which was not available in the control or N-ras cells. The coupling can depend on binding of GTP to the mutant p21 protein since the mutant cannot hydrolyze the GTP in contrast to N-ras which can be deactivated by conversion of GTP to GDP. It is not clear at this time how the activated GTP-ras can activate the Na^+/H^+ exchanger in response to electron transport. As one possibility, the conformational change produced by GTP binding to the p21 ras protein may be transmitted to the oxidase (Wittinghofer and Pai, 1991).

In rat liver or pineal cells transfected with the SV40 oncogene, the stoichiometry of proton release induced by oxidants is increased. This increased proton release is also sensitive to amiloride. The presence of the SV40 large T gene product in the membrane or SV40-induced change in metabolism appears to increase the efficiency of interaction between the redox system and the exchanger (Table 4).

GLYCOLYTIC ACIDS

In erythrocytes, external oxidants such as ferricyanide increase glycolysis (Harrison et al., 1991). Since these cells have no mitochondria, the increase in glycolysis could increase lactate or pyruvate formation, and excretion of these acids could be a basis for proton movement across the plasma membrane. In cells with mitochondria, the transmembrane electron transport decreases available NADH, so lactate formation would be decreased with consequent accumulation of pyruvate. The pyruvate

Table 4. Effect of SV40 Transformation on Oxidant-Induced H^+ Release

	H^+ Release Rate	$Fe(CN)_6$ Reduction	Ratio (H^+/e^-)
Pineal cells			
SV40 transformed	208 ± 21	65 ± 15	3.2
nontransformed	308 ± 22	192 ± 13	1.6
Rat liver cells			
SV40 transformed	552	36	15.3
nontransformed	588	88	6.7

Note: Sun et al., 1986a,b.

would be preferentially oxidized by the mitochondria, so it would not increase proton export. At this time there is no evidence that external oxidants increase export of organic acids from cells.

INHIBITORS OF PLASMA MEMBRANE ELECTRON TRANSPORT

Agents which inhibit plasma membrane electron transport also inhibit the associated proton release. The most notable of these are certain well-known antitumor drugs which are effective inhibitors of transplasma membrane electron transport, especially in transformed cell lines. The most effective drugs are adriamycin and related anthracyclines, bleomycin, and cis platin (Sun et al., 1992b). A second series of inhibitors of proton release activated by electron transport are agents which inhibit functional components of the electron transport. The best effects are seen with coenzyme Q analogs which can compete with coenzyme Q. The competitive aspect of the inhibition is revealed by reversal of the analog inhibition by increased coenzyme Q (Sun et al., 1992a). A remarkable effect of retinoic acid is stimulation of the Na^+/H^+ exchanger, but inhibition of exchanger activation by di-ferric transferrin (Sun and Crane, 1990) (Table 4).

HORMONE AND GROWTH FACTOR CONTROL

Few studies have been done on the combined effects of hormones and redox activation agents on proton transport. Since hormones and factors such as EGF, insulin, and bombesin which activate proton release through the Na^+/H^+ exchanger also activate transplasma membrane electron transport, part of their effect may be through activation of the oxidase (Brightman et al., 1992). If so, agents which inhibit the oxidase should decrease the effect of the hormones on the exchanger. This has not been examined. The effects of insulin on ferricyanide reduction, induction of proton release and stimulation of cell proliferation with HeLa cells are listed in Table 5. The action of di-ferric transferrin as a ligand to activate the oxidase is also a growth factor effect since di-ferric transferrin is an almost universal requirement for cell growth even in cells which derive iron from other sources (Barnes and Sato, 1980). It is clear that ferric transferrin can elicit significant proton release from many cells to contribute to its proliferative action (Crane et al., 1990).

A possible dual action of growth factors on kinases and on the oxidase makes definition of separate effects difficult. Further complication arises with cytokines which induce H_2O_2 production by receptive cells, since H_2O_2 itself may activate proton movement and cell proliferation (Meier et al., 1990; Shibanuma et al., 1990; Kaufman et al., 1993; Yuo et al., 1993).

Since binding of transferrin to its receptor is not known to activate a kinase, its effect on exchanger activation through the oxidase may be the best example of oxidase action uncomplicated by alternative pathways. An even less complicated

Table 5. Effects of Insulin on Cell Response to Ferricyanide and Bombesin on Response to Di-ferric Transferrin

	No Addition	Fe(CN)$_6$ (0.1 M)	Insulin (30 µg/ml)	Fe(CN)$_6$ + Insulin
HeLa cells				
Cell growth with cells/flask	0.8	1.6	12.4	24
Fe(CN)$_6$ reduction	0	94	0	175
H$^+$ release (nequiv min gww^{-1})	—	145	119	199
	No Addition	Fe$_2$Tf 10 µM	Bombesin 25 mM	Fe$_2$Tf + bombesin
Swiss 3T3 cells				
Cell growth	6	7.4	9.6	24.6
Fe$_2$Tf reduction	—	5.5	—	14.8
H$^+$ release (with Fe$_2$Tf)	—	217	—	391

Note: HeLa cells grown in αMEM with 10% FCS. Assay in absence of serum. Swiss 3T3 cells also assayed in absence of serum. Bombesin effects are discussed by Rozengurt (1986). Fe(CN)$_6$ is potassium ferricyanide and Fe$_2$Tf is di-ferric transferrin. Fe$_2$Tf and Fe(CN)$_6$ reduction rate is in nmole min^{-1} gww of cells^{-1}. Cell growth is measured after 48 hr as cells × 10^5/25 cm^2 flask. For proton release, cells are equilibrated first with insulin or bombesin to set a baseline, then oxidant is added. Insulin alone can activate the exchanger (Ives and Rector, 1984).

response can be ferric lactoferrin stimulation of K562 cell growth and proton release without contribution to iron uptake (Sun et al., 1991).

The additive effect of kinase-activating growth factors such as insulin and the oxidase further suggests that the oxidase may act through alternatives to kinase action. For example, the activation of electron transport with ferricyanide causes a rounding up of fibroblasts indicating cytoskeletal changes. It is known that cytoskeletal change through fibronectin also induces Na$^+$/H$^+$ exchanger activation (Schwartz et al., 1991), so the redox message may be transmitted through cytoskeletal change.

Table 6. Effect of Retinoic Acid on Proton Release from Pineal Cells in Absence and Presence of Di-ferric Transferrin

Retinoic Acid Concentration	Retinoic Acid-Induced H$^+$ Release	Ferric Transferrin-Induced H$^+$ Release: Increase Over Retinoic Acid (nequiv min^{-1} gww^{-1})	Total H$^+$ Release With Retinoic Acid and Transferrin
0	—	900	900
10^{-6}M	764	471	1235
10^{-5}M	3300	154	3454

Note: Proton release by untransformed pineal cells with temperature-sensitive SV40 grown at 40 °C nonpermissive temperature. The exchanger is not stimulated by retinoic acid in the SV40 transformed cells grown at the permissive 33 °C, so that SV40 is expressed but transferrin also activates the exchanger in transformed cells (Sun et al., 1987a; Sun et al., 1988).

CONCLUSION

Electron transport through oxidases in the plasma membrane contributes to, or controls, part of the proton release from the cell. The details of oxidase function and the mechanism of control remain to be elucidated. The NADPH oxidase of neutrophils is a special case in which proton transport is coupled to the cytochrome b_{557} electron carrier. This type of proton transport has its precedents in the well-characterized proton pumping through electron carriers in mitochondrial and chloroplast membranes and prokaryotic plasma membranes.

In most cases the NADH oxidase activates the Na^+/H^+ exchanger in animal cells. In plant cells, it appears to activate the H^+-ATPase. The mechanism for channel activation is analogous to channel activation by ligand receptors. Since the oxidase can control tyrosine kinase, activation of the exchanger is most likely through the MAP tyrosine kinase-activated serine kinase which phosphorylates the antiport.

Anomalous response to the NADH oxidase in animal cells, such as amiloride-insensitive proton transport, may be based on activation of the H^+-ATPase or direct electron transport-linked proton transfer. Further definition of the components of the NADH oxidase and the characteristics of electron transport are needed. In addition, the presence of a poorly characterized glutathione oxidase in the plasma membrane opens an alternative for oxidation-reduction control of proton transport. At this stage no evidence has been found for control of HCO_3^-/Cl^- exchange or organic acid transport by the plasma membrane oxidase.

REFERENCES

Barnes, H. & Sato, G. (1980). Serum-free cell culture: a unifying approach. Cell 22, 649–654.

Brightman, A.O., Wang, J., Min, R.K.-M., Sun, I.L., Barr, R., Crane, F.L., & Morré, D.J. (1992). A growth factor and hormone-stimulated NADH oxidase from rat liver plasma membrane. Biochim. Biophys. Acta 1105, 109–117.

Brzezinski, P. & Malmström, B.G. (1987). The mechanism of electron gating in proton pumping cytochrome c oxidase. Biochim. Biophys. Acta 894, 29–38.

Crane, F.L. (1990). In: Highlights in Ubiquinone Research (Lenaz, G., Barnabei, O., Rabbi, A., & Battino, M., eds.), pp. 3–20, Taylor and Francis, London.

Crane, F.L., Löw, H., Sun, I.L., Morré, D.J., & Faulk, W.P. (1990). Interaction between oxidoreductase, transferrin receptor and channels in the plasma membrane. In: Growth Factors: From Genes to Clinical Applications (Sara, V.R., Hall, K., Löw, H., eds.), pp. 129–139, Raven, New York.

Crane, F.L., Sun, I.L., Barr, R., & Löw, H. (1991). Electron and proton transport across the plasma membrane. J. Bioenerg. Biomemb. 23, 773–803.

Dancis, A., Roman, D.G., Anderson, G.J., Hinnebusch, A.G., & Klausner, R.D. (1992). Ferric reductase of Saccharomyces cerevisiae. Proc. Natl. Acad. Sci. USA 89, 3869–3873.

Harrison, M.L., Rathinavelu, P., Arese, P., Geahlen, R.L., & Low, P.S. (1991). Role of band three tyrosine phosphorylation in regulation of erythrocyte glycolysis. J. Biol. Chem. 266, 4106–4111.

Henderson, L.M. & Chappell, J.B. (1992). The NADPH oxidase associated H^+ channel is opened by arachidonate. Biochem. J. 283, 171–175.

Ibes, H.E. & Rector, F.C. (1984). Proton transport and cell function. J. Clin. Invest. 73, 285–290.

Kaufman, D.S., Goligorsky, M.S., Nord, E., & Graber, M.L. (1993). Perturbation of cell pH regulation by H_2O_2 in renal epithelial cells. Arch. Biochem. Biophys. 302, 245–254.

Krab, K. & Wikström, M. (1987). Principles of coupling between electron transfer and proton translocation with special reference to proton-translocation mechanisms in cytochrome oxidase. Biochim. Biophys. Acta 895, 25–39.

Lange-Carter, C.A., Pleiman, C.M., Gardner, A.M., Blumer, K.J., & Johnson, A.L. (1993). A divergence in the MAP kinase regulatory network defined by MEK kinase and Raf. Science 260, 315–319.

McCormick, F. (1993). How receptors turn ras on. Nature 363, 15–16.

Meier, B., Radeke, H.H., Selle, S., Habermehl, G.G., Resch, K., & Sies, H. (1990). Human fibroblasts release low amounts of reactive oxygen species in response to potent phagocyte stimulants, N-formylmethionylleucylphenylalenine, leukotriene B_4 or 12-O-tetradecanoyl phorbol-13-acetate. Biol. Chem. Hoppe Seyler 371, 1021–1025.

Mrkic, B., Forgo, J., Murer, H., & Helmle-Kolb, C. (1992). J. Mem. Biol. 130, 205–218.

Nanda, A., Grinstein J., & Curnette, J.T. (1993). Abnormal activation of H^+ conductance in NADPH oxidase-defective neutrophils. Proc. Natl. Acad. Sci. USA 90, 760–764.

Rozengurt, E. (1986). Early signals in the mitogenic response. Science 234, 161–166.

Sardet, C., Fafournoux, P., & Pouyssegur, J. (1991). αThrombin EGF and okadaic acid activate the Na^+/H^+ exchange by phosphorylating common sites. J. Biol. Chem. 266, 19166–19170.

Sasaki, S., Ishibashi, K., Nagai, T., & Marumo, F. (1992). Regulation mechanisms of intracellular pH in *Xenopus* oocytes. Biochim. Biophys. Acta 1137, 45–51.

Schwartz, M.A., Lechene, C., & Ingher, D.E. (1991). Insoluble: fibronectin activates the Na/H antiporter by clustering and immobilizing integrin $\alpha_5\beta_1$ independent of cell shape. Proc. Natl. Acad. Sci. USA 88, 7849–7853.

Shibanuma, M., Koroki, T., & Nose, K. (1990). Stimulation by hydrogen peroxide of DNA synthesis, competence family gene expression and phosphorylation of a specific protein in quiescent Balb/3T3 cells. Oncogene 5, 1025–1032.

Stahl, J.D. & Aust, S.D. (1993). Plasma membrane-dependent reduction of 2,4,6-trinitrotoluene by phanerochaete chrysosporium. Biochem. Biophys. Res. Commun. 192, 471–476.

Sun, E.E., Sun, I.L., & Crane, F.L. (1987a). The mechanism of retinoic acid-mediated effects on cell growth and differentiation. Proc. Indiana Acad. Sci. 97, 421–429.

Sun, I.L. & Crane, F.L. (1990). Interactions of antitumor drugs with plasma membranes. In: Oxidoreduction at the Plasma Membrane (Crane, F.L., Morré, D.J., & Löw, H., eds.), pp. 257–280, CRC Press, Boca Raton.

Sun, I.L., Crane, F.L., & Chou, J.Y. (1986a). Modification of transmembrane electron transport in plasma membranes of simian virus 40 transformed pineal cells. Biochim. Biophys. Acta 886, 327–336.

Sun, I.L., Navas, P., Crane, F.L., Chou, J.Y., & Löw, H. (1986b). Transplasmalemma electron transport is changed in simian virus 40 transformed liver cells. J. Bioenerg. Biomemb. 18, 471–486.

Sun, I.L., Garcia-Cañero, R., Liu, W., Toole-Simms, W., Crane, F.L., Morré, D.J., & Löw, H. (1987b). Diferric transferrin stimulates the Na^+/H^+ antiport of HeLa cells. Biochem. Biophys. Res. Commun. 145, 467–473.

Sun, I.L., Toole-Simms, W., Crane, F.L., Morré, D.J., Löw, H., & Chou, J.Y. (1988). Reduction of diferric transferrin by SV40 transformed pineal cells stimulates the Na^+/H^+ antiport activity. Biochim. Biophys. Acta 938, 17–23.

Sun, I.L., Crane, F.L., Morré, D.J., Löw, H., & Faulk, W.P. (1991). Lactoferrin activates plasma membrane oxidase and Na^+/H^+ antiport activity. Biochem. Biophys. Res. Commun. 176, 498–504.

Sun, I.L., Sun, E.E., Crane, F.L., Morré, D.J., Lindgren, A., & Löw, H. (1992a). Evidence for coenzyme Q function in plasma membrane electron transport. Proc. Natl Acad. Sci. USA 89, 11126–11130.

Sun, I.L., Sun, E.E., Crane, F.L., Morré, D.J., & Faulk, W.P. (1992b). Inhibition of transplasma membrane electron transport by transferrin-adriamycin conjugates. Biochim. Biophys. Acta 1105, 84–88.

Swallow, C.J., Grinstein, S., & Rothstein, O.D. (1990). A vacuolar type H^+-ATPase regulates cytoplasmic pH in murine macrophages. J. Biol. Chem. 265, 7645–7654.

Toole-Simms, W. (1988). Regulation of proton release from HeLa cells by ferric reductase. Ph.D. Thesis, 160 pp., Purdue University, West Lafayette, IN.

Viniegra, S., Cragoe, E.J. Jr., & Rabito, C.A. (1992). Heterogeneity of the Na^+/H^+ antiport systems in renal cells. Biochim. Biophys. Acta 1106, 99–109.

Wittinghofer, A. & Pai, E.F. (1991). Signal transduction through the ras p21 protein. Trends Biochem. Sci. 16, 382–387.

Ye, J. & Zadunaisky, J.A. (1992). H_2O_2 activates Na^+/H^+ exchange in plasma membrane vesicles from eye lens. Exp. Eye Res. 55, 251–260.

Yun, C.H., Gurubhagavatula, S., Levine, S.A., Montgomery, J.L.M., Brant, S.R., Cohn, M.E., Cragoe, E.J. Jr., Pouyssegur, J., Tse, C.M. & Donowitz, M. (1993). Glucocortical stimulated epithelial isoform of the Na^+/H^+ antiport. J. Biol. Chem. 268, 206–211.

Yuo, A., Kitagawa, S., Azuma, E., Natori, Y., Togawa, A., Saito, M., & Takaka, F. (1993). Tyrosine phosphorylation and intracellular alkalinization in human neutrophils stimulated by TNF, GMCSF, G-CSF. Biochim. Biophys. Acta 1156, 197–203.

Zhao, Z. & Willis, J.S. (1993). Cold activation of Na influx through the Na^+/H^+ antiport of guinea pig red cells. J. Memb. Biol. 131, 43–54.

FURTHER READINGS

Boron, W.F., Hogan, E., & Russell, J.M. (1988). pH sensitive activation of intracellular pH regulation system in squid axons by ATPγS. Nature 332, 262–265.

Cramer, W.A. & Knaff, D.B. (1990). Energy Transduction in Biological Membranes. 545 pp., Springer Verlag, Berlin.

Crane, F.L., Morré, D.J., & Löw, H. (eds.) (1988). Plasma Membrane Oxidoreductases in Control of Animal and Plant Growth. 443 pp., Plenum, New York.

Crane, F.L., Morré, D.J., & Löw, H. (eds.) (1990). Oxidoreduction at the Plasma Membrane: Relation to Growth and Transport, Vol. I, Animals. 318 pp., CRC Press, Boca Raton.

Dutton, P.L. (1991). The role of quinones in generating proton gradients. Ann. Rev. Biochem. 60, 325–368.

Forgac, M. (1992). Vacuolar type ATPases. J. Bioenerg. Biomemb. 24, 337–424.

Maly, K., Oberhuber, H., Doppler, W., Hoflacher, J., Jaggi, R., Groner, B., & Grunicke, H. (1988). Effect of H-ras on phosphatidyl inositide metabolism, Na^+/H^+ antiporter and intracellular calcium. Adv. Enzyme Regul. 27, 121–143.

Mirossay, L., DiGloria, Y., Chastre, E., & Emami, S. (1992). Pharmacological control of gastric acid secretion: molecular and cellular aspects. Bioscience Rep. 12, 319–368.

Mitchell, P. & Moyle, J. (1985). The role of ubiquinone and plastoquinone in chemiosmotic coupling between electron transfer and proton translocation in coenzyme Q. In: (Lenaz, G., ed.) pp. 145–163, Wiley, Chichester.

Rosenberg, S.V., Fadil, T., & Schuster, V.L. (1993). A basolateral lactate/H^+ transporter in kidney cells. Biochem. J. 289, 263–268.

Scott, D.R., Helander, H.F., Hersey, S.J., & Sachs, G. (1993). The site of acid secretion in the mammalian parietal cell. Biochim. Biophys. Acta 1146, 73–80.

Wigglesworth, J.M. (1991). Mechanisms of redox-linked proton translocations: an introduction. J. Bioenerg. Biomemb. 23, 701–702.

Chapter 7

Cell Volume Regulation

DIETER HÄUSSINGER

INTRODUCTION

Cell volume homeostasis is of crucial importance in overall cellular function. It is therefore not surprising that volume regulatory mechanisms have been found in

Principles of Medical Biology, Volume 4
Cell Chemistry and Physiology: Part III, pages 187–209.
Copyright © 1996 by JAI Press Inc.
All rights of reproduction in any form reserved.
ISBN: 1-55938-807-2

almost every cell studied so far. These mechanisms, however, are not only activated in order to compensate for cell volume deviations induced by anisoosmotic exposure or during substrate accumulation in the cell, but can also be activated in the resting state by hormones so that a new steady-state of cell volume is achieved. In this view, cell volume homeostasis does not imply absolute constancy of cell volume, but rather integration of those events allowing cell volume changes to act as a regulator of cellular function. Indeed, small alterations in cellular hydration, i.e., in cell volume, serve as an important signal which triggers a variety of cellular functions within minutes of the change. The liver is a prime example: hepatocyte volume can change considerably within minutes of a physiological alteration in the hormonal or substrate milieu, and such alterations in cell volume per se markedly influence a variety of metabolic pathways. It appears that cell swelling and shrinkage lead to certain opposite patterns of cellular metabolic function, which are triggered by hormones and amino acids wholly by altering cell hydration. Thus, alterations in cell hydration in response to different stimuli must be viewed as another second messenger triggering cellular function. However, the nature of the interplay between signal "hydration" and other established mechanisms of signal transduction remains to be elucidated.

ANISOOSMOTIC CELL VOLUME REGULATION

Most cell membranes are highly permeable to water and the net transmembrane movement of water is driven primarily by the osmotic gradient across the plasma membrane. Thus, any change in osmotic gradient induced by alterations in the activity of plasma membrane transport systems or substrate accumulation inside the cell would be expected to affect cell volume. For experimental purposes, exposure of cells to anisoosmotic fluid has been widely used to study cell volume regulation. This particular approach cannot be seen as an unphysiological tool for cell volume modification. For example, a hyperosmotic environment is created in the renal medulla during antidiuresis, which makes cell volume regulatory mechanisms mandatory not only in medullary cells but also in circulating blood cells when they pass through the vasa recta of the renal medulla. Further, during intestinal absorption, portal venous blood may become slightly hypo- or hyperosmotic. Liver swelling is found during intestinal absorption of water. In general, however, physiologically more important cell volume changes are known to occur in response to cumulative substrate uptake into the cells and under the influence of hormones (see p. 193).

When cells are suddenly suspended in a hypoosmotic medium they initially swell as if they were perfect osmometers, but a few minutes later they down-regulate their volume, thereby approaching the volume they had at the start. This behavior has been termed regulatory volume decrease (RVD). RVD, however, does not completely return the cell volume to normal (Figure 1). The extent of this remaining volume deviation, which is equivalent to a steady-state change in cell

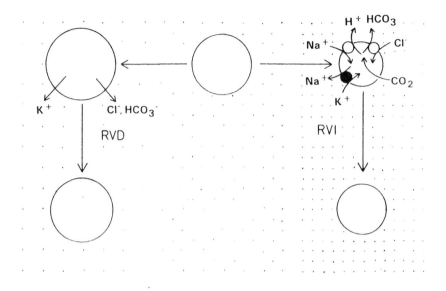

Figure 1. Volume regulatory mechanisms activated upon exposure of hepatocytes to hypoosmotic (left) or hyperosmotic (right) fluid. (From Häussinger and Lang, 1991.)

hydration, apparently acts as a potent signal which changes cellular function (see p. 196). RVD is achieved by a reduction in cytosolic osmolarity resulting from the exit of K^+ and Cl^- and nonelectrolytes (osmolytes), and by metabolic disposal of osmotically active substances.

When cells are suddenly suspended in a hyperosmotic medium, they initially shrink as if they were perfect osmometers but a few minutes later they up-regulate their volume, thereby returning the volume towards near-normal. This mechanism is termed regulatory volume increase (RVI). RVI, however, does not completely restore the cell volume to normal. RVI is achieved by an increase in cytosolic osmolarity due to net gain of K^+ and Cl^- and metabolic generation of osmotically active substances. Which mechanism is used for RVI and RVD, respectively, depends upon the cell type involved (for reviews see Chamberlin & Strange, 1989; Häussinger and Lang, 1991; Sarkadi and Parker, 1991; Lang and Häussinger, 1993).

Ionic Mechanisms

The ionic mechanisms involved in RVD and RVI reflect the most immediate regulatory response to challenges in cell volume. In many mammalian cells, RVD

Table 1. Ionic Mechanisms Involved in Regulatory Volume
Decrease (RVD) and Increase (RVI)

Regulatory Cell Volume Decrease (RVD)

Parallel activation of K^+ and Cl^- channels

  Lymphocytes, platelets, endothelial cells, fibroblast, hepatocytes, astrocytes, intestine, Madin-Darby canine kidney (MDCK) cells, tracheal ciliary epithelia, choroid plexus, eye lens, frog skin, frog urinary bladder, Ehrlich ascites tumor cells

Activation of KCl cotransport

  Erythrocytes from man, toadfish, birds, and sheep; Ca^{++}-depleted Ehrlich ascites tumor cells; Necturus gallbladder

Parallel Operation of K^+/H^+ exchange and Cl^-/HCO_3^- cotransport

  Amphiuma red cells, frog skin, cornea K^+/H^+ ATPase

  gastric epithelium

Na^+/Ca^{++} exchange and Ca^{++}-ATPase

  Dog erythrocytes

Regulatory Cell Volume Increase (RVI)

Parallel activation of Na^+/H^+ antiport and Cl^-/HCO_3^- antiport

  Erythrocytes from dog, rabbit, and Amphiuma; lymphocytes; osteoclasts; endothelial cells; parotis; pancreas, liver, gallbladder, proximal tubule, medullary thick ascending limb, and collecting duct; MDCK cells

Activation of Na^+-K^+-$2Cl^-$ cotransport

  Erythrocytes from humans, birds, and rat; Ehrlich ascites tumor cells; astrocytes; fibroblasts; C67 glioma cells; intestine; trachea; parotid; MDCK cells; retinal pigment epithelium; frog skin; HeLa cells; rabbit medullary thick ascending limb

Inhibition of K^+ and Cl^- channels

Hepatocytes, intestine, proximal tubule, cardiomyocytes, gallbladder, urinary bladder, frog skin, MDCK cells

is the result of cellular release of K^+, Cl^- and HCO_3^-. The primary mechanism leading to the release of these ions is dictated by the cell type (see Table 1). Depending on whether the resting membrane potential is closer to the chloride (e.g, hepatocytes) or to the potassium equilibrium potential (e.g., astrocytes and lymphocytes), RVD caused by activation of K^+ and Cl^- channels is associated with either hyperpolarization or depolarization of the plasma membrane, respectively. RVD involves different ionic mechanisms such as: (a) parallel activation of K^+ and Cl^- channels in the plasma membrane; (b) activation of KCl cotransport; (c) parallel activation of K^+/H^+ and HCO_3^-/Cl^- exchange; (d) parallel activation of Na^+/Ca^{2+} exchange and Ca^{2+}-ATPase; and (e) stimulation of electrogenic bicarbonate exit or activation of K^+/H^+-ATPase.

Different mechanisms are involved in the opening of volume-regulatory ion channels; they are cell type-dependent and involve direct channel activation by membrane stretch, alterations in intracellular free $[Ca^{2+}]$ or activation of membrane-bound signaling systems. For example, swelling of hepatocytes apparently opens stretch-activated nonselective cation channels, which allow passage of Ca^{2+} into the cell (Bear, 1990). Swelling in turn stimulates phospholipase C to produce inositol-1,4,5-trisphosphate, which in turn mobilizes Ca^{2+} from intracellular stores. The resulting increase in $[Ca^{2+}]_i$ may then activate Ca^{2+}-sensitive K^+ channels, thus triggering the volume-regulatory K^+ efflux (Bear and Petersen, 1987). However, Ca^{2+} may activate K^+ channels not only directly, but also indirectly by affecting other channel regulating processes, such as phosphorylation (McCarty and O'Neil, 1992). In the liver, Ca^{2+} activation of K^+ channels may not be the only mechanism allowing RVD; indeed, K^+ channels have also been described in isolated hepatocytes, which do not require an increase in intracellular calcium activity (Henderson et al., 1989) and are probably activated by cell membrane stretch, as also observed in rat kidney proximal tubules (Sackin, 1987). Studies of volume regulation in Ehrlich ascites tumor cells indicate that leukotrienes and prostaglandins play a role in triggering volume-regulatory ion fluxes (for review see Hoffmann et al., 1993). A completely different mode of RVD occurs in dog erythrocytes, in which intracellular Na^+ is high and K^+ is relatively low. These cells extrude Na^+ during RVD by activation of Na^+/Ca^{2+} exchange and the Ca^{2+} taken up is subsequently extruded by a Ca^{2+}-ATPase. In human erythrocytes, however, KCl release during RVD is independent of the membrane potential, suggesting activation of an electroneutral KCl symport system.

RVI following hyperosmotic exposure of cells is accomplished by an uptake of ions by either parallel activation of amiloride-sensitive Na^+/H^+ exchange and Cl^-/HCO_3^- exchange, or loop diuretic-sensitive Na^+-K^+-$2Cl^-$ cotransport (Table 1). In addition, loss of cellular ions is simultaneously minimized in most cell types by a reduction in membrane conductances. In liver, a sudden increase in extracellular osmolarity stimulates amiloride- and ouabain-sensitive K^+ uptake (Graf et al., 1988, 1993; Häussinger and Lang, 1991), whereas loop diuretics do not appreciably modify cell volume regulatory K^+ uptake. Thus, RVI in liver cells is apparently the result of activation of Na^+/H^+ exchange with subsequent extrusion of sodium in exchange for potassium by the Na^+-K^+-ATPase. Na^+-K^+-$2Cl^-$-cotransport does not seem to appreciably participate in hepatic RVI, even though the transporter exists in the cell membrane, and its activation, for example, by insulin is followed by an increase in cell volume.

The activation of Na^+/H^+ exchange or Na^+-K^+-$2Cl^-$ cotransport during RVI involves phosphorylation of these transporters, suggesting a cell volume-dependent shift in the activities of protein kinases and/or phosphatases (Grinstein et al., 1992; Minton et al., 1992; Bianchini and Grinstein, 1993). In line with this, okadaic acid, an inhibitor of protein phosphatases, increases cell volume in lymphocytes and hepatocytes, probably via phosphorylation-mediated activation of these transporters. Thus, alterations in cell hydration are associated with activity changes that have yet to be identified as involving protein kinases/phosphatases systems. These mechanisms may not only be involved in mediating volume-regulatory responses but may also trigger some of the metabolic changes occurring in response to cell volume changes (see p. 204).

Osmolyte Strategy

Organic compounds which accumulate in cells following cell shrinkage or which are released from the cells in response to a decrease in the ambient osmolarity (cell swelling) are called organic osmolytes. Osmolytes need to be nonperturbing solutes that do not interfere with protein function even when occurring in high intracellular concentrations (Yancey et al., 1982; Chamberlin and Strange, 1989). Such a prerequisite may explain why only a few classes of organic compounds, viz. polyols, such as inositol and sorbitol (e.g., astrocytes and renal medulla), methylamines, such as betaine and α-glycerophosphorylcholine (renal medulla), and certain amino acids, such as taurine (e.g., Ehrlich ascites tumor cells) have evolved as osmolytes in living cells. Osmolytes in the renal medulla are important, since medullary fluid osmolarity can increase up to 3,800 mosmol/l during antidiuresis and decrease to 170 mosmol/l during diuresis (Graf et al., 1993). In the antidiuretic state (high extracellular osmolarity in renal medulla), intracellular osmolarity increases in renal medullary cells as the result of the accumulation of inositol and betaine, which are taken up via Na^+-dependent transporters, and as the result of increased synthesis of sorbitol and α-glycerophosphorylcholine. This process of intracellular osmolyte accumulation takes hours–days, but produces intracellular organic osmolyte concentrations of several hundred mmol/l. The enhanced synthesis of sorbitol from glucose by aldose reductase under these conditions involves an increased expression of the enzyme due to activation of the encoding gene (Uchida et al., 1989). Likewise, the Na^+-dependent transporters for inositol and betaine are induced upon hyperosmotic exposure (Kwon et al., 1992). Transition from the antidiuretic to the diuretic state and accordingly from a high to a low ambient osmolarity leads, within minutes, to a dramatic increase in the permeability of medullary cells to organic osmolytes. This facilitates the rapid efflux of osmolytes from the cells in response to the decline of the extracellular osmolarity. Here, osmolyte-specific transport systems (so-called permeases) and non-specific ion channels are thought to be involved in this response, but the mechanisms underlying their regulation are far from clear.

Exocytosis

There is some evidence that cell swelling stimulates exocytosis, which plays a role in RVD, e.g., in liver cells (van Rossum et al., 1987). Indeed, hypoosmotic cell swelling transiently increases the plasma membrane surface. This may well involve microtubule-dependent exocytosis at the canalicular and basolateral membrane of the hepatocyte (Pfaller et al., 1993; Häussinger et al., 1993a; Graf et al., 1993). It is tempting to speculate that these exocytotic mechanisms bring about the insertion into the plasma membrane of K^+ and Cl^- channels, as well as bile acid exporter molecules, thereby augmenting the release of osmotically active solutes during RVD.

PHYSIOLOGICAL MODULATORS OF CELL VOLUME

Cumulative Substrate Transport

One of the most important challenges for cell volume homeostasis is the cumulative uptake of osmotically active substances, such as amino acids, by specific Na^+-dependent transport systems (for review see Kilberg and Häussinger, 1992). These transport systems can create intra/extracellular amino acid concentration gradients across the plasma membrane of up to 20 by utilizing the energy of the transmembrane electrochemical Na^+ gradient. The accumulation of amino acids and K^+ in cells leads to cell swelling, which in turn triggers a volume regulatory K^+ efflux. This is similar to that activated during hypoosmotic exposure (Kristensen and Folke, 1984). As illustrated in Figure 2, infusion of glutamine into isolated perfused rat liver creates in a matter of about 12 minutes an intra/extracellular glutamine concentration gradient of about 12. As a result of this accumulation of glutamine via the Na^+-coupled transporter, together with Na^+ (the Na^+ in turn is exchanged for K^+ by the Na^+-K^+-ATPase system which accounts for the net K^+ uptake during the first 2 min of glutamine infusion shown in Figure 2), hepatocytes *in situ* are found to swell by about 10%. It is important to note that this increase in hepatocellular hydration is maintained as long as glutamine is present, despite subsequent activation of a volume-regulatory net K^+ efflux. The latter mechanism does not restore liver cell volume; it only prevents cell swelling from becoming excessive as would otherwise be predicted from the continuing accumulation of glutamine inside the cell. During the rapid washout of glutamine following its withdrawal from the perfusate, the swollen hepatocytes shrink rapidly and a secondary volume-regulatory net K^+ uptake, which lasts for more than 10 minutes, restores cell volume to its original value (Figure 2).

Liver cell swelling occurs under the influence of not only glutamine, but also alanine, proline, serine, glycine, aminoisobutyrate, phenylalanine, hydroxyproline, and bile acids. This is not the case with glucose or leucine, i.e., compounds which are not concentrated by liver cells. Two points here are noteworthy. First, amino acid-induced cell swelling and volume-regulatory responses occur upon exposure of hepatocytes to amino acids in physiological concentrations. And second, physiological fluctuations in portal vein amino acid concentration in response to the feeding/starvation cycle are accompanied by parallel alterations in liver cell volume. The degree of amino acid-induced cell swelling seems largely related to the existing steady state intra/extracellular amino acid concentration gradients. Further, the degree of amino acid-induced cell swelling is modified by hormones, as well as the nutritional state by means of three mechanisms: (a) regulation of expression of the concentrating transport systems in the plasma membrane; (b) modification of the electrochemical Na^+ gradient as a driving force for Na^+ coupled transport;

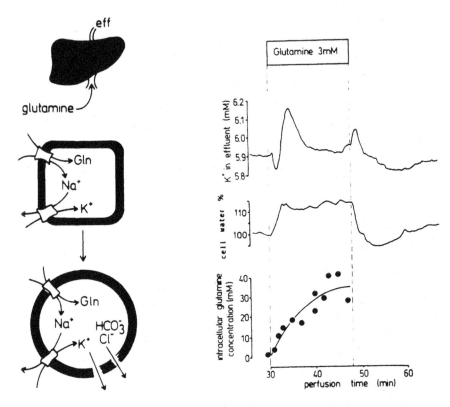

Figure 2. Effect of glutamine (3 mM) addition to perfusate of isolated, single-pass perfused rat liver on intracellular glutamine accumulation, cell volume, and volume-regulatory K+ fluxes. Addition of glutamine to portal perfusate leads to rapid cell swelling due to cumulative, Na+-dependent uptake of glutamine into liver cells. The initial net K+ uptake is explained by exchange of cotransported Na+ for K+ through the Na+-K+-ATPase system. Glutamine-induced cell swelling during the first two minutes of glutamine infusion activates volume regulatory K+ (plus Cl− and HCO_3^-) efflux. This volume-regulatory response prevents further cell swelling despite continuing glutamine accumulation inside the cell until a steady-state intracellular glutamine concentration of about 35 mM is reached. However, the liver cell remains in a swollen state as long as glutamine is infused. This degree of cell swelling modifies cellular function. (From Häussinger and Lang, 1991.)

and (c) alteration of intracellular amino acid metabolism. The third mechanism attenuates the transmembrane osmotic gradient created by the amino acid transporters when products are generated that rapidly equilibrate across the plasma membrane, but enhances cell swelling when poorly permeant products are formed and then accumulate in the cells.

Hormones

Among other actions, hormones are known to modify the activities of membrane ion transporters and ion channels, affect the membrane potential, and modulate Na^+-driven substrate transport (for review see Moule and McGivan, 1990). These actions in turn are expected to modify cell volume. Thus, for example, hormones are now recognized as potent modulators of liver cell volume (Häussinger and Lang, 1992). Whereas anisoosmotic exposure and amino acid accumulation in hepatocytes primarily lead to cell swelling with secondary activation of volume-regulatory ion transporters, hormones primarily affect the activity of these volume-

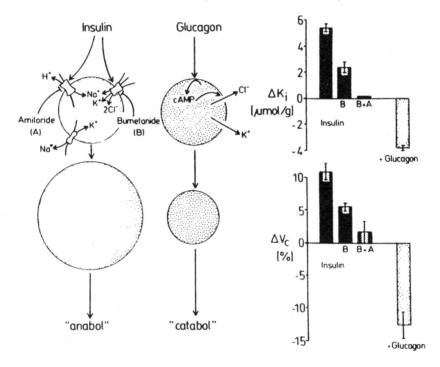

Figure 3. Modulation of liver cell volume and cellular K^+ balance by insulin and glucagon. Insulin activates amiloride-sensitive Na^+/H^+ exchange, bumetanide-sensitive Na^+-K^+-$2Cl^-$ cotransport as well as Na^+-K^+-ATPase, resulting in cellular accumulation of K^+ (ΔK_i), Na^+ and Cl^-, and cell swelling (ΔV_c). Both cellular K^+ accumulation and cell swelling by insulin are abolished in presence of amiloride (A) plus bumetanide (B). Conversely, glucagon and cAMP deplete cellular K^+ by activating Cl^- channels and quinidine/Ba^{2+}-sensitive K^+ channels, thereby inducing cell shrinkage. The hormone-induced cell volume alterations act like another signal which participates in mediating hormonal effects on hepatic metabolism: cell swelling triggers an anabolic pattern of cellular function, while cell shrinkage triggers a catabolic pattern. (From Häussinger and Lang, 1991, 1993.)

regulatory transport systems. In liver, for example, insulin stimulates Na^+/H^+ exchange, Na^+-K^+-$2Cl^-$-cotransport and the Na^+-K^+-ATPase, i.e., transport systems which when turned on bring about RVI in a variety of tissues (Table 1). The overall outcome of insulin action is accumulation of K^+, Na^+, and Cl^- in cells, and consequently, cell swelling. In sharp contrast, glucagon activates Na^+-K^+-ATPase, and simultaneously decreases internal K^+ in isolated perfused rat liver by opening Ba^{2+}- and quinidine-sensitive K^+ channels. Thus, glucagon action leads to depletion of cellular Na^+, K^+, and Cl^-. The result is cell shrinkage (Figure 3). Indeed, in both perfused liver and isolated hepatocytes, insulin increases while glucagon decreases cell volume within minutes of application of the hormone (Figure 3). Modulation of cell volume by insulin or glucagon is not a pharmacological phenomenon because half-maximal effects are observed with physiological concentrations. It is also noteworthy that other hormones can induce hepatocellular shrinkage (e.g., adenosine, vasopressin, and serotonin) or swelling (e.g., bradykinin and α-adrenergic compounds).

Other Cell Volume Effectors

Hypoxia reduces the activity of membrane Na^+-K^+-ATPase; this results in accumulation of Na^+ (and Cl^-) inside the cell, and consequently, cell swelling. However, the duration of hypoxia required for appreciable cell swelling can be quite variable while swelling itself depends on the rate of Na^+ entry and exit, and the extent of compensatory exocytosis. An increase in the extracellular K^+ concentration resulting from reduced Na^+ pump activity or depolarization favors the occurrence of cell swelling caused by an impairment of Cl^- extrusion. In regard to oxidative stress due to an increased load of hydrogen peroxide, hepatocyte shrinkage occurs as the result of the opening of K^+ channels (Hallbrucker et al., 1993). In such circumstances, it is the equilibrium between hydrogen peroxide-generating and disposing reactions inside the hepatocyte which determines cellular K^+ balance, and accordingly, cell volume (Saha et al., 1993).

REGULATION OF CELLULAR FUNCTIONS BY THE INTERNAL HYDRATION STATE

It is thus evident that cell volume is a dynamic parameter which seems to change within minutes in response to a physiological load of substrates or hormones and anisotonicity. Present evidence suggests that alterations in cellular hydration lead to profound effects on cellular function (Table 2). The significance of this stems from the conclusion that cell volume changes represent another principle of metabolic control (Häussinger and Lang, 1991, 1992, 1993), and that substrates and hormones, e.g., insulin and glucagon, exert their influence on metabolism in part by changing cellular hydration. It also stems from the conclusion that cell swelling stimulates the *de novo* synthesis of protein and glycogen, and simultane-

Table 2. Liver Cell Swelling as an Anabolic Signal

Cell Swelling	
Increases	*Decreases*
Protein synthesis	Proteolysis
Glycogen synthesis	Glycogenolysis
Lactate uptake	Glycolysis
Amino acid uptake	
Glutaminase	Glutamine synthesis
Glycine oxidation	
Ketoisocaproate oxidation	
Acetyl-CoA carboxylase	
Urea synthesis from amino acids	Urea synthesis from NH_4^+
Glutathione (GSH) efflux	GSSG release
Taurocholate excretion into bile	
Actin polymerization	
Microtubule stability	
Lysosomal pH	
mRNA levels for c-jun, β-actin, tubulin, ornithine decarboxylase	mRNA levels for phosphoenol-pyruvate carboxykinase
	Viral replication

ously inhibits proteolysis and glycogenolysis (Table 2), and that cell shrinkage leads to the opposite metabolic pattern.

This is apparent from the metabolic changes listed in Table 2 which are in accord with the notion that cell swelling serves as a proliferative anabolic signal, while cell shrinkage serves as a catabolic signal. Thus, alterations in cellular hydration in response to physiological stimuli are an important but only recently recognized signal which facilitates the adaptation of cell metabolism to alterations in the immediate environment, e.g., substrates, tonicity, and hormones. One far-reaching consequence of this is that the role of Na^+-dependent amino acid transport systems in the plasma membrane can no longer be merely identified with amino acid translocation; rather, these transporters act as if they are a transmembrane signaling system triggering cellular function by altering cell hydration in response to substrate delivery (Häussinger and Lang, 1991, 1993). Such a signaling role may shed new light on the long-known heterogeneity of transport systems among different cell types, and their different expression during development (for review see Kilberg and Häussinger, 1992). Likewise, transmembrane ion movements under the influence of hormones are an integral part of hormonal signal transduction mechanisms with alterations in cell hydration acting as another second messenger of hormone action (Häussinger and Lang, 1992). However, the exact step of hormone-induced cell volume changes in the hierarchy of post-receptor events in intracellular signaling, and its interplay with other known hormone-activated messenger systems, remains to be established.

The role of internal hydration as a signal for cellular function may be recognized by analogy to metabolic regulation by $[Ca^{2+}]_i$ or pH_i. Cells possess efficient mechanisms that keep the intracellular calcium and proton concentrations within narrow limits, otherwise cells would not survive. These homeostatic mechanisms can also be used to produce small physiological changes in cell hydration, which, in turn, regulate cell function.

Although regulation of cell function by cell hydration has been studied most extensively in liver and renal cells, there is growing evidence that cell volume also determines cell function in other cell types. Moreover, alterations in cell hydration affect the volume of intracellular organelles, thereby modifying their function. For example, there is a close relationship between mitochondrial volume changes and mitochondrial function (see Halestrap, 1989, 1993).

Cellular Hydration and Protein Turnover

Hepatic proteolysis is known to be under the control of amino acids and hormones, e.g., insulin and glucagon, but the nature of the underlying mechanism remains obscure (Seglen and Gordon, 1984; Mortimore and Pösö, 1987). More recently, evidence has come to light that hypoosmotic cell swelling inhibits proteolysis in liver, whereas hyperosmotic cell shrinkage stimulates protein break-down under conditions where the proteolytic pathway is not already fully activated (Häussinger et al., 1991; vom Dahl et al., 1991). This is in line with the view that internal hydration is a major determinant of proteolysis. The known antiproteolytic effect of insulin and that of several amino acids, such as glutamine and glycine, can be ascribed to the accompanying cell swelling, whereas stimulation of proteolysis by glucagon can be ascribed to cell shrinkage (Häussinger et al., 1991; vom Dahl et al., 1991). This is borne out by the fact that the effects of glutamine, glycine, insulin, and glucagon on proteolysis are quantitatively mimicked when the cell volume changes observed in response to these effectors are induced to the same degree by anisoosmotic exposure. Further, when insulin-induced cell swelling is stopped by inhibitors of the Na^+/H^+-exchanger and the Na^+-K^+-$2Cl^-$ cotransporter, the antiproteolytic activity of the hormone disappears. In liver, there is a close relationship between the proteolytic rate and cell hydration, regardless of whether cell hydration is modified by hormones, glutamine, glycine, bile acids, the K^+ channel blocker Ba^{2+}, or anisoosmotic exposure (Figure 4). In liver from fed rats, the antiproteolytic effect of glycine is only about one third that found in liver from starved rats. This is explained by an almost three-fold increase in the swelling potency of glycine resulting from greater activity of the glycine transport system during starvation. It is also worth noting that the antiproteolytic effect of amino acids other than glutamine and glycine can only be partly explained by cell volume changes. Apparently, other mechanisms involving proteolytic control must play a role here. The mechanism by which cell hydration regulates proteolysis is poorly

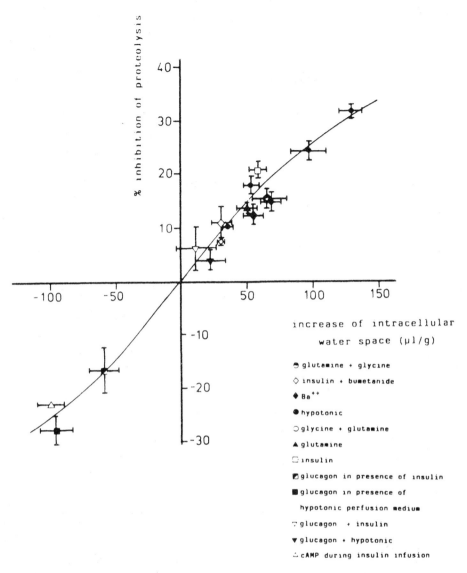

Figure 4. Relationship between cell volume and proteolysis in liver.
Cell volume in perfused liver was determined as intracellular water space and
proteolysis was assessed as [³H]leucine release in the effluent perfusate from perfused
rat livers, which were prelabelled *in vivo* by intraperitoneal injection of [³H]leucine
16 h prior to the perfusion experiment. Cell shrinkage stimulates proteolysis, whereas
cell swelling inhibits it. It should be noted that proteolysis is already maximally
activated in the absence of hormones and amino acids, and cannot be further
stimulated by hyperosmotic or glucagon-induced cell shrinkage. The proteolysis-
stimulating effect of these cell-shrinking maneuvers, however, becomes apparent
when proteolysis is preinhibited by either amino acids or insulin. Cell volume changes
were induced by insulin, cAMP, glucagon, amino acids, Ba^{2+}, or anisoosmotic
exposure. (From Häussinger et al., 1991.)

understood. Nonetheless, it appears that microtubules play a role. This is strongly suggested by evidence that the disruption of microtubules with colchicine eliminates the sensitivity of proteolysis to cell hydration changes (Häussinger et al., 1994). Furthermore, studies of hepatocytes using acridine orange fluorescence show that an increase in cell hydration is accompanied by rapid alkalinization of internal acidic compartments (Völkl et al., 1993). Since autophagic proteolysis, which constitutes about 70% of total hepatic proteolysis, is critically dependent on a low autophagosomal pH (e.g., ≈ 5), the possibility must be considered that cell swelling reduces the activity of vacuolar type H^+-ATPases, and that this, in turn, leads to impaired proteolysis. Although different steps in hepatic proteolysis are known to be sensitive to small alterations in ATP concentration (Plomp et al., 1987), it is not yet clear whether inhibition of proteolysis following an increase in cell hydration can be attributed to a decrease in cytosolic ATP concentration.

Liver cell swelling not only inhibits proteolysis, but also stimulates protein synthesis. In sharp contrast, cell shrinkage stimulates proteolysis but inhibits protein synthesis (Stoll et al., 1992). That cell hydration affects both protein degradation and synthesis in opposite directions is a finding that has a direct bearing on the problem of protein-catabolic states in disease (see p. 204).

Cellular Hydration and Carbohydrate Metabolism

In rat liver, cell swelling reduces whereas cell shrinkage increases lactate, pyruvate, and glucose release from the liver (Graf et al., 1988; Lang et al., 1989). Specifically, in hepatocytes isolated from fed rats glycogen phosphorylase *a* activity increases or decreases following hypoosmotic or hyperosmotic exposure, respectively. Although glycogenolysis is inhibited by cell swelling, glycogen synthesis is stimulated. Stimulation cannot be ascribed to metabolites formed from amino acids, e.g., glutamine (Katz et al., 1976) but seems due to amino acid-induced cell swelling (Baquet et al., 1990). Similarly, stimulation of glycogen synthesis by quinine, choline, or mercaptopicolinate can be ascribed to agonist-induced cell swelling (Baquet et al., 1990). How cell swelling, or an increase in cell hydration, elicits stimulation of glycogen synthesis is not yet clear but it could be due to a lowering in intracellular $[Cl^-]$, which leads to deinhibition of glycogen synthase phosphatase (Meijer et al., 1992). Increases in cell hydration also markedly increase glucose flux through the pentose phosphate shunt (Saha et al., 1992). The underlying mechanism is also unclear, but increased NADPH generation caused by swelling-induced activation of the pentose phosphate shunt may at least in part explain the diminished loss of oxidized glutathione from hepatocytes, as well as the diminished susceptibility of swollen cells to damage by oxidative stress (Saha et al., 1992).

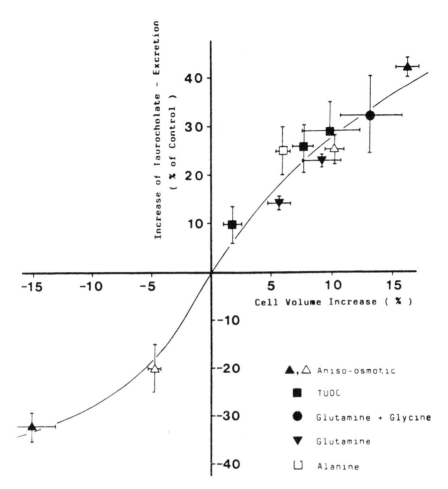

Figure 5. Relationship between taurocholate excretion (into bile) and cell volume. Livers were single-pass-perfused with taurocholate (100 μM). Taurocholate excretion into bile under normotonic perfusion conditions in the absence of hormones and amino acids was 185±9 nmol/min/g liver (n = 15); i.e., about 53% of the portal taurocholate load. Data on taurocholate excretion under the influence of various effectors on cell volume are given as percentage thereof. TUDC, tauroursodeoxycholate. (From Häussinger et. al., 1992.)

Hepatocellular Hydration and Bile Acid Excretion

Cellular hydration markedly affects epithelial transcellular transport, as exemplified by transcellular bile acid transport in the liver. In the hepatocyte, conjugated bile acids are taken up at the sinusoidal (basolateral) side by an Na^+-dependent carrier and are excreted at the canalicular (apical) membrane by means of a specific transport ATPase (Boyer et al., 1992). There is general agreement that this canalicular secretion step is rate-controlling for overall transcellular bile acid transport. In rat hepatocytes, transcellular taurocholate transport is strongly dependent upon cellular hydration: cell swelling stimulates, whereas cell shrinkage inhibits canalicular bile acid secretion (Figure 5), regardless of whether cell volume is modified by anisoosmoticity, amino acids, or insulin. Increments or decrements in cell hydration increase or decrease the V_{max} for canalicular bile acid excretion within minutes: a 10% increase in cell hydration is sufficient to achieve a doubling of V_{max} for transcellular taurocholate transport in perfused rat liver. This effect is sensitive to colchicine (an inhibitor of microtubules), and is unaccompanied by significant alterations in the driving forces involved in bile acid excretion (Häussinger et al., 1992, 1993). Furthermore, hepatocellular swelling leads to a rapid exocytotic process across the basolateral and canalicular membranes (Boyer et al., 1992; Pfaller et al., 1993; Graf et al., 1993). These observations suggest that an increase in hepatic hydration stimulates canalicular bile acid excretion within minutes by exocytotic insertion into the canalicular membrane of previously intracellularly stored bile acid transporter molecules. They also show that small alterations in cell hydration induce rapid changes in membrane transport, as in the case of basolateral amino acid transport (Bode and Kilberg, 1991).

Cellular Hydration and Gene Expression

Cellular hydration also modifies cell metabolism on a long-term basis by modifying gene expression. Consider the following few examples. In liver, hyperosmotic cell shrinkage increases the mRNA levels for the gluconeogenic enzyme phosphoenolpyruvate carboxykinase (PEPCK) by about five-fold due to an enhanced transcription of the gene. In sharp contrast, the levels decrease following hypoosmotic cell swelling (Newsome et al., 1994). Thus, hyperosmotic cell shrinkage mimicks the inductive effect of cAMP, whereas hypoosmotic cell swelling mimicks the known repressive effect of insulin on PEPCK mRNA levels. However, the mechanisms underlying these anisoosmotic effects on PEPCK mRNA levels have not yet been elucidated. It is noteworthy that hepatocyte levels for β-actin and tubulin mRNA increase in response to cell swelling, regardless of whether swelling is induced by insulin, hypoosmotic exposure, or glutamine. Ornithine decarboxylase (ODC) is induced in liver following hypoosmotic exposure or addition of amino acids. The effect turns out to be due to both an increased synthesis and decreased degradation of ODC (Tohyama et al., 1991), and to be sensitive to inhibition by okadaic acid. Further, hypoosmotic cell swelling increases the mRNA levels of c-jun, but not of c-fos (Finkenzeller et al., 1994), whose gene products are components of the AP-1 transcription factor. Moreover, there is evidence that hyperosmotic exposure of renal medullary cells increases the expression of aldose reductase. The signal by which hyperosmolarity triggers the increase in aldose reductase activity most likely involves an increase in intracellular ionic strength (Uchida et al., 1989). In MDCK cells (dog kidney cell line), hyperosmotic exposure increases mRNA coding for the protein of transport systems for betaine and myo-inositol (Kwon et al., 1992). In cultured human fibroblasts, amino acid deprivation leads to a cycloheximide-sensitive adaptive increase in the activity of the amino acid transport system. This adaptive increase is potentiated by hyperosmotic cell shrinkage and counteracted by hypoosmoticy-induced cell swelling (Gazzola et al., 1991). The mechanisms by which cell volume changes affect gene expression are still largely obscure, but changes in cytosolic ionic composition, the cytoskeleton, and levels of protein phosphorylation are likely candidates. Interestingly, viral replication depends upon host cell hydration. For example, hypoosmotic swelling of cultured duck hepatocytes inhibited replication of duck hepatitis B virus, whereas hyperosmotic shrinkage stimulated its replication (Offensperger et al., 1994).

Cell differentiation and growth under the influence of growth factors and insulin are coupled to activation of amiloride-sensitive Na^+/H^+ exchange and bumetanide-sensitive Na^+-K^+-$2Cl^-$ cotransport, which leads to cell swelling. An important role of cell swelling in cell proliferation is suggested by several findings including: (a) in mouse 3T3 cells, amiloride together with bumetanide completely block the growth factor-induced exit from the G_0/G_1-phase, and entry into the S-phase (Panet et al., 1986); and (b) hyperosmotic cell shrinkage inhibits cell proliferation in SV3T3 cells (Petronini et al., 1992).

Alterations in gene expression may also in turn affect cell volume. This was shown to be the case by an increase in the resting cell volume by about 30% following expression of ras-oncogene, e.g., in NIH fibroblasts (Lang et al., 1992). The growth factor-independent proliferation of the ras oncogene expressing cells is sensitive to amiloride and furosemide, that is, to blockers of Na^+/H^+ exchange and Na^+-K^+-$2Cl^-$ cotransport, suggesting that cell swelling induced by activation of these transporters plays a role in cell proliferation. In lymphocytes, further, mitogenic signals activate these transporters and may shift the set-point of cell volume regulation to higher resting values (Bianchini and Grinstein, 1993). This cell volume increase may be an important prerequisite for cellular proliferation.

Signals Linking Cell Hydration to Cellular Function

This issue is far from being settled. However, several potential mechanisms linking alterations in cell hydration to specific functional changes have already been described. Clearly, no single mechanism can be expected to account for all the diverse metabolic effects occurring in response to cellular hydration changes. Alterations in hydration are expected to influence membrane stretch, membrane-bound signaling systems, the cytoskeleton, and, in turn, to alter the ionic interior of the cell, as well as the extent of macromolecular crowding in the cytosol. The relative importance of these potential mechanisms may vary, depending on the specific metabolic pathway. For example, cell swelling leads to rapid polymerization of actin, an increase in the mRNA levels for β-actin and tubulin, and stabilization of microtubules, e.g., in liver (Theodoropoulos et al., 1992; Häussinger et al., 1994). These cytoskeletal alterations following an increase in cell hydration appear to play an important role in swelling-induced stimulation of exocytosis, transcellular bile acid transport, and inhibition of proteolysis, but not stimulation of pentose phosphate shunt activity or glycine oxidation following hepatocyte swelling. On the other hand, alterations in Na^+, K^+ and Cl^- concentrations inside the cell, accompanied by changes in cell hydration, may be a main signal for expression of osmoregulatory genes (e.g., aldose reductase in renal medulla) or regulation of glycogen synthesis in liver. Cell swelling is accompanied by membrane stretching; this may not only open stretch-activated ion channels, but also increase the formation of inositol-1,4,5-trisphosphate and produce a rise in $[Ca^{2+}]_i$ in many cell types. However, it seems doubtful that these transients triggering metabolic effects as indicated in Table 2 are as important as protein phosphorylation/dephosphorylation reactions in the mechanism of response to changes in cell volume. This seems clear in the case of activation of Na^+/H^+ exchange and Na^+-K^+-$2Cl^-$ cotransport during RVI since phosphorylation/dephosphorylation of these transporters involves protein kinases/phosphatases (Sarkadi and Parker, 1991; Grinstein et al., 1992). In line with this, a model has been presented postulating that the extent of macromolecular crowding, i.e., cytosolic protein concentration, determines the tendency of intracellular macromolecules to associate with the plasma membrane and determines enzymatic activity (Minton et al., 1992). Conceivably, alterations in cell hydration could interfere with the activity of protein kinases and phosphatases, which are not only involved in the regulation of volume-regulatory responses, but also in the regulation of cell metabolism in general. Indeed, a swelling-induced activation of mitogen-activated protein kinases (MAP-kinases) has already been identified (Schliess et al., 1995), which transduces some of the swelling induced metabolic effects, such as stimulation of taurocholate excretion into bile.

PATHOPHYSIOLOGICAL AND CLINICAL ASPECTS

Contemporary clinical medicine pays careful attention to the hydration state of the extracellular space, but not enough to cellular hydration probably because of the lack of routinely applicable techniques for the assessment of cell volume in patients. However, it should be kept in mind that cell hydration is determined primarily by the activity of ion and substrate transporting systems in the plasma membrane, and, to a minor extent, by the hydration state of the extracellular space. The role of cell hydration in regulating protein turnover is an important one, partly because it has a direct bearing on the problem of the pathogenesis of protein-catabolic states in the severely ill. As emphasized earlier, a decrease in cell hydration inhibits protein

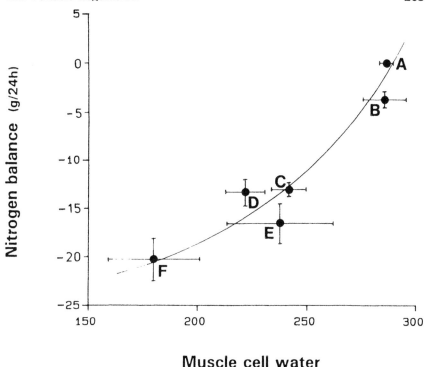

Muscle cell water
(ml/100 g fat-free dry weight)

Figure 6. Whole body nitrogen balance and cellular hydration of skeletal muscle. Data were obtained from humans. A: healthy subjects (n = 17); B: liver tumors (n = 5), C: polytrauma day 2 and D: day 9 after trauma (n = 11); E: acute necrotizing pancreatitis (n = 6); F: burn patients (n = 4). (From Häussinger et. al., 1993.)

synthesis and stimulates protein breakdown, e.g., in liver. This may also be true of skeletal muscle. Indeed, a close relationship between cell hydration in skeletal muscle and whole body nitrogen balance has been established in patients, irrespective of the underlying disease (Figure 6). The working hypothesis put forward is that cell shrinkage in liver and skeletal muscle triggers the protein-catabolic state found in various diseases (Häussinger et al., 1993). Although this implies that the degree of cell hydration determines the extent of nitrogen wasting, the pathogenetic mechanisms leading to cell shrinkage may well be multifactorial and could involve disease-specific components. Ironically, the physician interfering empirically with the cellular hydration state is also trying to overcome protein-catabolic states by the infusion of amino acids.

Another problem relates to the use of ursodesoxycholic acid to treat cholestatic diseases, such as primary biliary cirrhosis. The mechanism of its action remains poorly understood. However, it is significant that tauroursodesoxycholic acid, which is its taurine conjugate, is rapidly formed *in vivo* and responsible for liver cell swelling which apparently stimulates the excretion of endogenous bile acids (see Figure 5).

Last, there is some new evidence that hepatic encephalopathy is a primary disorder of astroglia (Norenberg et. al., 1992). Ammonia induces glial swelling as the result of glutamine accumulation in astrocytes. Both glial swelling and brain edema are recognized as major events leading to brain dysfunction following acute ammonia intoxication or acute fulminant liver failure. This raises the possibility that astrocyte swelling may also be of pathophysiogical significance in chronic hepatic encephalopathy. This seems likely in view of evidence that ammonia toxicity is reduced by inhibitors of astroglial glutamine synthetase (Hawkins et al., 1993), and that this in turn prevents cell swelling. Since myo-inositol serves as an astrocyte osmolyte, any ammonia-induced glial swelling would be expected to reduce its cellular content. Such a reduction can be demonstrated in man *in vivo* by proton-NMR spectroscopy. Chronic hepatic encephalopathy is found to be associated with a disappearance of the inositol signal (Kreis et. al., 1992).

SUMMARY

Cells possess potent mechanisms that keep their volume relatively constant within narrow limits. Major challenges for cellular volume are cumulative substrate uptake, alterations in the ambient osmolarity, and the effects of hormones. Thus, cell volume is not regarded as a constant but rather as a dynamic parameter; in fact, recent evidence suggests that alterations in cellular hydration represent a new, not yet fully recognized principle of metabolic control. That is to say, they act like second messengers, mediating metabolic hormone and amino acid effects. Accordingly, concentrative amino acid transport systems and hormone-modulated ion transporters lying in the plasma membrane can be viewed as an integral part of transmembrane signaling systems.

REFERENCES

Baquet, A., Hue L., Meijer, A.J., van Woerkom, G.M., & Plomp, P.J.A.M. (1990). Swelling of rat hepatocytes stimulates glycogen synthesis. J. Biol. Chem. 265, 955–959.

Bear, C.E. & Petersen, O.H. (1987). L-alanine evokes opening of single Ca^{2+}-activated K^+ channels in rat liver cells. Pflügers Arch. Physiol. 410, 342–344.

Bear, C.E. (1990). A nonselective cation channel in rat liver cells is activated by membrane stretch. Am. J. Physiol. 258, C421–C428.

Bianchini, L. & Grinstein, S. (1993). Regulation of volume-modulating ion transport systems by growth promoters. In: Interaction of Cell Volume and Cell Function (Lang, F. & Häussinger, D., eds.), pp. 249–277, Springer-Verlag, Heidelberg.

Bode, B. & Kilberg, M. (1991). Amino acid dependent increase of hepatic system N activity is linked to cell swelling. J. Biol. Chem. 266, 7376–7381.

Finkenzeller, G., Newsome, W.P., Lang, F., & Häussinger, D. (1994). Increase of c-jun mRNA upon hypoosmotic swelling of rat hepatoma cells. FEBS Lett. 340, 163–166.

Gazzola, G.C., Dall'Asta, V., Nucci, F.A., Rossi, P.A., Bussolati, O., Hoffmann, E.K., & Guidotti, G.G. (1991). Role of amino acid transport system A in the control of cell volume in cultured human fibroblasts. Cell. Physiol. Biochem. 1, 131–142.

Graf, J., Haddad, P., Häussinger, D., & Lang, F. (1988). Cell volume regulation in liver. Renal Physiol. Biochem. 11, 202–220.

Graf, J., Guggino, W.B., & Turnheim, K. (1993). Volume regulation in transporting epithelia. In: Interactions of Cell Volume and Cell Function (Lang, F. & Häussinger, D., eds.), pp. 67–117, Springer-Verlag, Heidelberg.

Grinstein S., Furuya, W., & Bianchini, L. (1992). Protein kinases, phosphatases, and the control of cell volume. News Physiol. Sci. 7, 232–237.

Häussinger, D., Hallbrucker, C., vom Dahl, S., Decker, S., Schweizer, U., Lang, F., & Gerok, W. (1991). Cell volume is a major determinant of proteolysis control in liver. FEBS Lett. 283, 70–72.

Häussinger, D. & Lang, F. (1992). Cell volume and hormone action. Trends Pharmacol. Sci. 13, 371–373.

Häussinger, D., Hallbrucker, C., Saha, N., Lang, F., & Gerok, W. (1992). Cell volume and bile acid excretion. Biochem. J. 288, 681–689.

Häussinger, D., Roth, E., Lang, F., & Gerok, W. (1993). Cellular hydration state: An important determinant of protein catabolism in health and disease. Lancet 341, 1330–1332.

Häussinger, D., Saha, N., Hallbrucker, C., Lang, F., & Gerok, W. (1993b). Involvement of microtubules in the swelling-induced stimulation of transcellular taurocholate transport in perfused rat liver. Biochem. J. 291, 355–360.

Häussinger, D., Stoll, B., vom Dahl, S., Theodoropoulos, P.A., Markogiannakis, E., Gravanis, A., Lang, F., & Stournaras, C. (1994). Microtubule stabilization and induction of tubulin mRNA by cell swelling in isolated rat hepatocytes. Biochem. Cell. Biol. 72, 12–19.

Halestrap, A.P. (1989). The regulation of the matrix volume of mammalian mitochondria *in vivo* and *in vitro* and its role in the control of mitochondrial metabolism. Biochim. Biophys. Acta 973, 355–382.

Halestrap, A.P. (1993). The regulation of organelle function through changes of their volume. In: Interactions of Cell Volume and Cell Function (Lang, F. & Häussinger, D., eds.), pp. 279–307, Springer-Verlag, Heidelberg.

Hallbrucker, C., Ritter, M., Lang, F., Gerok, W., & Häussinger, D. (1993). Hydroperoxide metabolism in rat liver. K^+ channel opening, cell volume changes and eicosanoid formation. Eur. J. Biochem. 211, 449–458.

Hawkins, R.A., Jessy, J., Mans, A.M., & De Joseph, M.R. (1993). Effect of reducing brain glutamine synthesis on metabolic symptoms of hepatic encephalopathy. J. Neurochem. 60, 1000–1006.

Henderson, R.M., Graf, J., & Boyer, J.L. (1989). Inward-rectifying potassium channels in rat hepatocytes. Am. J. Physiol. 256, G1028–G1035.

Hoffmann, E.K., Simonsen, L.O., & Lambert, I.H. (1993). Cell volume regulation: Intracellular transmission. In: Interactions of Cell Volume and Cell Function (Lang, F. & Häussinger, eds.), pp. 187–248, Springer-Verlag, Heidelberg.

Katz, J., Golden, S., & Wals, P.A. (1976). Stimulation of hepatic glycogen synthesis by amino acids. Proc. Natl. Acad. Sci. USA 73, 3433–3437.

Kilberg, M. & Häussinger, D. (eds.) (1992). Mammalian Amino Acid Transport: Mechanisms and Control. Plenum Press, New York.

Kreis, R., Ross, B.D., Farrow, N.A., & Ackerman, Z. (1992). Metabolic disorders of the brain in chronic hepatic encephalopathy detected with ^{1}H-NMR spectroscopy. Radiology 182, 19–27.

Kristensen, L.O. & Folke, M. (1984). Volume-regulatory K^+ efflux during concentrative uptake of alanine in isolated rat hepatocytes. Biochem. J. 221, 265–268.

Kwon, M.H., Yamauchi, A., Uchida, S., Preston, A.S., Garcia-Perez, A., Burg, M.H., & Handler, J.S. (1992). Cloning of the cDNA for a Na^+/myo-inositol cotransporter, a hypertonicity stress protein. J. Biol. Chem. 267, 6297–6301.

Lang, F., Stehle, T., & Häussinger, D. (1989). Water, H^+, lactate and glucose fluxes during cell volume regulation in perfused rat liver. Pflügers Arch. Physiol. 413, 209–216.

Lang, F., Ritter, M., Wöll, E., Weiss, H., Häussinger, D., Maly, K., & Grunicke, H. (1992). Altered cell volume regulation in ras oncogene expressing NIH fibroblasts. Pflügers Arch. Physiol. 420, 424–427.

Lang, F. & Häussinger, D. (eds.) (1993). Interaction of Cell Volume and Cell Function. Springer-Verlag, Heidelberg.

McCarty, N.A. & O'Neil, R.G. (1992). Calcium signaling in cell volume regulation. Physiol. Rev. 72, 1037–1061.

Meijer, A.J., Baquet, A., Gustafson, L., van Woerkom, G.M., & Hue, L. (1992). Mechanism of activation of liver glycogen synthase by swelling. J. Biol. Chem. 267, 5823–5828.

Minton, A.P., Colclasure, G.C., & Parker, J.C. (1992). Model for the role of macromolecular crowding in regulation of cellular volume. Proc. Natl. Acad. Sci. USA 89, 10504–10506.

Mortimore, G.E. & Pösö, A.R. (1987). Intracellular protein catabolism and its control during nutrient deprivation and supply. Ann. Rev. Nutr. 7, 539–564.

Moule, S.K. & McGivan, J.D. (1990). Regulation of the plasma membrane potential in hepatocytes—mechanism and physiological significance. Biochim. Biophys. Acta 1031, 383–397.

Newsome, W.P., Warskulat, U., Noe, B., Wettstein, M., Stoll, B., Gerok, W., & Häussinger, D. (1994). Modulation of phosphoenolpyruvate carboxykinase mRNA levels by the hepatocellular hydration state. Biochem. J. 304, 555–560.

Norenberg, M.D., Neary, J.T., Bender, A.S., & Dombro, R.S. (1992). Hepatic encephalopathy: A disorder in glial-neuronal communication. Progress Brain Res. 94, 261–269.

Offensperger, W.B., Offensperger, S., Stoll, B., Gerok, W., & Häussinger, D. (1994). Effects of anisotonic exposure on duck hepatitis B virus replication. Hepatology 20, 1–7.

Panet, R., Snyder, D., & Atlan, H. (1986). Amiloride added together with bumetanide completely blocks mouse 3T3-cell exit from G_0/G_1 phase and entry into S-phase. Biochem. J. 239, 745–750.

Petronini, P.G., De Angelis, E.M., Borghetti, P., Borghetti, A.F., & Wheeler, K. (1992). Modulation by betaine of cellular responses to osmotic stress. Biochem. J. 282, 69–73.

Pfaller, W., Willinger, C., Stoll, B., Hallbrucker, C., Lang, F., & Häussinger, D. (1993). Structural reaction pattern of hepatocytes following exposure to hypotonicity. J. Cell. Physiol. 154, 248–253.

Plomp, P.J.A.M., Wolvetang, E.J., Groen, A.K., Meijer, A.J., Gordon, P.B., & Seglen, P.O. (1987). Energy dependence of autophagic protein degradation in isolated rat hepatocytes. Eur. J. Biochem. 164, 197–203.

Sackin, H. (1987). Stretch-activated potassium channels in renal proximal tubule. Am. J. Physiol. 253, F1253–1262.

Saha, N., Stoll, B., Lang, F., & Häussinger, D. (1992). Effect of anisotonic cell volume modulation on glutathione-S-conjugate release, t-butylhydroperoxide metabolism and the pentose-phosphate shunt in perfused rat liver. Eur. J. Biochem. 209, 437–444.

Saha, N., Schreiber, R., vom Dahl, S., Lang, F., Gerok, W., & Häussinger, D. (1993). Endogenous hydroperoxide formation, cell volume and cellular K^+ balance in perfused rat liver. Biochem. J. 296, 701–707.

Schliess, F., Schreiber, R., & Haussinger, D. (1995). Activation of extracellular signal related kinases Erk-1 and Erk-2 by cell swelling in H4 IIE hepatoma cells. Biochem. J. 309, 13–17.

Seglen, P.O. & Gordon, P.B. (1984). Amino acid control of autophagic sequestration and protein degradation in isolated rat hepatocytes. J. Cell. Biol. 99, 435–444.

Stoll, B., Gerok, W., Lang, F., & Häussinger, D. (1992). Liver cell volume and protein synthesis. Biochem. J. 287, 217–222.

Theodoropoulos, T., Stournaras, C., Stoll B., Markogiannakis, E., Lang, F., Gravani, A., & Häussinger, D. (1992). Hepatocyte swelling leads to rapid decrease of G-/total actin ratio and increases actin mRNA levels. FEBS Lett. 311, 241–245.

Tohyama, Y., Kameji, T., & Hayashi, S. (1991). Mechanisms of dramatic fluctuations of ornithine decarboxylase activity upon tonicity changes in primary cultured rat hepatocytes. Eur. J. Biochem. 202, 1327–1331.

Uchida, S., Garcia-Perez, A., Murphy, H., & Burg, M. (1989). Signal for induction of aldose reductase in renal medullary cells by high external NaCl. Am. J. Physiol. 256, C614–C620.

van Rossum, G.V.D., Russo, M.A., & Schisselbauer, J.C. (1987). Role of cytoplasmic vesicles in volume maintenance. Curr. Top. Membr. Transp. 30, 45–74.

Völkl, H., Friedrich, F., Häussinger, D., & Lang, F. (1993). Effect of cell volume on acridine orange fluorescence in hepatocytes. Biochem. J. 295, 11–14.

vom Dahl, S., Hallbrucker, C., Lang, F., Gerok, W., & Häussinger, D. (1991). Regulation of liver cell volume and proteolysis by glucagon and insulin. Biochem. J. 278, 771–777.

Yancey, P.H., Clark, M.E., Hand, S.C., Bowlus, R.D., & Somero, G.N. (1982). Living with water stress: Evolution of osmolyte systems. Science 217, 1214–1222.

RECOMMENDED READINGS

Chamberlin, M.E. & Strange, K. (1989). Anisosmotic cell volume regulation: A comparative view. Am. J. Physiol. 257, C159–C173.

Häussinger, D. & Lang, F. (1991). Cell volume in the regulation of hepatic function: A new mechanism for metabolic control. Biochim. Biophys. Acta 1071, 331–350.

Sarkadi, B. & Parker, J.C. (1991). Activation of ion transport pathways by changes in cell volume. Biochim. Biophys. Acta 1071, 407–427.

Chapter 8

Some Aspects of Medical Imaging

H.K. HUANG

Principles of Medical Biology, Volume 4
Cell Chemistry and Physiology: Part III, pages 211–239.
Copyright © 1996 by JAI Press Inc.
All rights of reproduction in any form reserved.
ISBN: 1-55938-807-2

INTRODUCTION

Medical imaging is the study of human functions and anatomy through pictorial information. In order to generate this pictoral information, multidisciplinary knowledge, including biology, anatomy, physiology, chemistry, computer science, optical science, radiological science, electrical engineering, mathematics, and physics are required (Huang, 1993a). Generally speaking, medical imaging investigates methods and procedures of:

1. Converting a conventional medical image, or synthesizing some biological, anatomical, or physiological information, to a digital image.
2. Analyzing the digital image according to a specific application or clinical need.
3. Extracting key results and casting them into a format suitable for presentation, archiving, and decision making.

Some successful medical imaging applications in the early 1970s were the blood cell analyzer (Pressman and Wied, 1979) and the gamma camera in nuclear medicine (NM). The development of the computed tomography (CT) scanner resulted in the award of the Nobel Prize in Medicine to Allan M. Cormack and Godfrey N. Hounsfield in 1979. Major medical imaging developments in the 1980s were electron microscopy (EM), laser microscopy (LM), digital subtraction angiography (DSA), magnetic resonance imaging (MRI), positron emission tomography (PET), computed radiography (CR), Doppler ultrasound, and picture archiving and communication systems (PACS) (Huang, 1981a; Huang et al., 1990). EM can reveal minute details in biological infrastructures as small as a few angstroms in

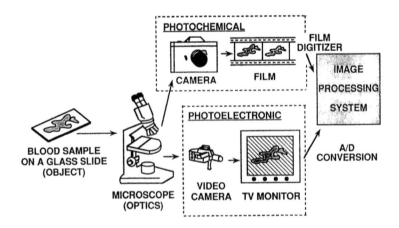

Figure 1. An example of biomedical image detectors and recorders.

size. LM yields thin serial images providing three-dimensional morphology of living cells. DSA allows real-time subtraction to enhance the vascularities in millimeters. Without the use of ionizing radiation, MRI reveals high-contrast images of anatomical structures in any plane of the body. MRI is the method of choice for neuroradiological, vascular, and musculoskeletal diagnosis.

PET provides chemical and physiological images of the human body that complement anatomical images obtained using MRI and CT (Hawkins et al., 1992). The registration of MRI or CT with PET head images provides an insight into the specific function of various parts of the brain (Valentino et al., 1991). CR allows an X-ray image to be recorded directly as a digital image, opening new avenues for using digital image processing as an aid in medical diagnosis (Kangarloo et al., 1988).

PACS is a novel concept for medical image management and communication (Huang et al., 1988). When fully implemented, the system will revolutionize medical practice (Huang et al., 1993b). PACS storage technology includes parallel transfer disks and optical disk libraries. In the former, a conventional X-ray image of 8 megabytes can be stored or retrieved from the parallel transfer disks within one second. In the latter, an optical disk library that occupies a footprint of no more than 3×6 feet allows the storage of one terabyte of information, equivalent to about two years worth of all MR and CT examinations conducted in a large teaching hospital.

In communication components, fiber optic systems with specially designed fiber optic transmitters and receivers can transmit images at a rate up to 1 gigabit per second. A conventional 8-megabyte X-ray image can be transmitted between two points in about one second (Huang et al., 1992). For display, $2,000 \times 2,000$ pixel monitors are readily available that display a conventional X-ray image without loss of diagnostic quality (Huang et al., 1993b). Three-dimensional display stations are used in various clinical applications (Udupa et al., 1993).

MEDICAL IMAGE FUNDAMENTALS

Medical Image Detectors and Recorders

Medical image detection and recording methods can be categorized as being either photochemical or photoelectronic. An example of a photochemical method is the phosphorous screen and silver halide film combination system used for X-ray detection. The television camera and display monitor used in fluorography is a photoelectronic technique. The photochemical method is a direct process; it has the advantage of combining image detection and image recording in a single step. The screen/film system simultaneously detects and records the attenuated X-rays. A photoelectronic system, on the other hand, usually involves a two-step process; the image is detected first and then recorded in a subsequent step.

In the case of DSA, an image intensifier tube (II) is used as the X-ray detector instead of a screen/film combination. The detected X-rays are converted first into light photons and then into electronic signals that are recorded by a video camera. The images from the video camera can be displayed on a television monitor, and the video signal can be digitized to form a digital image. The photoelectronic system, while clearly more complicated, has one important advantage: the output information can be converted to digital format for image processing.

Figure 1 shows an example of medical image detectors and recorders. In this case, an image of blood cells from a blood sample on a glass slide is to be recorded. The glass slide is first placed under the microscope. If a 35 mm camera is attached to the microscope, the image of the blood cells can be recorded on film. On the other hand, if a television or CCD (charge couple device) camera is attached to the microscope, then the blood cells are seen as an electronic image on a television monitor. In either case, the recorded image is in analog form. For a digital computer to process these images, they must first be converted to digital form, a step called analog to digital conversion.

Digital Images

A digital image $P(x,y)$ is defined as an integer function of two variables x,y such that

$$0 \leq P(x,y) \leq N \text{ where } 1 \leq x \leq m, 1 \leq y \leq n$$

and x, y, m, n, and N are positive integers. For simplicity, we let $m = n$ (i.e., $P(x,y)$ is a square image). Given (x,y), $P(x,y)$ is called a picture element, or *pixel*. The computer memory requirement for storing image $P(x,y)$ is $n \times n \times k$ bits where $k = \log_2 (N + 1)$. Thus, $n \times n \times k$ means that the image has n lines, each line has n pixels, and each pixel can have a discrete gray-level value that ranges from 0 to $2^k - 1$ (Udupa et al., 1993). A typical microscopic digital image has 512×512 pixels and each pixel can have values from 0 to 255.

Spatial and Density Resolution

Once an object of interest has been digitally recorded, we would like to know the image quality. Image quality is characterized by three parameters: spatial resolution, density resolution, and signal to noise ratio. Spatial resolution is a measure of the number of pixels used to represent the object, and density resolution is the total number of discrete gray level values in the digital image. It is apparent that n and N are proportional to spatial resolution and density resolution, respectively. A high signal-to-noise ratio means that the image is very pleasing to the eye and hence is a better quality image. Figure 2 demonstrates the concept of spatial and density resolution of a digital image of a lymphocyte. The left-hand column in

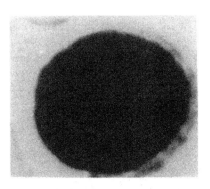

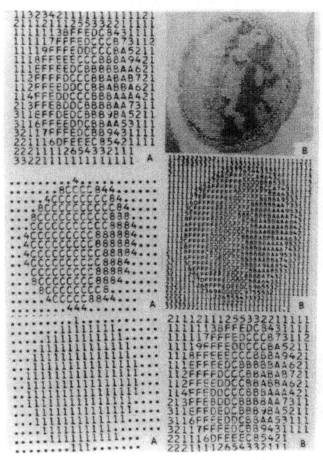

Figure 2. Illustration of spatial and density resolution, using a lymphocyte image as an example. (**A**) Fixed spatial resolution, variable density resolution: 16, 4, and 2 grey levels. (**B**) Fixed density resolution (16 levels), variable spatial resolutions. The 16 levels are represented by: •, 1, 2, 3, 4, 5, 6, 7, 8, 9, A, B, C, D, E, and F.

Figure 2 shows digitized images of the lymphocyte with a fixed spatial resolution (21 × 15) and variable density resolutions (from top to bottom: 16, 4, and 2 gray levels). The right-hand column depicts the digital representation of the same analog image with a fixed density resolution (16) and variable spatial resolutions (from top to bottom: high, medium, low). It is clear from this example that the upper right corner digital image has the best quality (highest spatial and density resolutions), whereas the lower left corner image has the lowest spatial and density resolutions. Depending on the application requirement, the spatial resolution, density resolution, and signal-to-noise ratio of the image should be adjusted properly during image acquisition. A high-resolution image requires a larger memory capacity for storage and a correspondingly longer time for image processing than a lower-resolution image.

Sources of Medical Images

By far, the richest sources of medical images is in radiology; sometimes we call those images macroscopic to differentiate them from microscopic and infrastructural images. In radiology, about 70% of the examinations, including those that involve skull, chest, abdomen, and bone, produce images that are acquired and stored on X-ray film. These images have a spatial resolution of about 5 lp/mm. Line pair per millimeter (lp/mm) is a measure of spatial resolution; one line pair represents two pixels. These films can be converted to digital format using a film digitizer. Among various types of digitizers, the laser scanning digitizer is considered superior because it can best preserve the density and spatial resolutions of the original analog image. A laser film scanner can digitize a 14" × 17" X-ray film to 4,000 × 5,000 pixels (about 5 lp/mm), with 12 bits per pixel. At this spatial and density resolution, the quality of the original analog image and the digitized image is essentially equivalent. In clinical practice, however, we digitize an X-ray film to 2,000 × 2,500 pixels. CR, which uses a laser stimulable luminescence phosphor imaging plate as a detector, is gradually replacing the screen/film combination as the image detector. In this case, a laser beam is used to scan the imaging plate that contains the latent X-ray image. The latent image is excited and emits light photons that are detected and converted to digitized electronic signals forming a direct digital X-ray image.

The other 30% of radiological examinations—those that involve CT, ultrasonography (US), MRI, PET, and DSA—produce images that are already in digital format. A CT, US, MRI, PET, and DSA image has sizes of $512 × 512 × 12$, $512 × 512 × 8$, $256 × 256 × 12$, $128 × 128 × 12$, and $512 × 512 × 8$ bits, respectively. These techniques use different energy sources and detectors to generate images and are complementary in their clinical applications to each other. CT uses X-rays as an energy source and gas or scintillating crystals as detectors. US uses an ultrasonic transducer both as the energy source and detector. MR uses two energy sources, magnetic fields and radio-frequency electromagnetic waves, and a radio-frequency receiver as the detector. DSA uses X-rays as an energy source and an image intensifier tube as the detector. Conventional X-ray examinations and DSA produce a projectional image, whereas CT, US, PET, and MRI give sectional images. Sectional images can be stacked to form a three-dimensional image set which represents the true three-dimensional object. All radiologic images are monochromatic except NM, PET, and Doppler US in which pseudo colors are used as an image enhancement tool.

Other medical images sources used in anatomy, biology, and pathology are light and electron microscopes. Images from these sources are collected with a video or CCD camera and then digitized to a 512 × 512 × 8 bit image. Light microscopy produces true color images, using red, green, and blue filters for color separation. Thus, a color image after digitization yields three digital images, the combination of which produces a true-color digital image encoded at 24 bits/pixel. Figure 3a–j shows some examples of these medical images.

Image Processing Systems

After a medical image is formed, it is analyzed by an image processing system. The architecture of an image processing (IP) system consists of three major components: image processor(s), image memories, and video processor(s). They are connected by internal computer buses to form an integrated system. Figure 4 shows the general block diagram of an integrated image processing system. For this particular system, only the system controller is connected to the computer host bus. The image processor is a high-speed array processor. It is composed of arithmetic-logic units, multipliers and shifters, comparators, and look-up tables. The image memories can be partitioned into various sizes for efficient storage of image data. The video processor takes the images from the image memories and selectively displays them on video monitors.

An IP system requires extensive software support. The trend in IP software development is towards portability. Figure 5 shows the general organization of IP software. Portability is preferred in the three higher levels of software so that they can be used again when the system hardware is upgraded in the future. The two lower levels are machine dependent and have to be rewritten for every new hardware architecture. IP functions include pixel, local, global, and statistical operations, and also consist of image database manipulation and image display. In the past, Fortran was used in most IP software development, but C programming language running under UNIX operation system is now standard.

In medical imaging applications, contour extraction of an object of interest is important because it leads to quantitative measurements. Despite many years of software research and development, soft-tissue segmentation in radiologic images, and differentiation of objects of interest in histology specimens, are still a very

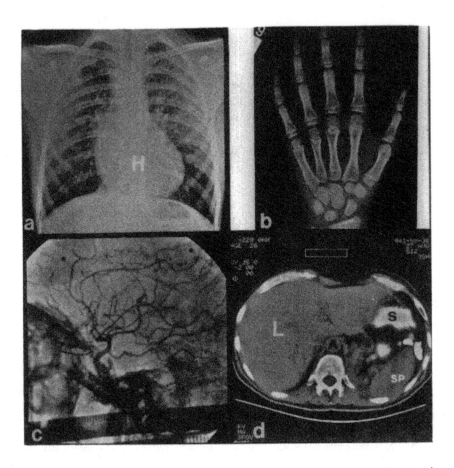

(continued)

Figure 3. Examples of medical images obtained from various sources. (**a**) CR (computed radiography) image of the chest, the image is 2,000 × 2,500 pixels and 10 bits/pixel. Excellent delineation of the blood vessels behind the heart (H). (**b**) CR image of the hand. Both soft tissue and the detail of the bones are seen very clearly. (**c**) DSA (digital subtraction angiography) of the brain showing contrast enhanced blood vessels. (Courtesy of E. Pietka.) (**d**) CT (computed tomography) image of the upper abdomen. Contrast media is shown in the stomach (S). L, liver; SP, spleen; A, descending aorta. (**e, f, g**) MR (magnetic resonance) images of the head from the same patient in the transverse, sagittal, and coronal plane. Images show fine structures of the brain. (Courtesy of S. Sinha.) (**h**) Mapping of the brain function to anatomy. The grey level image is from the MR, the color is from the PET (positron emission tomography) of the same patient. Red color shows high metabolic rates. Registration of these two images required sophisticated mathematics and computer programming. (Courtesy of D. Valentino.) (**i**) A longitudinal section Doppler ultrasound image of the abdomen. Red color shows the blood flow, arrow indicates that flow in the portal vein is hepatopetal. (Courtesy of E. Grant.) (**j**) Three-dimensional reconstruction of the lumber spine from sectional CT images.

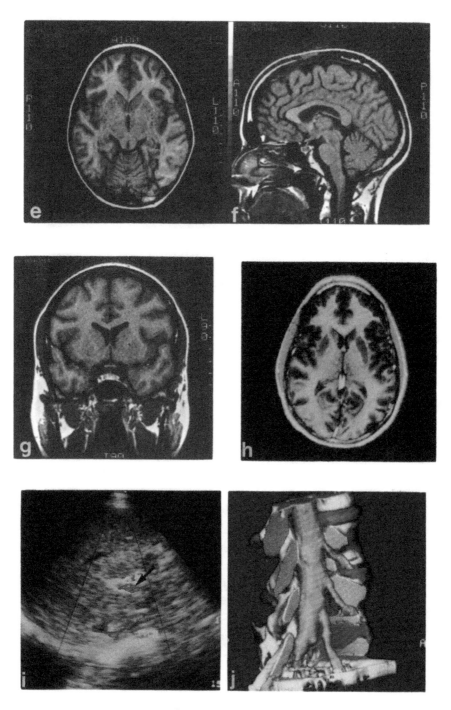

Figure 3. (Continued)

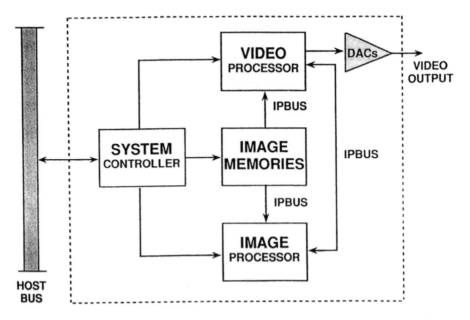

Figure 4. A general architecture of an integrated image processing system. Only the system controller is connected to the computer host bus. IP: Image Processing.

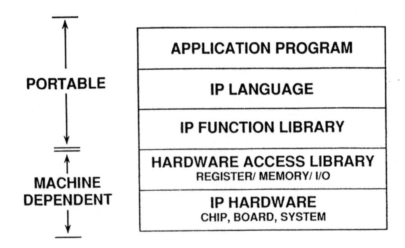

Figure 5. Organization of image processing software. The three higher-level software should be portable so that they can be used for any future hardware architecture.

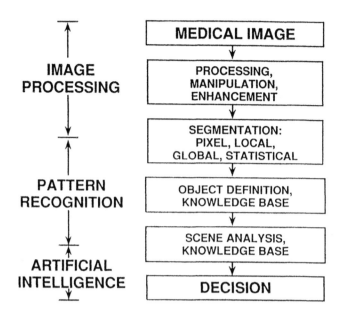

Figure 6. Levels of sophistication in medical image processing. Medical database, pattern recognition and artificial intelligence will be major research topics in the 1990s.

difficult task. Advances in medical image processing remain largely in the domain of quantization. Figure 6 shows the levels of sophistication in medical image processing.

DIGITAL MICROSCOPY

Instrumentation

Digital microscopy is used to extract quantitative information from microscopic slides. A digital microscopic imaging system consists of the following six components (Huang, 1981b):

- a compound microscope with proper illumination for specimen input,
- a vidicon (or CCD) camera for scanning microscopic images,
- TV monitors for displaying the image,
- an analog to digital (A/D) converter, and
- a computer (or image processor) to process the digital image.

Figure 7 shows the block diagram and the physical setup of the instrumentation.

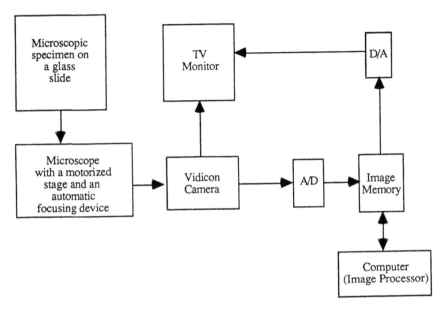

Figure 7. A block diagram showing the digital microscopy instrumentation.

Resolution

The resolution of a microscope is defined as the minimum distance between two objects in the specimen which can be resolved by the microscope. Three factors control the resolution of a microscope (Burrells, 1971; Rochow and Rochow, 1978).

1. The angle subtended by the object of interest in the specimen and the objective lens; the larger the angle, the higher the resolution.
2. The medium between the front lens and the coverslip of the glass slide; the higher the refractive index of the medium, the higher the resolution.
3. The wavelength of light employed; the shorter the wavelength, the higher the resolution.

These three factors can be combined into a single equation (Ernst Abbe 1840–1905):

$$s = \frac{\lambda}{2(NA)} = \frac{\lambda}{2n \sin i}$$

where s is the distance between two objects in the specimen that can be resolved; the smaller the s, the greater the resolution. λ is the wavelength of the light employed; n is the refractive index of the medium; i is the half-angle subtended by the object at the objective lens; and NA is the numerical aperture commonly used for defining the resolution; the larger the NA, the higher the resolution. Therefore, in order to obtain a higher resolution for a microscope, use an objective lens with an oil immersion lens (large n) with a large angular aperture and select a shorter wavelength of light source for illumination.

Contrast

Contrast is the ability to differentiate various components in the specimen with different intensity levels. There are black-and-white contrast and color contrast. Black-and-white contrast is equivalent to the range of the grey scale; the larger the range, the better the contrast. The color contrast is an important parameter in microscopic image processing: in order to bring out the color contrast from the image, various color filters have to be used with the illumination. It is clear that the spatial and density resolutions of a digital image are limited by the resolution and contrast of a microscope, respectively. In order to do effective quantitative analysis with the microscope, two additional attachments to the microscope are necessary: a motorized stage assembly and an automatic focusing device.

Motorized Stage Assembly

The purpose of a motorized stage assembly is for rapid screening and locating the exact position of objects of interest for subsequent detailed analysis. The motorized stage assembly consists of a high precision x-y stage with a specially designed holder for the slide to minimize the vibration due to transmission when the stage is moving. Two stepping motors are used for driving the stage in the x and the y directions. A typical motor step is about 2.5 micron with an accuracy and repeatability to within ±1.25 microns. The motors can move the stage in either direction with a maximum speed of 650 steps, or .165 cm, per second. The two stepping motors can either be controlled manually or automatically by the computer.

Automatic Focusing Device

The purpose of the automatic focusing device is to assure that the microscope is focusing all the time when the stage is moving from one field to another by the stepping motors. It is essential to have the microscope in focus before the vidicon camera starts to scan. Two common methods for automatic focusing are using a third stepping motor in the z-direction or an air pump. The principle of using a stepping motor in the z-direction for automatic focusing is as follows: the stage is moved up and down with respect to the objective lens by the z-direction motor. The z movements are nested in large +z and −z values initially and then gradually to smaller +z and −z values. After each movement, a video scan of the specimen through the microscope is made and some optical parameters are derived from the scan. A focused image is defined as the scan with these optical parameters above certain threshold values. Since this method requires nested upward and downward movements of the stage in the z-direction, though automatic, it is time consuming.

The use of an air pump for automatic focusing is based on the assumption that in order to have automatic focusing, the specimen lying on the upper surface of a glass slide has to be on a perfect horizontal plane with respect to the objective lens all the time. One reason why the slide is not focused is that the glass slide is not of uniform thickness. When it rests on the horizontal stage, the lower surface of the slide will form a horizontal plane with respect to the objective but not the upper surface. If an air pump is used to create a vacuum from above, such that the upper surface of the slide is suctioned from above to form a perfect horizontal plane with respect to the objective, then the slide will be focused all the time. Using an air pump for automatic focusing does not require additional time during operation; however, it requires precision machinery.

Vidicon (CCD) Camera and Scanning

Once the specimen is focused under the microscope, a vidicon or a CCD camera can be attached to the microscope tube to detect the light emitted from the specimen within the microscopic field of view. The camera scans the specimen point by point from left to right, top to bottom and forms a light image of the specimen on the photosensitive face of the camera. The brightness $B(x, y)$ of each pixel is converted into an electrical voltage (video signal) which is transmitted to the display monitor. This voltage is used to control the brightness of a corresponding spot on the fluorescent screen of the monitor. These spots reconstruct a video image of the specimen on the television screen. In order to have the camera perform satisfactorily for microscopic imaging, the following specifications should be met:

Gamma (γ) of the tube: .65 or less
Dynamic range: 200:1
Resolution: The MTF (modulation transfer function derived by plotting the video amplitude versus the number of lines) should be comparable to that of an ideal Gaussian Spot with diameter 1/500 of the image width.
Linearity: $\pm 1/2\%$ of the image height for all pixels.

Analog to Digital Conversion

There are two methods for digitizing the microscopic image formed by the vidicon camera: the real-time digitizing with a fast A/D converter, usually in the

10 MHz range. It converts the video signal B(x, y) into a digital number P(x, y) and sends it to the (x, y) location of the image memory. A complete TV frame (512 × 512 pixels with eight bits/pixel) can be digitized in 1/30 second. Because of the high speed A/D conversion, the signal to noise ratio of the real-time digitized image tends to be low. In order to have a better signal to noise ratio, it is common to digitize the same frame many times and take the average value for each pixel. The high-resolution digitizer uses a slower but better signal-to-noise ratio A/D converter; the digital image thus obtained is of better quality. Since the A/D conversion rate is much slower than the TV scanning rate, the same microscopic field has to be scanned many times in order to have a complete digital image.

Image Memory

The purpose of the image refresh memory is for storage and display of the digitized microscopic image. Once the image is digitized and stored in the memory, it is continuously refreshed in synchronism with the video scan. The refresh memory is not a component of the main computer and should be considered as a very fast peripheral storage device. Once the image is stored, it can be accessed by the computer for image processing. The refresh memory is generally organized into memory planes, with each plane having the storage capacity to refresh a 512 × 512

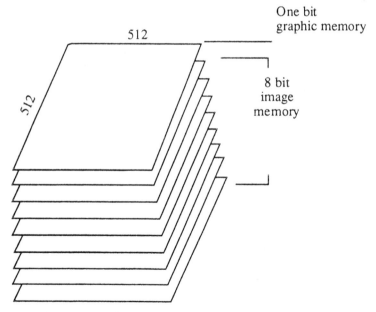

Figure 8. Organization of a 512 × 512 × 8 image memory and a 512 × 512 × 1 graphic memory. For black-and-white image processing, one eight-bit image memory is sufficient; for real color image processing, three image memories are necessary.

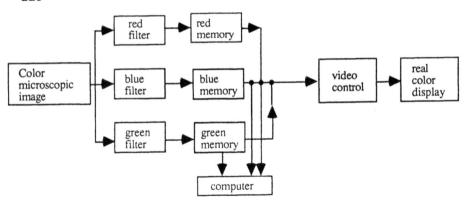

Figure 9. Color image processing block diagram. Three filters (red, blue, and green) are used to filter the image before digitization, respectively. The three digitized filtered images are stored in the red, blue, and green memories. The real color image can be displayed back on the color monitor from these three memories through the composite video control.

one-bit grey-level image. The most commonly used refresh memory for imaging has 8 or 12 memory planes which give 256 to 4,096 grey levels. In addition, there should be one extra memory plane for graphic overlay on top of the image memory for interactive image processing. Figure 8 shows the architecture of the image memory.

If real color microscopic image is used, the color specimen is generally digitized in three steps, each time with a red, blue, and green filter, respectively. The three color-filtered images are then stored separately in three refresh memories, the blue, red, and green, each of which with eight planes. The computer will treat each of the eight planes refresh memory as an individual microscopic image and process them separately. Thus, a true color image has 24 bits/pixel. The real color digital microscopic image can be displayed back on a color monitor from these three memories through a color composite video control. Figure 9 shows the block diagram of the real color microscopic imaging.

Computer

The computer, usually a personal computer with necessary peripherals, serves as a control, as well as performing image analysis in the system. The following are the major control functions:

Movement of the x-y stepping motors
Control of the automatic focusing
Control of the color filtering

Control of the digitization of the microscopic image
Image processing

ANALYSIS OF MICROSCOPIC IMAGES

For reviews, see Castleman, 1979; Onoe et al., 1980; Huang, 1981a, 1982.

Microscopic Glass Slide

The specimen under consideration is first prepared with proper staining and put on the glass slide. A thin cover slip is then placed on top of the specimen. The slide is positioned on the microscope stage and an origin is selected with respect to the motorized stage. The slide is now ready to be analyzed. For automatic analysis of the slide, the following nomenclature is commonly used (Huang, 1982). (See Figures 10 and 11 for illustration.)

The total scan area (TSA) is that portion of the slide to be scanned by the vidicon camera. The dimensions of the TSA cannot exceed the size of the cover slip (15 mm × 15 mm). For example, 10 mm × 10 mm is a commonly used dimension. A strip is a lateral (or vertical) portion of the TSA. For example, if a 20× objective lens is used, a strip would have the dimensions of about 10 mm × 200 μ. A field represents a portion of a strip; using the same example, it would have the dimensions of about 200 μ × 200 μ. This field represents a square area to be scanned by the vidicon camera and when it is digitized it will consist of 512 × 512 pixels. The

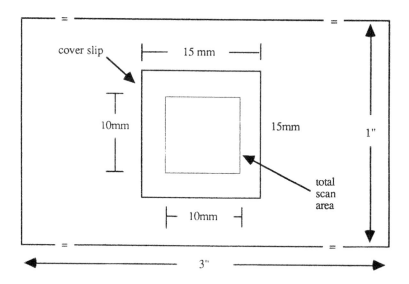

Figure 10. The dimensions of a sample glass slide.

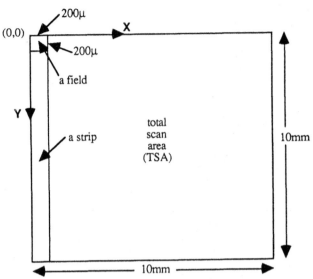

Figure 11. The subdivision of the total scan area (TSA). The numerical values used in this drawing are based on a 100× oil immersion objective lens. There is some overlapped area between adjacent fields.

origin is at the left upper corner of the TSA. The positive x-axis is taken as the line starting from the origin and extending horizontally (towards the right) along the upper edge of the TSA. The positive y-axis is taken as the line starting from the origin moving longitudinally along the left edge of the TSA. The step is the smallest increment the x, y stepping motors can advance, for example, 2.5 microns.

Search for Objects of Interest

The first step in analyzing a microscopic slide is to search for objects of interest from the TSA using a lower magnification objective lens. For example, in blood karyotyping, the objects of interest would be metaphase blood cells. Since different objects of interest have different characteristics, there is no general search algorithm for every biological specimen. However, some guidelines can be used to facilitate the search. Consider the chromosome karyotyping as an example to illustrate the search technique.

In chromosome karyotyping, the goals are to search for isolated metaphase cells under low magnification and karyotype these metaphase cells using higher magnification. The blood cell sample is first prepared with proper staining. A low magnification objective (for example, 20×) is used to scan each field in the TSA. Since only low resolution is needed to identify metaphase cells, two grey level and every other scan line from the image are sufficient. Let us define:

An object in a binary image (two grey level) consists of a string which has n consecutive ones separated from other groups of ones by at least n zeros on either side. The parameter, n, is chosen as the optimal choice between the width of a chromosome and the spacing between two chromosomes in a metaphase cell under this magnification. Thus, for example, the string

$$0000001\ 1\ 1\ 1000000$$

is an object with $n = 4$. The location of the object is the order pair consisting of the coordinates of the left most "one" and right most "one." If n is too small, the object could be dirt in the field and if n is too large, the object could be a line through a non-metaphase cell. In both cases, the object should be discarded.

A cover is a horizontal string of at least p such objects, where the distance between two neighboring objects must be less than q points (p and q are parameters typical of a metaphase cell). The starting coordinates of a cover (x_s, y_s) is the left-hand coordinates of its left most object, and the ending coordinate of this cover (x_e, y_e) is the right-hand coordinates of its right most object, and

$$s > (x_e - x_s) > \gamma$$

where γ, s are parameters typical of a metaphase cell. For example, the string

$y = 1$:   000 1 1 1 0000 1 1 1 1 00000000 1 1 1 0000

        object1   object2   object3

        First Cover   Second Cover

consists of three objects, if $q = 5$, then object 1 and 2 form a cover with $p = 2$ and object 3 forms a second cover with $p = 1$. In the first cover, $x_s = 4$ and $x_e = 14$, and the second cover, $x_s = 23$ and $x_e = 25$.

A chromosome spread is a collection of at least two longitudinal proximate covers; its size must lie within a square of certain dimensions which is determined by the magnification. The center of the spread is determined as the arithmetic means of the minimum and maximum coordinates of all the covers.

A search algorithm can then be developed to look for chromosome spreads based on these hierarchical structures starting from objects, to cover, and to chromosome spreads. The algorithm can continue searching from field to field, strip to strip, until the complete TSA is scanned. With a carefully prepared specimen, and a 20× objective, an average of 100 good metaphase cells can be located. The center of these spreads can be recorded and the computer can move the stage back to all these centers for subsequent analysis with a higher magnification objective. Figure 12 shows the relative magnification of a partial chromosome spread using a low (20×) and a high (100×) objective. Figure 13 shows some covers and chromosome spreads found by this algorithm.

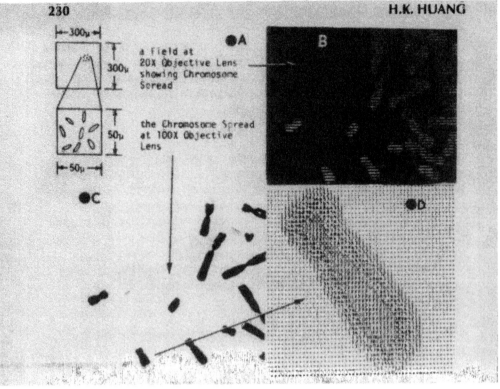

Figure 12. (A) Relative magnification of metaphase cells under a low and a high objective lens; (B) a partial chromosome spread under a low objective lens digitized in two grey-level and low resolution; (C) the same partial chromosome spread under 100× objective; (D) a chromosome spread digitized in 16 levels.

Analysis: Boundary Determination

After objects of interest have been located, a higher magnification objective lens, for example a 100× oil immersion, can be used for analysis. The first step is to find the boundary or boundaries of the objects. The simplest is the histogram method. The histogram method generates the histogram of the field which includes objects of interest as well as the background. The algorithm defines the grey-levels at the trough(s) of the histogram as the boundary cut-off level of objects. Once a cut-off grey level is known, the coordinates of the boundary of objects can be obtained by a program searching through the digital picture.

Figure 14 shows an example of the histogram technique (Huang et al., 1983). Figure 14a is a bone biopsy microradiograph obtained from a cross section of the rib of a dog. This image is as seen directly from the television monitor through the camera attached to the microscope focusing on the microradiograph. The goal of this study is to determine the total bone area, area of the trabecular bone (T), area

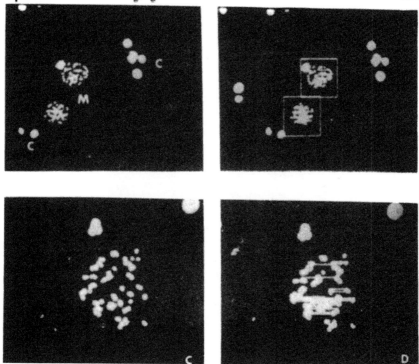

Figure 13. (A) A field showing two metaphase chromosome spreads (M) and some blood cells (C); (B) search algorithm locating the two spreads based on the hierarchal structure of objects, covers, and spreads; (C) another metaphase spread with a higher magnification; (D) horizontal lines showing the covers found during each line scan.

of marrow cavity (M), area of the cortical bone (C), and area of holes (arrows) in the cortical bone from this microradiograph. In order to extract these parameters, the boundaries of these areas must be determined first. The histogram technique is used to determine these boundaries. Figure 14b shows the histogram, h, of the complete image including the bone and the background, where

$$h = h(g) \qquad 0 \le g \le 255$$

where g is the grey-level value, and h is the grey-level count. Both the image and the graphic memory are used to display the result. The histogram can be smoothed by using a low pass digital filter in the frequency domain as follows: the Fourier Transform, H or h, is first performed and its spectrum |H| is shown in Figure 14c. A low pass filter, F, is used to smooth H such that

$$S = FH$$

where S is the transform of the histogram to be smoothed. The inverse transform, s of S, is then the smoothed histogram. The smoothed histogram, using a second

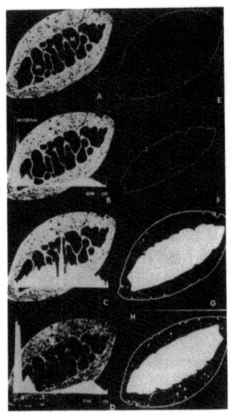

Figure 14. An example demonstrating the boundary detection technique. A bone biopsy microradiograph from a cross section of the rib of a dog is shown in (**A**). The objective is to find the boundaries of the cortical bone (c), trabecular bone (T) and holes (arrows) in the cortical bone. Histogram technique is used in the spatial domain for boundary detection. The histogram is smoothed in the frequency domain. See text for description. M, marrow.

order Butterworth low pass filter on H, is shown in Figure 14d. Compare the original histogram shown in Figure 14b with that of 14d. From this smoothed histogram, the boundary of the bone can be automatically traced by a standard boundary trace program using the grey level value at trough T (Figure 14d) as the cut-off value. The automatically traced bone boundary is shown in Figure 14e. The interface between the cortical and the trabecular bone is difficult to determine automatically since there is no clear-cut boundary between them. An interactive graphic technique has to be used which requires judgment from the operator. The interactively traced boundary between the cortical and the trabecular bone is shown in Figure 14f. The grey level value at trough B in Figure 14d is used as the cut-off value for automatically tracing the holes in the cortical bone areas; some holes are being automatically traced at the lower left corner of the bone in Figure 14g. Figure 14h shows the final result after all the holes have been traced and counted, only the graphic memory is shown in this figure to highlight the tracing. Thus, in this microscopic image, all the boundaries of objects of interest have been identified and their coordinates recorded. The next step is to perform measurements from these coordinates.

Analysis: Geometric and Density Parameters

Geometric Parameters

In black and white microscopic imaging, two types of parameters can be extracted from boundary coordinates for quantitative analysis: geometric parameters and density parameters. Geometric parameters (Huang, 1981a) only measure the size and shape of the object under consideration whereas density parameters measure both the geometry and the grey level distribution of the object.

Some commonly used descriptive geometric parameters are height, width, major diameter, minor diameter, elongation (major/minor), perimeter, convex perimeter, irregularity of outline (perimeter/convex perimeter), total area, inner area, outer area, features with holes (outer area/inner plus outer), circularity (area) (4π)/Perimeter^2), goodness of fit to square (perimeter/4·minor diameter), coil packing (perimeter/2·major diameter), orientation, location, intercept, and nearest neighbor distance. Figure 15 illustrates these features along with their definitions. Thus, the areal measurements in the microradiograph example previously described is only one of these many parameters.

Mathematically, geometric parameters can be described by using the Fourier Series in polar coordinates (ρ, θ). Thus, given a set of boundary points (x, y) from an object of interest, they can be transformed into the polar coordinates with respect to its geometric center (x, y). A curve fitting technique in polar coordinates can be used to fit this set of points into a Fourier Series such that any point $\rho(\theta)$ on this boundary can be expressed by

$$\rho(\theta) = A_o + \sum_{n=1}^{\infty} (A_n \cos n\theta + B_n \sin n\theta)$$

where A_o, A_n, and B_n, coefficients from the fitting. Each coefficient and each term in this expression has a certain meaning in terms of the shape of the boundary. For example, the plot $A_o + A_1 \cos\theta$ versus θ represents the error in locating the centroid; the A_2 term is the elongation; the A_3 term represents the triangularity; the A_4 term indicates the degree of squareness. Also, the higher the

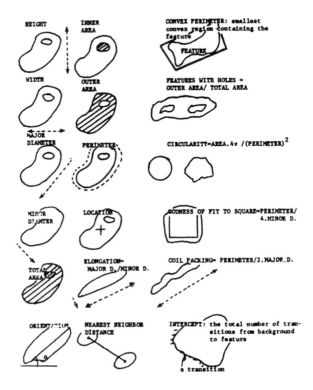

Figure 15. Some commonly used descriptive geometric parameters.

order of the term, the finer the morphological detail it represents. Figure 16 illustrates the concept of using the Fourier Series to describe geometric parameters.

Density Parameters

Density parameters take into consideration both the geometry and the density distribution within the object. Commonly used density parameters are the total mass, the center of gravity, and the second moment. Mathematically, they can be expressed as follows:

$$\text{Total mass: } M = \sum_{x,y} p(x,y)\Delta A$$

$$\text{Center of gravity: } X = \sum_{x,y} p(x,y)x \sum_{x,y} p(x,y)$$

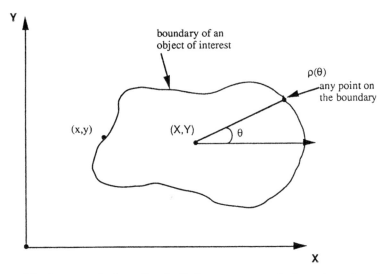

Figure 16. Concept of using a Fourier Series to represent the boundary of an object. The (x,y) coordinates of the boundary points of an object is transformed to polar coordinates. Each point on the boundary $\rho(\theta)$ can be expressed by a Fourier Series obtained from curve fitting of the boundary points. (X,Y) are the coordinates of the center of gravity.

$$Y = \sum_{x,y} p(x,y)y \sum_{x,y} p(x,y)$$

Second moments: $I_{xx} = \sum_{x,y} p(x,y)(y - Y)^2$

$$I_{yy} = \sum_{x,y} p(x,y)(x - X)^2$$

$$I_{xy} = \sum_{x,y} p(x,y)(x - X)(y - Y)$$

where $p(x,y)$ is the pixel value located at (x,y), and ΔA is the size of a pixel.

Scan and Count

In some specimens, the objects of interest in a microscopic image are not well-defined. In this case, boundary detection technique will not be effective. Quantization can only be done using a scan and count technique which produces a density distribution map of the image according to some preassigned conditions,

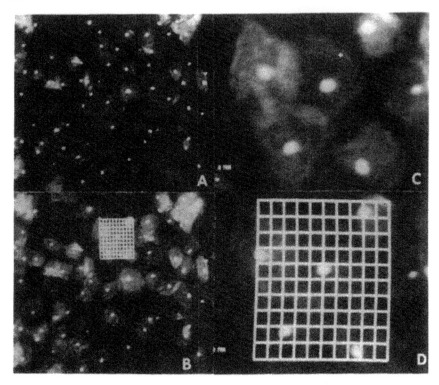

Figure 17. An example using the scan and count technique to count exfoliative cells in a Pap smear; see text for description.

for example, within a square mesh. This technique is best explained with an example. Figure 17a shows some exfoliative cells as seen in a Pap smear; bright dots are nuclei of the cells. Suppose it is required to find a quick count on the number of cells in a certain subregion. A square mesh from the graphic memory can be superimposed onto the image memory (Figure 17b) and the average density of each square can be evaluated. Since nuclei have higher density than the cytoplasm and the background, a proper choice threshold for the average density in each square will yield an approximate count of number of nuclei in the area of interest. Figure 17c and 17d are the zoom images of the corresponding area.

Color Parameters

Color parameters are extremely important in characterizing objects of interest in a microscopic slide. The method of separating the three primary colors, red, blue, and green, from a microscopic image to three image memories has been previously described. Various parametric analyses using a set of two images (red and blue, blue and green, and red and green) can be performed to isolate the color of interest, the detail of which is beyond the scope of this chapter and will not be treated here

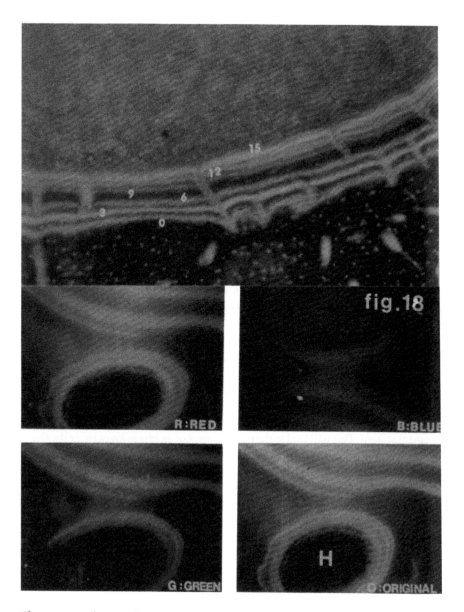

Figure 18. A bone cell in a fluorochromic image with tetracycline label shown in orange red. The objective is to quantify the tetracycline label. **a.** A fluorochromic image form the tibia of a rat bone biopsy with five different tetracycline labels. Each label has its own color characteristics: 0-day, oxytetracycline label; 3-day, DCAF; 6-day, xylenol orange, 90mg/kg; 9-day, hematoporphin, 300mg/kg (did not stain); 12-day, doxycycline; 15-day, alizarin red 5. The dose is 20 mg/kg except otherwise specified. **b.** A partial osteonal unit depicting the osteoid and the Haversian canal (H). One tetracycline label is shown in orange red, the inside ring immediately adjacent to H. The three color images (red, blue, and green) are also shown (the blue image was accidentally flipped during the photographic process).

(Huang et al., 1983). However, as an example, consider a bone cell in a fluorochromic image with tetracycline label shown in Figure 18. The objective of the study is to identify the tetracycline label (orange red) from the cell and quantify its intensity. If black-and-white imaging is used to perform the analysis, it would be extremely difficult because of the lack of its color components. However, if color parameters are used with parametric analysis, it is possible to identify the tetracycline label fairly easily. Figure 18a shows a fluorochromic image from the tibia of a rat bone biopsy with five different tetracycline labels; each label has its own color characteristic. Figure 18b depicts a fluorochromic image of a partial osteonal unit showing the osteoid and the Haversian canal. The image is decomposed to red, green, and blue images.

SUMMARY

We have discussed the fundamentals of medical imaging and have covered several matters including some definitions, image detectors and recorders, sources of medical images, and image processing systems.

The use of medical imaging in the radiological sciences is expected to increase about 40% in the next five years (Huang, 1987). New methods producing medical images will not progress drastically, but image quality from existing imaging modality will continue to improve. Traditionally, medical imaging is used only for diagnostic purposes, but we see the trend that they are also being used for therapeutic applications as well. PACS will become a vital image database management system. This will lead to the development of an image knowledge database that will require new IP hardware and software. The leading candidate in hardware design for medical image processing will be a modified parallel processing architecture that will shorten the time required for interprocessor communication. Mathematical advances may provide a new approach for image segmentation (Trambert, 1986). Fractal analysis shows promise for image feature extraction and object definition. Neutral networks may prove useful for medical pattern recognition, and other artificial intelligence techniques may bring medical imaging to the threshold of a mature science.

With regard to digital microscopic image processing, it requires fundamental knowledge of microscopy, television scanning, and digital image processing. In microscopy, specimen preparation is the most important step, since a good preparation can account for almost 50% of the success in designing a digital microscopic imaging system. The motorized x-y stage and automatic focusing are two important components for screening and quantitative analysis of microscopic images. The television camera used for scanning should have the proper specifications so that its resolution is adequate for imaging the objects of interest. In digital imaging technology, an image memory (or memories if color processing is necessary) and a graphic memory are essential in addition to the computer memory.

lens is used during the search and a high power objective with oil immersion is used for the feature extraction. The starting point in searching is using a two grey level and low resolution image. And the development of a search algorithm depends on the characteristics of the objects of interest. Three types of parameters should be considered in feature extraction: geometric, density, and color. All these parameters are essential in the success of performing quantitative microscopic imaging.

REFERENCES

Burrells, W. (1971). In: Microscope Technique. John Wiley & Sons, New York.

Castleman, K.R. (1979). In: Digital Image Processing. Prentice-Hall, New Jersey.

Hawkins, R.A., Hoh C., Glaspy, J., Choi, T., Dahlbom, M., Rege, S., Messa, C., Nietszche, E., Hoffman, E., Seeger, L., Moddahi, J. & Phelps, M.E. (1992). The role of positron emission tomography in oncology and other whole body applications. Sem. Nucl. Med., XXII (No. 4), 268–284.

Huang, H.K. (1981a). Biomedical image processing. CRC Critical Revs. Bioeng. 4, 185–271.

Huang, H.K. (1982). Introduction to digital microscopy. In: Application of Computers in Medicine (Schwarz, M. D., ed.), IEEE Engineering in Med. and Biol. TH0095–0; 177–186.

Huang, H.K. (1987). In: Elements of Digital Radiology. Prentice Hall, Englewood Cliffs, N.J.

Huang, H.K. (1993a). In: Medical Imaging Encyclopedia of Computer Sciences, 3rd ed. (Ralston, A. & Reilly, E.D. eds.), pp. 842–847, Van Norstrand, New York.

Huang, H.K., Aberle, D.R., Lufkin, R., Grant, E.G., Hanafee, W.N., & Kangarloo, H. (1990). Advances in medical imaging. Ann. Int. Med. 112, 203–220.

Huang, H.K., Mankovich, N.J., & Altmaier, J. (1981b). A macroscopic and microscopic image processing system. Proc. Fifth Ann. Symp. Comp. Applic. Med. Care, pp. 580–585, Washington, D.C.

Huang, H.K., Mankovich, N.J., Taira, R.K., Cho, P.S., Stewart, B.K. Ho, B.K.T., Chan, K.K., & Ishimitsa, Y. (1988). Picture archiving and communication systems for radiological images: State-of-the-art. CRC Critical Revs. Diagn. Imag. 28, 383–427.

Huang, H.K., Meyers, G.L., Martin, R.K., & Albright, J.P. (1983). Digital processing of microradiographic and fluorochromic images from bone biopsy. Comp. Biol. Med. 13, 27–47.

Huang, H.K., Taira, R.K., Lou, S.L., Wong, A.W.K., et al. (1993b). Implementation of a large-scale picture archiving and communication system. Comp. Med. Imag. & Graph. 17, 1–11.

Huang, H.K., Wong, W.K., Lou, S.L., & Stewart, B.K. (1992). Architecture of a comprehensive radiologic imaging network. IEEE Trans. SA Commun. 10, 1188–1196.

Kangarloo, H. Boechat, M.I., Barboric, Z.,Taira, R.K., Cho, P.S., Mankovich, N.J., Ho, B.K.T., Eldredge, S.L., & Huang, H.K. (1988). Two-year clinical experience with a computed radiography system. Am. J. Radiol. 151, 605–608.

Onoe, M., Preston, K., & Roxenfeld, A., (eds.) (1980). In: Real-Time Medical Image Processing, Plenum Press, New York.

Pressman, N.J. & Wied, G.I. (1979). The automation of cancer cytology and cell image analysis. Ed., Proc. Second Intl. Conf., Chicago, IL.

Rochow, T.G. & Rochow, E.G. (1978). In: An Introduction to Microscopy by Means of Light, Electrons, X-rays, or Ultrasound. Plenum Press, New York.

Udupa, J.K., Goncalves, R.J., Iyer, K., Narendula, S., et al. (1993). Imaging transforms for 3D biomedical imaging: An open, transportable system (3DVIEWNIX) approach. In: Computer Assisted Radiology CAR '93 (Lemke, H.U., Inamura, K., Jaffe, C.C., & Felix, R., Eds.), pp. 370–383, Springer-Verlag, Berlin.

Trambert, M.A. (1986). Cellular logic implementation on a generalized image processor applied to biomedical image processing. Anal. Quant. Cytol. Histol. 8, 131–137.

Valentino, D.J., Mazziota, J.C., & Huang, H.K. (1991). Volume rendering of multimodal images: Applications to MRI and PET imaging of the human brain. IEEE Trans. Med. Imaging 10, 554–562.

Chapter 9

Nuclear Magnetic Resonance Studies of Cell Metabolism *In Vivo*

RAINER CALLIES and KEVIN M. BRINDLE

INTRODUCTION

The phenomenon of nuclear magnetic resonance (NMR) was discovered independently by Purcell et al. (1946) and Bloch et al. (1946) and subsequently led to Purcell and Bloch being awarded the Nobel Prize for Physics in 1952. Early studies *in vivo* used proton NMR. However, due to an intrinsic lack of sensitivity, it was

Principles of Medical Biology, Volume 4
Cell Chemistry and Physiology: Part III, pages 241–269.
Copyright © 1996 by JAI Press Inc.
All rights of reproduction in any form reserved.
ISBN: 1-55938-807-2

not until the early 1970s and the introduction of higher field magnets and more sensitive instrumentation and techniques that NMR spectroscopy on living samples became more commonplace. Among the first of the studies on living tissues was a ^{13}C-NMR study of yeast cells (Eakin et al., 1972). This was quickly followed by ^{31}P (Moon and Richards, 1973) and ^{15}N (Llinás et al., 1975) NMR studies on a variety of isolated cell preparations. The technique is now a well-established tool for the noninvasive measurement of metabolite levels and fluxes in systems ranging from isolated cells (Szwergold, 1992), through perfused organs (Styles et al., 1987), to the intact animal (Koretsky and Williams, 1992), including man (Radda, 1986).

This chapter will deal predominantly with studies on mammalian cells, although some examples of studies on plant cells, fungi, and bacteria will also be included. In the context of this chapter, the term *in vivo* refers to the intact cell. However, we have also included a discussion of NMR measurements on extracts of cells and cell growth media, particularly NMR measurements of isotopic labeling, as these illustrate the power of the NMR technique for the investigation of metabolic processes.

Given the facility of NMR to make noninvasive measurements on an intact animal, why bother with measurements on isolated cells? There are a number of possible reasons for this approach. An isolated cell preparation can be studied under relatively well defined and controlled conditions. Thus, detailed studies can be made on the effect of individual factors on cellular metabolism, free of the influence of humoral factors and changes in blood supply which complicate measurements in the intact animal. However, care must be exercised in this approach, as it has been shown that the media on which the cells are grown can have a profound influence on the metabolism of the cells and on the levels of metabolites observed in NMR experiments (Macdonald et al., 1993). Studies on clonal cell lines offer the opportunity to study a relatively homogeneous population of cells, whereas in the animal, NMR studies of a single cell type require localization techniques so that the signal is acquired from a specific tissue, e.g., liver, muscle, etc. Using modern localization techniques with the nucleus most receptive to NMR detection, the proton, the smallest volumes from which a signal can be obtained, in a reasonable time, are of the order of 5 μl or more in animals (Bourgeois et al., 1991) and more than 1 ml in humans (Frahm et al., 1990). The greater sensivity in animals reflects the fact that, in general, they are smaller and thus can be examined in relatively small bore, higher field magnets. Thus, studies on man and animals are often hampered by signal contamination from regions outside the volume of interest. Even if localization were specific for a particular tissue, this would still not guarantee study of a homogeneous cell population. Tissues are invariably composed of more than one cell type or alternatively there may be regional heterogeneity in the metabolism of the tissue, e.g., the muscle wall of the heart (Robitaille et al., 1989) or the periportal and perivenous regions of the liver (Häussinger, 1990). Probably the best cell line to use is a primary cell line derived from the tissue of interest, e.g., hepatocytes from liver or astrocytes from brain (Shulman et al., 1979; Brand et al., 1992; Urenjak et al., 1993). However, this can create problems with supply of material as an NMR experiment on the cells themselves can require gram quantities of material. The use of established, tumor-derived cell lines relieves this problem to a certain extent, although the metabolism of the cells may be significantly different from their tissue of origin. Cell immortalization is an alternative, but again the metabolism of the cells may be altered (McLean, 1993).

In conclusion, studies on isolated cells allow measurements on homogeneous cell populations under well-defined conditions and can aid in the interpretation of metabolic changes seen by NMR in the same or similar cells in the intact animal. In some cases the cells may be a valid system for study in their own right. For example, there have been several NMR studies of commercially important mammalian cell lines which are used in the biotechnology industry for the production of various monoclonal antibodies and recombinant proteins with therapeutic or analytical applications (Mancuso et al., 1990; Gillies et al., 1991).

METHODOLOGY

Physical Principles of NMR

An in-depth description of the physical basis of NMR is beyond the scope of this chapter and we will give only a very brief outline. More detailed descriptions can be found in the books by Derome (1987), Sanders and Hunter (1987), and Gadian (1982), among others.

Nuclei with nonzero nuclear spin possess a magnetic moment. In the presence of an applied magnetic field these nuclei assume a number of different energy levels determined by their spin quantum number. For nuclei with a spin quantum number of 1/2, which includes those nuclei most frequently used in NMR experiments, the spins have only two allowed energy levels in which the nuclear magnetic moments are aligned with (for nuclei with a positive gyromagnetic ratio this is the lower energy level) or against (higher energy) the applied field. The energy difference (ΔE) between these two levels is given by:

$$\Delta E = h \, \nu$$

where h is Planck's constant and ν is the Larmor frequency (in hertz). The Larmor frequency is dependent on the gyromagnetic ratio of the nucleus (γ) and the strength of the applied magnetic field (B_0). The gyromagnetic ratio is a proportionality constant relating frequency to the applied field, i.e.,

$$\nu = \frac{|\gamma| B_0}{2\pi}$$

At thermal equilibrium the energy states are populated according to the Boltzmann distribution. As the energy difference is very small, the energy levels are nearly equally populated, with a slight excess of spins in the lower energy level. Application of an oscillating magnetic field, which has the same frequency as, or is resonant with, the Larmor frequency, induces transitions between these energy levels. A net transfer of spins from the lower to the higher energy level results in absorption of energy and it is this which is detected by the NMR spectrometer. As the population difference is very small, the strength of the NMR signal is weak, and hence the technique is very insensitive. The information obtained in the NMR experiment is displayed in the form of a spectrum, which has axes of intensity versus frequency (see Figure 1).

The resonance frequencies in NMR are in the radiofrequency range. While the nuclei of different elements resonate at frequencies differing by several tens of

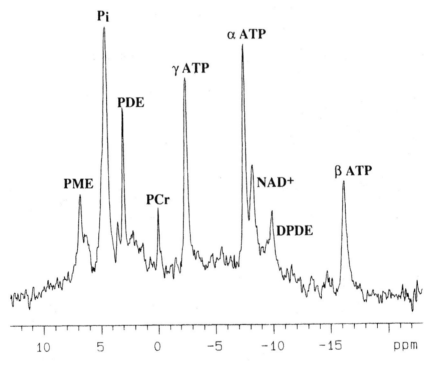

Figure 1. ^{31}P NMR spectrum of Chinese hamster ovary cells (CHO K1) growing in a hollow-fiber bioreactor (Callies, Jackson, and Brindle, unpublished observations). The assignments are: PME, phosphomonoesters; P$_i$, inorganic phosphate (predominantly extracellular); PDE, phosphodiesters (predominantly glycerophosphocholine); PCr, phosphocreatine; ATP, adenosine-triphosphate (γ-, α-, and β-phosphates); NAD, nicotinamide adenine dinucleotide; DPDE, diphosphodiesters.

megahertz, according to their gyromagnetic ratios, the utility of the technique in chemistry and biology resides in the fact that each nucleus of a particular element resonates at a slightly different frequency, depending on its chemical environment. The magnetic field experienced by each nucleus is modified slightly by the surrounding electrons. Currents of electrons induced by the external field create a small magnetic field which opposes the applied field. The extent of this shielding depends on the electron density and thus provides a characteristic fingerprint of the chemical environment of a nucleus. Differences in frequency due to this effect are of the order of a few kilohertz, and can be detected in a single experiment (see Figure 1). Since resonance frequency depends linearly on the strength of the applied magnetic field and thus varies with the spectrometer used, it is usual to express the frequency of each peak or resonance in parts per million (ppm) of the operating frequency of the spectrometer, yielding a value which is independent of the magnetic field strength, B_0. It is called the chemical shift (δ),

$$\delta = \frac{(\nu - \nu_{ref}) \times 10^6}{\nu_{ref}}$$

where ν_{ref} is the operating frequency of the spectrometer and ν is the frequency of a resonance. The intensity of a peak in the spectrum is proportional to the number of the nuclei giving rise to the peak. However, the relative intensities are also influenced by the relaxation times of the resonances. Following excitation of the NMR signal, the spins relax back to equilibrium. If a resonance is excited again before equilibrium has been reestablished, then the signal can become saturated and its intensity in the spectrum reduced. The time constants of the relaxation processes are dependent, among other things, on the motion of the nuclei.

Signals from different elements cannot normally be acquired in a single NMR experiment since they require changes in spectrometer hardware. However, by using an NMR probe which is tuned to several different resonance frequencies, it is possible with modern spectrometers, to acquire signals concurrently from more than one nuclear isotope, e.g., ^{31}P, ^1H, and ^{13}C (Styles et al., 1979), and thus markedly increase the information content of the experiment. It is also possible to detect one or more nuclei indirectly via another nucleus (see below).

Observable Nuclei

The NMR properties of the most commonly used nuclei in *in vivo* NMR are listed in Table 1.

^1H

The hydrogen nucleus is the most sensitive to NMR detection in absolute terms, although less sensitive, in relative terms, than tritium. The relative sensitivity

Table 1.

Isotope	Spin	Natural Abundance (%)	Relative Sensitivity	Absolute Sensitivity	Frequency at 9.4 T (MHz)
^{1}H	1/2	99.98	1.00	1.00	400.0
^{31}P	1/2	100	6.6×10^{-2}	6.6×10^{-2}	161.9
^{13}C	1/2	1.11	1.6×10^{-2}	1.8×10^{-4}	100.6
^{15}N	1/2	0.37	10^{-3}	3.8×10^{-6}	40.5
^{23}Na	3/2	100	9.3×10^{-2}	9.3×10^{-2}	105.8
^{19}F	1/2	100	0.834	0.834	376.3

Note: NMR properties of the most commonly used nuclei in *in vivo* NMR. The relative sensitivity derives from the physical properties of the nucleus, whereas the absolute sensitivity is the product of the relative sensitivity and the natural abundance of the isotope.

derives from the physical properties of the nucleus, whereas the absolute sensitivity is the product of the relative sensitivity and the natural abundance of the isotope. The ubiquity of the proton in molecules of biological interest, such as lipids, carbohydrates, proteins, and various metabolites, makes ^{1}H NMR a fundamental tool for the biological spectroscopist. However, the narrow chemical shift range (~10 ppm), which compresses these signals into a narrow frequency range, has meant that techniques have had to be devised which either improve signal resolution or suppress unwanted signals. These techniques are discussed further below. A particular problem in living systems arises from the huge concentration of water, the ^{1}H signal from which can swamp the signals from metabolites present in only millimolar concentrations. This problem is exacerbated, particularly at low magnet field strengths, by the very small frequency dispersion of the spectra. In early studies on cells this problem was overcome by exchanging ^{1}H$_2$O for ^{2}H$_2$O (Brown et al., 1977). However, there are now a variety of sophisticated and very effective NMR techniques for the suppression or elimination of the water signal (Freeman, 1990). The effectiveness of one such sequence is illustrated by the ^{1}H NMR spectra of cells shown in Figure 2.

^{31}P

The only naturally abundant isotope of phosphorus, ^{31}P, is NMR detectable with good sensitivity. ^{31}P NMR has been widely used to investigate cellular energy metabolism since it can be used to monitor, noninvasively, the cellular concentrations of ATP and P_i. Determination of the free intracellular P_i concentration is particularly useful. Measurements of its level using conventional assays on cell extracts can yield overestimates due to artefactual hydrolysis of organic phosphates during the extraction procedure. The chemical shift of the P_i resonance is sensitive to pH changes in the physiological range and therefore it has often been used as a pH indicator *in vivo* (Szwergold, 1992). The concentration of ADP is not usually

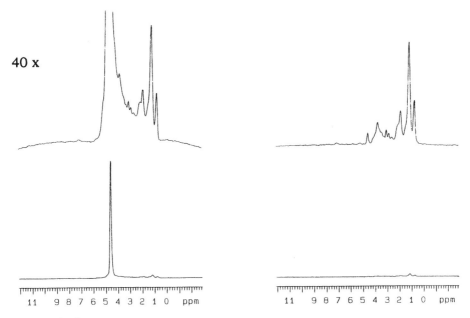

Figure 2. ^{1}H NMR spectra of mammalian cells growing in a hollow-fiber bioreactor without (left) and with (right) suppression of the water signal (4.7 ppm) using the CHESS pulse sequence (Haase et al., 1985). The spectra were acquired using the STEAM sequence with two, three-pulse CHESS cycles (Callies, Jackson, and Brindle, unpublished observations) (Moonen and Van Zijl, 1990).

directly measurable from ^{31}P NMR spectra of tissues. This is because either the concentration is too low to detect by NMR or because the bulk of the nucleotide is bound to cellular macromolecules, e.g., actin in muscle (Balaban, 1984, and references cited therein); or it is sequestered in some way, in mitochondria for example (Balaban, 1984, and references cited therein). The lack of ADP visibility when bound to a macromolecule could arise because of a decrease in the molecule's mobility leading to a decrease in the spin-spin relaxation time for the phosphorus nuclei. This would result in an increase in their resonance linewidths such that they could be broadened beyond detection. However, in this context, it is worth noting that resonances of ADP can be detected when the nucleotide is bound to an enzyme (Rao et al., 1978). Why ADP should not be visible when sequestered in mitochondria is not clear, given that resonances from ADP, ATP, and P$_i$ can be detected in isolated mitochondria (Hutson et al., 1989) and there is evidence that intramitochondrial P$_i$ can be observed in ^{31}P NMR spectra of perfused hearts (Garlick et al., 1992). In those tissues possessing creatine kinase, such as heart, muscle, brain, and some tumor cells, the free cytosolic ADP concentration can be calculated from ^{31}P NMR measurements of the near-equilibrium concentrations of the enzyme's substrates (Matthews et al., 1982). ^{31}P NMR is used to measure the concentrations of

PCr and ATP. The H^+ concentration is determined from the chemical shift of the P_i resonance and the free Mg^{2+} concentration from the chemical shifts of the resonances of ATP, which binds Mg^{2+} (Gupta and Gupta, 1984). The total cellular creatine plus phosphocreatine concentration can be determined using conventional assay techniques on tissue extracts or by ^1H NMR. These measurements allow calculation of the free ADP concentration according to the following equation:

$$[\Sigma ADP] = \frac{[\Sigma ATP][\Sigma creatine]}{K_{eq}[\Sigma phosphocreatine][H^+]}$$

where K_{eq} is the equilibrium constant for the reaction and the concentrations are the total measurable concentrations of the substrates. The equilibrium constant is dependent on the free Mg^{2+} concentration. However, the effect of the Mg^{2+} concentration on this constant is easily calculated from the known binding constants of the nucleotides for this cation and the H^+ ion (Lawson and Veech, 1979).

^{31}P NMR measurements of the free cytosolic ADP concentration have proved to be very powerful in monitoring cellular energy status and in investigating the control of mitochondrial oxidative phosphorylation in $vivo$ (Chance et al., 1982; Balaban et al., 1986; Brindle et al., 1989). The range of tissues on which this measurement can be made has been extended by using molecular genetic techniques to introduce creatine kinase into yeast cells (Brindle et al., 1990) and the livers of transgenic mice (Koretsky et al., 1990), two systems which do not normally contain the enzyme.

The ^{31}P NMR spectrum also gives information on the levels of various metabolites involved in phospholipid metabolism, such as the phosphomonoesters, phosphocholine, and phosphoethanolamine, and the phosphodiesters such as glycerophosphocholine and glycerophosphoethanolamine. There are also resonances from diphosphodiesters such as UDP-sugars. These include molecules involved in glycogen synthesis and protein glycosylation. A number of unusual phosphorylated compounds have also been detected in tumor cells (Ruiz-Cabello and Cohen, 1992). A ^{31}P NMR spectrum of mammalian cells growing in a hollow-fiber bioreactor is shown in Figure 1.

^{13}C and ^{15}N

Carbon and nitrogen are detectable by NMR as the isotopes ^{13}C and ^{15}N. These have a low natural abundance (1.1% and 0.37%, respectively) and thus are suitable for labeling studies. The naturally abundant carbon isotope, ^{12}C, is not detectable as it has zero spin. The naturally abundant nitrogen isotope, ^{14}N, has spin but is not detectable in a ^{15}N experiment as it has a different gyromagnetic ratio. Furthermore, ^{14}N is not as useful as ^{15}N for NMR spectroscopy since, with a few exceptions, it gives much broader resonances. Both ^{13}C and ^{15}N show a broad chemical shift range which aids signal resolution and identification.

As with radioisotopes, ^{13}C and ^{15}N can be used as labels for following metabolic fluxes *in vivo*. Although less sensitive than radioisotope techniques, NMR detection of these stable isotopes has a number of important advantages. The label can be detected noninvasively *in vivo* in anything from an isolated cell preparation (Shulman et al., 1979) to the human brain (Rothman et al., 1992). In some cases it may be advantageous to monitor the label distribution in deproteinized extracts prepared from the tissue as these samples are magnetically more homogeneous and give rise to better resolved spectra. However, even in this situation, NMR detection of the label has some significant advantages. The NMR spectrum not only shows which molecule is labeled but also which position in the molecule is labeled. Contrast this with radiotracer techniques where the identification of the labeled molecule(s) first requires chromatographic separation of the cellular metabolites followed by radioactive counting. Identification of the labeled position(s) in the molecule requires degradation of the molecule, followed by further chromatography and counting. Mass spectrometry can be used for stable isotope detection and distribution within a molecule, but even here some chromatographic pretreatment of the sample may be required. Another important feature of the NMR experiment, which has been very important in ^{13}C NMR, is the facility to detect adjacent labels, i.e., the signal from one ^{13}C nucleus in the molecule can indicate whether an adjacent carbon in the molecule is labeled or not. This labeling of an adjacent carbon or carbons is observed as splitting of the peak into a multiplet structure due to spin-spin coupling. Analysis of these splitting patterns can be used to identify the different isotope isomers of a molecule (so-called "isotopomer" analysis). The application of isotopomer analysis in ^{13}C-labeling studies has been enormously informative in terms of quantitating the relative fluxes through various metabolic pathways *in vivo* (Walker et al., 1982; Jeffrey et al., 1991).

NMR detection of ^{15}N-labeling has an added advantage when compared to radiotracer methods in that the radioisotope of nitrogen, ^{13}N, has a very short half-life (<10 minutes) which has limited its use. Although ^{15}N NMR is particularly insensitive and requires long signal acquisition times in order to obtain a spectrum, it has been shown to be useful in studies on bacteria, fungi, and plants for studying nitrogen uptake. The sensitivity of ^{15}N label detection, however, can be improved by detecting the nucleus indirectly via ^{1}H NMR (see below).

[23]Na

The only naturally abundant isotope of sodium, ^{23}Na, is detectable by NMR with a sensitivity which is slightly greater than that of [31]P. The physiological importance of this ion has ensured that there have been a number of *in vivo* NMR studies (Boulanger and Vinay, 1989). The principal problem with ^{23}Na NMR measurements *in vivo* has been to resolve the signal from intracellular sodium from the signal from extracellular sodium, which is present at about an 80-fold higher concentration. There are two basic techniques that have been used to separate the signals. One involves the addition of agents which shift the frequency of the extracellular signal. However, these can be toxic to the tissue and difficult to introduce. Another, more sophisticated technique, involves the use of multiple quantum coherence methods (Jelicks and Gupta, 1989). These exploit differences in the relaxation properties of the intra- and extracellular sodium resonances and therefore have no effect on the tissue. However, it has been shown that the techniques might not be completely effective at suppressing the extracellular signal (Jelicks and Gupta, 1989).

^{19}F

The ^{19}F nucleus, which is 100% naturally abundant, is almost as receptive to NMR detection as the proton. In general, there is very little fluorine in biological systems and therefore the nucleus has been used for labeling studies. The metabolism of fluorine-containing chemotherapeutic agents has been followed, for example, using ^{19}F NMR (Malet-Martino and Martino, 1992). ^{19}F-labeled calcium chelating agents have been used as probes of the intracellular free Ca^{2+} concentration in isolated cells and perfused hearts in much the same way that similar fluorescence-labeled probes have been used (Gupta and Gupta, 1984; Tsien and Poenie, 1986; Kirschenlohr et al., 1988). There are also ^{19}F NMR probes of pH and oxygen tension (Taylor and Deutsch, 1988). Recently specific proteins have been fluorine-labeled *in vivo* by inducing enzyme synthesis in the presence of a fluorine-labeled amino acid. These proteins have then been detected in the intact cell using ^{19}F NMR (Brindle et al., 1989). The resonances from these proteins give information on their ligand binding properties and mobility (Williams et al., 1993).

There are a number of other nuclei which can be detected by NMR *in vivo*. However, these have been used to a much lesser extent (see Szwergold, 1992, and references cited therein).

Techniques

The development of NMR spectroscopy has been propelled not only by improved instrumentation but also by the introduction of new and powerful techniques. A detailed description of these techniques is beyond the scope of this chapter. However, we have selected a few techniques which have been, or are likely to be, the most important in the area of NMR studies of cellular metabolism. We exclude NMR imaging and localized spectroscopy techniques which are not directly relevant to studies on isolated cells. The reader is referred to chapter 14 by Huang in this book and detailed discussions elsewhere (Mansfield and Morris, 1982; Leach, 1988; Koretsky and Williams, 1992).

Two-Dimensional NMR

The introduction of two-dimensional (2D) NMR techniques (Jeener, 1971) heralded a revolution in NMR spectroscopy, most notably in the determination of protein structure (Wüthrich, 1986). In a 2D NMR experiment, frequency information is spread over two axes as opposed to one in the conventional NMR experiments discussed so far. The spread of data over two axes (even three or four using more recently developed techniques, e.g., Clore and Gronenborn, 1991) can considerably enhance signal resolution and provide additional information which aids in resonance assignment. The spectra are often presented in the form of a contour plot (see Figure 3), where the peaks on the plot in some way correlate the signals on the two frequency axes. The nuclei giving rise to these signals may be connected to each other in a number of ways; for example, they may be closely linked to each other via covalent bonds and thus show spin-spin coupling. In the example shown in Figure 3, the conventional one-dimensional spectrum is represented by the peaks on the diagonal. The off-diagonal peaks, or cross peaks, correlate resonances in the one-dimensional spectrum which are spin-spin coupled to each other. The plot is symmetrical about the diagonal. A conventional one-dimensional spectrum is shown at the top of the plot. Although 2D techniques can be applied *in vivo* (May et al., 1986), the usually long data acquisition times can limit their usefulness. They are most useful when applied to body fluids, cell extracts, etc., where they can be used to resolve and assign resonances in complex and crowded spectra, usually ^{1}H spectra (Nicholson and Wilson 1989; Sze and Jardetzky, 1990a,b; Kriat et al., 1992).

Indirect Detection of Isotope Labels

Although ^{13}C and ^{15}N NMR are powerful techniques for monitoring isotope label redistribution and metabolic fluxes *in vivo* (see above), the utility of these methods can be limited, in some cases, by their insensitivity. The sensitivity of label detection, however, can be radically improved by detecting the ^{15}N or ^{13}C label indirectly via the more sensitive ^{1}H nucleus. In principle, in the case of ^{15}N, indirect detection via ^{1}H NMR can enhance sensitivity by nearly 1,000-fold (Live et al., 1984), although gains of 100–300-fold are more likely (Summers et al., 1986). The details of these techniques need not concern us here but they rely on the fact that in these multi-pulse, multi-nuclear experiments, where both the ^{1}H and ^{13}C or ^{15}N magnetizations are excited, the evolution of the ^{1}H magnetization is affected depending on whether it is spin-coupled to an ^{15}N or ^{13}C nucleus. The magnetization of a spin-coupled ^{1}H resonance effectively reports on whether an adjacent carbon or nitrogen in a molecule is labeled. A 2D version of an indirect detection experiment is shown in Figure 4. Only ^{1}H signal was acquired in this experiment. The ^{15}N spectrum was reconstructed from these ^{1}H data. As well as enhancing

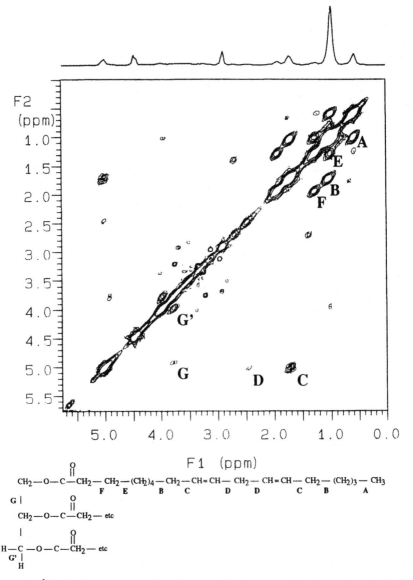

Figure 3. ¹H NMR spectrum of myeloma cells that contained cytoplasmic lipid droplets. The protons in the triglycerides which give rise to the cross peaks in the spectrum are indicated. The conventional one-dimensional spectrum is shown at the top of the figure. From Callies et al., 1993, with permission.

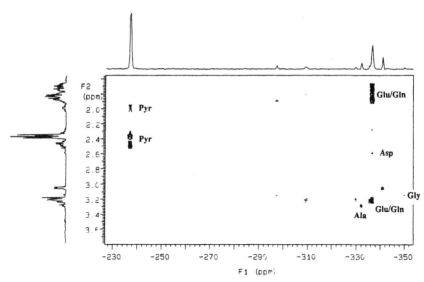

Figure 4. Two-dimensional heteronuclear multiple quantum coherence spectrum of a medium sample taken from a culture of HeLa cells which had been grown for 48 hours in the presence of 2 mM L[2-^{15}N]glutamine. In this type of experiment the ^{15}N label (F1 axis) is detected indirectly via spin-coupled protons (F2 axis). The peaks labeled in the contour plot arise from alanine (Ala), glutamate (Glu), glutamine (Gln), glycine (Gly), aspartate (Asp), and pyrollidone carboxylic acid (Pyr). From Street et al., 1993, with permission.

sensitivity these indirect detection methods also give fractional labeling or, in radioisotope terms, specific activity, since both the labeled and unlabeled species are detected in the ^{1}H spectrum (Brindle and Campbell, 1987). These techniques have been used, for example, to monitor ^{13}C-labeling of glutamate in the human brain (Rothman et al., 1992) and ^{15}N labeling of cellular metabolites in cultures of mammalian cells (Street et al., 1993) (see Figure 4). A very early application was in ^{1}H/^{2}H labeling studies, which were designed to examine the activities of individual enzymes in the human erythrocyte (reviewed in Brindle and Campbell, 1987). The ^{2}H label was detected indirectly, either by substitution of an observed ^{1}H for ^{2}H or by the effect of ^{2}H substitution on the magnetization of an adjacent spin-coupled ^{1}H.

Magnetization Transfer Measurements of Chemical Exchange

NMR has the capability, using magnetization transfer techniques, to measure very rapid metabolic fluxes *in vivo*, of the order of 10^{-3} M s^{-1}. A magnetization transfer experiment, in which exchange between ATP and P$_i$ was measured in yeast

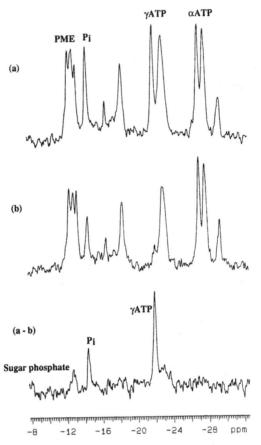

Figure 5. ³¹P NMR magnetization transfer measurements of ATP turnover in immobilized yeast cells. A control spectrum is shown in (**a**). Saturation of the γ-phosphate resonance of ATP and transfer of this magnetic label through chemical exchange results in a decrease in the intensity of the P_i and sugar phosphate resonances (**b**). This is most clearly seen in the difference spectrum (**a–b**). The magnitude of the decrease in the P_i resonance can be used to calculate the flux between P_i and ATP and, hence, the rate of ATP turnover (see Brindle, 1988a,b).

cells is illustrated in Figure 5. The experiment is analogous in some respects to an isotope exchange experiment except that in this case the label is introduced noninvasively in the form of a short-lived nuclear polarization. In the experiment shown in Figure 5, the γ-phosphate resonance of ATP was labeled. This is manifest as a loss of the γ-phosphate resonance from the spectrum (compare Figures 5a and 5b). Transfer of this label to the P_i resonance via chemical exchange causes a decrease in the P_i signal intensity (this is clearly shown in the difference spectrum,

Figure 5a–b). The magnitude of this decrease can be used to quantitate the exchange flux. The technique has been widely used in ^{31}P NMR to measure ATP turnover and flux in the creatine kinase reaction (Brindle, 1988a) in systems ranging from isolated mammalian cells (Neeman et al., 1987) to the human brain (Cadoux-Hudson et al., 1989). There have also been measurements of membrane transport using the technique, most notably of a fluorinated glucose analogue in erythrocytes (Potts et al., 1990). Although the number of fluxes measurable by this technique are very few, the flux between P_i and ATP is arguably one of the most important in the cell. This is a measurement which is unique to NMR and which is inaccessible to other techniques.

Spin Echoes and Diffusion

Conventional ^{1}H NMR spectra of tissues usually contain resonances from cellular metabolites which are largely overlapped by broad and intense signals from cell proteins and lipids. However, these crowded and ill-resolved spectra can be edited using spin-echo techniques. These remove the broad protein resonances and leave signals from small molecules and very mobile regions of proteins and lipids (Brown et al., 1977; Brindle and Campbell, 1984, 1987). The broad signals from proteins are lost due to their relatively short spin-spin relaxation times. Spin-echo techniques constitute the basis of most ^{1}H NMR experiments *in vivo* as they permit ready observation of metabolite resonances. The intensities of the resonances in these spectra are also modulated by the diffusion of the molecules, as well as by their relaxation times. In the presence of a magnetic field gradient, diffusion of a molecule leads to loss of its resonance intensities. Gradients may be endogenous (Brindle et al., 1979) or imposed externally using gradient coils (van Zijl et al., 1991). In these diffusion-weighted experiments, molecules outside cells, which have larger diffusion coefficients or longer diffusion pathlengths, give signals of reduced intensity compared to molecules inside cells, which have much lower diffusion coefficients and/or undergo restricted diffusion. These techniques can be used to edit out intracellular signals and to monitor transport and uptake of metabolites into cells (Brindle et al., 1979; van Zijl et al., 1991).

Cell Perfusion Systems

The inherently low sensitivity of NMR has necessitated the development of methods for maintaining cells at relatively high cell densities under near-physiological conditions within the NMR spectrometer (sample diameters are usually between 10–25 mm). This involves supplying the cells with nutrients, particularly oxygen, removing waste products, and preventing cell sedimentation. Although some studies have been done with suspensions of cells, in which the cells are stirred and oxygenated by bubbling oxygen through the suspension, these are usually short-term experiments and most studies now employ some form of cell entrapment and superfusion.

The choice of cell perfusion system will depend, to a certain extent, on whether the cells are capable of growing in suspension or if they need to adhere to a solid support (anchorage-dependent cells). At present there are three main types of cell perfusion system, for use with NMR. These are discussed briefly below; for more detailed discussions see Gillies et al. (1986) and Seguin et al. (1992).

Anchorage-dependent cells can be grown on solid or macroporous microcarrier beads and these can then be superfused with nutrient medium in the spectrometer. The advantages of the system include direct contact between the perfusate and the cells (see below) and growth of the cells at rates comparable to those found in tissue culture dishes. A disadvantage, with solid beads, is the relatively low cell density obtainable due to the large sample volume occupied by the beads.

Cell entrapment within a polymer matrix can be used with both anchorage-dependent and suspension cells. The earliest version of this method involved mixing the cells with molten low-gelling temperature agarose and then allowing this mixture to solidify in the form of fine threads (Foxall and Cohen, 1983). There are also variants of this technique in which the agarose is solidified in the form of beads or in which alginate or a basement membrane preparation is used as the immobilizing agent (Daly et al., 1988). The threads or beads containing the cells can then be perfused in the spectrometer. An advantage of the technique is that it is relatively easy to perform and it permits study at very high cell densities. However, it is not clear that the cells can divide in this system and therefore the properties of the cells may be altered in some way.

Another system which can be used with both anchorage-dependent and suspension cells is the hollow-fiber cartridge. This usually consists of a closed tube through which passes a bundle of semi-permeable hollow fibers. Nutrient medium pumped through these fibers exchanges nutrients for waste products, with cells growing in the extracapillary space. These may be anchorage-dependent cells, which grow on the fibers themselves or on a solid support, or suspension culture cells. The advantage of the system is that the cells can be grown to very high densities, which are comparable to those found in tissues (Minichiello et al., 1989). The disadvantage is that nutrient supply is via diffusion through the walls of the fibers. This can lead to insufficient supply and/or concentration gradients of nutrients within the extracapillary space (Drury et al., 1988). The homogeneity of the preparation can, therefore, be compromised.

A more rarely used system, which can serve as a good model for tumors *in vivo*, involves growing the cells as spheroids. These are clumps of cells, with diameters of the order of several hundred microns, which can be superfused in the spectrometer (Lin et al., 1987).

AREAS OF STUDY

Energy Metabolism

The facility of ^{31}P NMR to monitor the energy status of tumors *in vivo* and the potential importance of these measurements for predicting and monitoring the response of tumors to therapy has led to numerous studies on isolated tumor cells. These have been used as relatively well defined, homogeneous and controlled models of intact tumors (Evanochko et al., 1984). Most solid tumors, for example, contain some hypoxic cells, which can make them resistant to radiotherapy. A correlation between oxygen supply and energy status would, therefore, make ^{31}P NMR measurements *in vivo* a valuable tool for assessing the potential response of a tumor to radiotherapy (Dunn et al., 1989). However, studies on isolated tumor cells of the relationship between nutrient supply and energy status have sometimes failed to show a direct relationship (Freyer et al., 1991; Szwergold, 1992). A comparative study of a rat glioma cell line and its normal counterpart, rat cerebellar astrocytes in primary culture, showed that both cell types could maintain their ATP levels during a prolonged anoxic period (more than one hour). In two mice tumors, with substantially different hypoxic cell fractions, ^{31}P NMR measurements on the tumors *in vivo* showed no significant differences in energy status. However, studies on the cells derived from these tumors showed that *in vitro*, the cells were different in their response to oxygen deprivation, one being substantially more refractory to the energy changes induced by hypoxia than the other (Gerweck et al., 1992). The situation can be complicated, however, by the methods used to examine the cells *in vitro*. When human adenocarcinoma cells, which were retained in the NMR tube inside a dialysis sac, were perfused anaerobically there was a loss of NTP, an increase in P_i, a decrease in phosphocreatine, and a decrease in intracellular pH (Desmoulin et al., 1986). However, when the same cells were perfused on micro-carrier beads, 100 minutes of anoxia and even 40 minutes of total ischemia were without effect on the levels of ATP (Fantini et al., 1987).

There have been several studies on the importance of glycolysis as an energy source in cultured cells. In an early study, Ugurbil et al. (1981) used ^{31}P NMR to examine the dependence of energy production on glycolysis and oxidative phosphorylation in normal and X-ray transformed mouse embryo fibroblasts. The cells were grown and examined on microcarrier beads. Inhibition of respiration in these cells using cyanide had little effect on ATP levels. Inhibition of glycolysis, however, with 2-deoxyglucose, resulted in a decrease in ATP concentration to 40% of control levels. Sri-Pathmanathan et al. (1989) used ^{31}P NMR to examine the effects of glucose and glutamine as sole carbon sources on energy status in immobilized HeLa cells. Glutamine provides energy via oxidative phosphorylation whereas glucose is used initially in glycolysis. Both substrates maintained cellular energy status in terms of the levels of ADP and ATP, although there was a significantly higher P_i concentration in the cells superfused with glutamine. In other cells, glucose deprivation has been shown to result in a rapid loss of ATP (Desmoulin et al., 1986). However, Pianet et al. (1991) showed that in glioma cells, starvation for glucose caused no changes in the concentrations of ATP or P_i, as measured in cell extracts, but there were significant decreases in the ATP signals and increase in the P_i signal in the spectra of the cells. These changes in signal intensities *in vivo* were shown to be due to changes in the relaxation times of the ATP and P_i resonances and were proposed to reflect changes in the distribution of these metabolites within the cells. A notable feature of this study was the similar behavior of malignant glioma cells and normal astrocytes in primary culture under conditions of anoxia, ischemia, and glucose starvation.

Nutrient Metabolism

The utilization of substrates containing carbon and nitrogen can conveniently be followed by using ^{13}C and ^{15}N NMR to follow the fate of the ^{13}C- and ^{15}N-labeled substrates, respectively. ^{13}C NMR was used to monitor glycerol metabolism in rat hepatocytes by noninvasive measurements on cells (Cohen et al., 1979, 1981) and in rabbit renal cells by measurements on cell extracts and media supernatants (Jans and Willem, 1988). While in hepatocytes there was considerable gluconeogenic activity, this was much more restricted in renal cells where there was significant production of lactate and amino acids, mainly glutamate and glutamine. There have been a series of ^{13}C NMR studies on renal cells, both on primary cultures and on established cell lines, in which fluxes into the TCA cycle have been examined using labeled pyruvate (Jans and Leibfritz, 1989) and amino acids (Jans and Leibfritz, 1988; von Recklinghausen et al., 1991). A comparative study of fluxes into the TCA cycle in neuronal and glial cell lines and their normal counterparts was performed by Brand et al. (1992). Glial cells showed greater flux through pyruvate carboxylase than neural cells, reflecting their role in CO_2 fixation in the brain. In glial cells the TCA cycle was used largely for biosynthesis, whereas in the neuronal cells it appeared to be used preferentially for energy metabolism. Post et al. (1992) used ^{13}C NMR to investigate glucose metabolism in two human leukemic cell lines, one of which was sensitive and the other resistant to the toxic effects of the synthetic glucocorticoid, dexamethasone. Drug resistance did not correlate with changes in glucose utilization, which was in contrast to findings in drug-resistant human breast cancer cells (Cohen and Lyon, 1987). In the latter cells it was proposed that drug-resistance was conferred by the expression of an ATP-requiring membrane transporter, P-glycoprotein, which actively pumps the drug out of the cell. This leads to an increased ATP requirement and thus glycolytic flux. In the leukemic cell lines most of the glucose was consumed in glycolysis with only a minor flux through the pentose phosphate pathway (ca. 10%) and an even smaller flux into the TCA cycle.

Since the first *in vivo* ^{15}N NMR spectrum was recorded in 1975 (Llinás et al., 1975) there have been a number of studies of nitrogen metabolism using ^{15}N NMR, principally in microorganisms. Amino acid biosynthesis was investigated in *Neurospora crassa* using ^{15}N NMR measurements on suspensions of intact mycelia incubated with [15]NH$_4^+$ (Legerton et al., 1981; Kanamori et al., 1982). The biosynthetic pathways for alanine and glutamine were shown to compete for the common precursor, glutamate. In cells with increased glutamine synthesis, the increase correlated with the increase in glutamine synthetase activity measured in cell-free extracts (Kanamori et al., 1982). In photoautotrophic organisms, *in situ* ^{15}N NMR measurements of ^{15}N pulse-labeling have been used in conjunction with enzyme-specific inhibitors to investigate pathways involved in ammonia assimilation (Callies et al., 1992). Studies in higher organisms have been limited by the insensitivity of ^{15}N NMR, although there have been several studies in animals or perfused animal organs (e.g., Gründer et al., 1989; Farrow et al., 1990; Kanamori et al., 1993). There have been very few studies on cultured mammalian cells. Fernandez et al. (1988) used ^{15}N NMR measurements on concentrated cell supernatants to monitor the utilization of ^{15}N amine- or amide-labeled glutamine in cultures of hybridoma cells. Street et al. (1993) carried out similar experiments in a range of cultured mammalian cells which were incubated with amine- and amide-labeled glutamine and labeled ammonium ion. In this study the sensitivity of label detection was enhanced by using inverse detection via ^{1}H NMR. The patterns of metabolite labeling were correlated with the activities, measured in cell extracts, of the enzymes involved in label redistribution.

Lipid Metabolism

The observation of phospholipid metabolites in the ^{31}P NMR spectra of cells and cell extracts (see Figure 1) has led to several studies designed to investigate the control of phospholipid metabolism. These metabolites include phosphocholine (PC), phosphoethanolamine (PE), glycerophosphocholine (GPC), and glycerophosphoethanolamine (GPE). The phospholipids, themselves, are not observable *in vivo*, except at low field strengths, due to their relative lack of mobility and consequently broad signals (Murphy et al., 1989). They are observable, however, in nonaqueous cell extracts (Ronen et al., 1990). A primary motivation for these studies has been the observation of altered levels of these intermediates in tumor cells (see below).

Phosphatidylcholine is synthesized via the Kennedy pathway (Kennedy and Weiss, 1956) in three consecutive steps catalyzed by the enzymes choline kinase, phosphocholine cytidyltransferase, and phosphocholine transferase. Phosphatidylethanolamine is synthesized via the corresponding pathway for ethanolamine or by decarboxylation of phosphatidylserine (Borkenhagen et al., 1961) or by base exchange (Bjerve, 1973). In immobilized breast cancer cells, flux in these pathways was investigated by adding choline, ethanolamine, and a choline kinase inhibitor to the perfusate (Daly et al., 1987). This study indicated that the cytidyltransferase enzymes are rate-limiting as PC and PE both accumulated, but there was no accumulation of CDP-choline or CDP-ethanolamine. This finding has also been made in other cell types (reviewed by Ruiz-Cabello and Cohen, 1992).

Other nuclei can also be used in NMR investigations of phospholipid metabolism. In an early ^{1}H NMR study, differentiation of Friend leukemia cells was shown to be accompanied by a four-fold rise in PC concentration (Agris and Campbell, 1982). ^{13}C NMR has been used in conjunction with ^{13}C-labeling to investigate the metabolism of triacylglycerols in adipocytes (Sillerud et al., 1986) and phospholipid metabolism in human breast cancer cells (Ronen et al., 1992). In the latter study, ^{13}C NMR was used to measure turnover in metabolites such as PC, while ^{31}P NMR was used to measure their total concentration.

Characterization of Tumor Cells

NMR has been widely used in assessing the response of tumors *in vivo* to various therapeutic regimes. Studies on isolated tumor cells have formed an important part of this work as they offer the opportunity of studying a relatively well-defined and homogeneous cell population (reviewed in Evanochko et al., 1984) in which it can be easier to relate the observed changes in their NMR spectra to the underlying biochemistry. As well as differences in energy metabolism (see above), tumor cells also show changes in phospholipid metabolism which can be monitored by ^{31}P NMR. For example, a consistent observation has been the presence of a relatively large phosphomonoester peak in tumor cells due to the presence of high concentrations of phosphocholine and phosphoethanolamine (Daly and Cohen, 1989). However, this feature is not necessarily diagnostic of a tumor cell. For example, in the regenerating rat liver, which is a very rapidly growing tissue, there is a marked increase in the phosphomonoester peak due to an increase in the concentration of phosphoethanolamine (Murphy et al., 1992). Thus changes in the intensities of the phosphomonoesters need not indicate malignant transformation per se, but may be a consequence of rapid cell growth. ^{1}H NMR spectra also show features which are characteristic of tumor cells. For example, high levels of lactate have been observed in the ^{1}H NMR spectra of tumors (Negendank, 1992) and ^{1}H NMR spectra of gliomas show a much reduced N-acetylaspartate signal compared to normal brain (Le Fur et al., 1993). The ^{1}H NMR spectra of isolated tumor cells frequently contain intense and relatively narrow signals from neutral lipids, predominantly triglycerides. These signals arise from very mobile lipid molecules and thus are unlikely to come from the plasma membrane. A model has been proposed in which these lipid signals are thought to arise from neutral lipid domains associated with the plasma membrane (Mountford and Wright, 1988). However, it seems likely that the majority of these signals arise from cytoplasmic lipid droplets (Callies et al.,

1993). As with the increase in phosphomonoester signal, these changes in the ^{1}H spectra of lipids are not necessarily indicative of a malignant cell. For example, mitogen-stimulated lymphocytes accumulate cytoplasmic lipid droplets (Tanaka et al., 1963) and show the high resolution ^{1}H lipid signals (Mountford et al., 1982).

Drug Metabolism

Chemotherapy plays a prominent role in the clinical management of cancer. The absorption, distribution, metabolism, and excretion of an anticancer drug are important parameters to determine when assessing its efficacy. In recent years it has become feasible to use NMR as an analytical technique in pharmacokinetic studies.

Measurements in vivo

Noninvasive NMR measurements of drugs *in vivo* have important advantages over more conventional methods, such as chromatographic techniques which cannot be applied *in vivo* and radiotracer techniques, which do not directly yield information on the metabolism of the drug. Studies of anticancer drugs using NMR have been reviewed by Malet-Martino and Martino (1992) and Newell et al. (1992). The former review includes a detailed discussion of ^{19}F NMR investigations of fluorinated drug metabolism. ^{19}F NMR is well-suited to this work because of its relatively high sensitivity and the absence of background fluorine signals in biological systems.

Measurements on Body Fluids

As *in vivo* NMR measurements are often hampered by a lack of spatial resolution and sensitivity, a number of studies of drug metabolism have employed NMR measurements on biological fluids and isolated cells (often referred to as *in vitro* studies). High resolution NMR measurements on biofluids (blood plasma, urine, bile, cerebrospinal fluid, etc.) have important advantages over conventional methodologies in that they do not require extensive sample preparation and are nonselective in terms of the type of molecule detected, i.e., any molecule above a certain threshold concentration should give rise to detectable signals in a ^{1}H spectrum. There is no necessity, as is often the case with chromatographic methods, to select for a particular type or types of molecule. With the facility to detect a variety of low molecular weight compounds and macromolecules at the same time, it can provide a considerable amount of biochemical and chemical information in a relatively short period of time (Nicholson and Wilson, 1989; Bell, 1992). These techniques are making a significant impact in the field of clinical diagnosis and are established methods in pharmacology and toxicology (Vion-Dury et al., 1992).

Applications of these techniques to drug metabolism are reviewed by Nicholson and Wilson (1989), Preece and Timbrell (1990), and Malet-Martino and Martino (1992). Their application to the study of plasma composition in cancer has been reviewed by Vion-Dury et al. (1993).

Measurements on Isolated Cells

NMR studies of drug-resistant and drug-sensitive human breast cancer cells have been used to investigate pleiotropic drug resistance in terms of its effects on cellular energy metabolism (Cohen et al., 1986; Lyon et al., 1988) (see above). These cells were also used to test 2-deoxyglucose as a potential chemotherapeutic agent for drug-resistant cancer cells (Kaplan et al., 1990). The biotransformation of a cytotoxic agent has been studied using ^{31}P NMR in mice leukemia cells (Sonawat et al., 1990) and fluoropyrimidine metabolism has been studied using ^{19}F NMR in murine and human tumor cell lines. The effects of growth-inhibitory bioflavinoids on the metabolism of human leukemic cells have been studied by ^{13}C NMR (Post and Varma, 1992).

CONCLUDING REMARKS

The isolated cell preparation is an important system for the *in vivo* NMR spectroscopist in terms of gaining a better understanding of the NMR spectra obtained in the intact animal. For example, ^{1}H spectra of isolated neuronal and glial cells may help us to determine the contribution that these two cell types make to the ^{1}H spectra of intact brain. The response of tumor cells to anoxia, ischemia, etc. may aid interpretation of the spectra obtained from similar cells in a solid tumor in an animal or human patient following certain clinical interventions, such as chemotherapy or alteration of tumor perfusion with vasoactive drugs.

In some studies the cells may be a valid system for study in their own right. In this situation the merits of the NMR methods must be considered in relation to other competing techniques. For example in ^{31}P NMR, the chemical shift of the inorganic phosphate resonance is frequently used as a pH indicator. While this technique has no competition when measuring regional pHs in the human brain (Radda, 1992), there are some very sensitive alternative techniques for measuring pH in isolated cells. For example, fluorescent probe techniques can not only be used to measure pH in individual cells, they can also be used to measure pH heterogeneity within these cells (Tsien and Poenie, 1986). However, ^{31}P NMR has the advantages that it can also be used to monitor, in the same experiment, the energy status of the cell (from the levels of ATP, P_i, and, when present, phosphocreatine), the levels of intermediates involved in phospholipid metabolism (phosphocholine, glycerophosphocholine, etc.), and the concentrations of intermediates involved in glycogen synthesis and protein glycosylation (UDP-sugars). The merits of ^{13}C and ^{15}N NMR and indirect ^{1}H NMR methods for isotope label detection have already been discussed and apply equally well in isolated cell preparations as in the intact animal.

What an investigator gets out of an NMR experiment will depend very much on his or her perspective. For example, for a clinician involved in treating cancer, it may be the determination of changes in the NMR spectra which are characteristic of disease progression or which indicate response to therapy. For a physiologist it may be changes in pH or metal ion concentrations in the cell in response to an external stimulus. For a biochemist it may be an understanding of events manifest at the whole cell or whole organ level in terms of the underlying processes taking place at the molecular level. In this context, the combination of molecular genetic techniques with NMR methods for investigation appears a particularly attractive approach (Brindle et al., 1994). Noninvasive NMR measurements of changes in metabolite levels and metabolic fluxes in response to molecular genetic manipulation of the concentrations or properties of specific enzymes or membrane transporters promises to be a valuable approach to studying the control of metabolism *in vivo*. These techniques can be applied in any system, from the isolated cell to the transgenic animal.

To a newcomer to the technique, once you have gotten over the disappointment of its relative insensitivity and you have invested a little time and effort in understanding the basic principles, you cannot fail to be impressed by its power and versatility. NMR will inevitably have applications in your field of study, whether it is structure determination on your favorite protein or peptide or measurements on the intact biological system in which it functions.

ACKNOWLEDGMENTS

We would like to thank the BBSRC, the Wellcome Trust, and the Royal Society for supporting research in K.M.B.'s laboratory.

REFERENCES

Agris, P.F. & Campbell, I.D. (1982). Proton nuclear magnetic resonance of intact friend leukemia cells: Phosphorylcholine increase during differentiation. Science 216, 1325–1327.

Balaban, R.S. (1984). The application of nuclear magnetic resonance to the study of cellular physiology. Am. J. Physiol. 246, C10–C19.

Balaban, R.S., Kantor, H.L., Katz L.A., & Briggs, R.W. (1986). Relation between work and phosphate metabolite in the *in vivo* paced mammalian heart. Science 232, 1121–1123.

Bell, J.D. (1992). In: Magnetic Resonance Spectroscopy in Biology and Medicine (Bovée, W.M.M.J., de Certaines, J.D., & Podo, F., eds.), pp. 529–557, Pergamon Press, Oxford.

Bjerve, K.S. (1973). The Ca^{2+}-dependent biosynthesis of lecithin, phosphatidylethanolamine and phosphatidylserine in rat liver subcellular particles. Biochim. Biophys. Acta 296, 549–562.

Bloch, F., Hansen, W.W., & Packard, M. (1946). Nuclear induction. Phys. Rev. 69, 127.

Borkenhagen, L.F., Kennedy, E.P., & Fielding, L. (1961). Enzymatic formation and decarboxylation of phosphatidylserine. J. Biol. Chem. 236, PC28–PC30.

Boulanger, Y. & Vinay, P. (1989). Nuclear magnetic resonance monitoring of sodium in biological tissues. Can. J. Physiol. Pharmacol. 67, 820–828.

Bourgeois, D., Remy, C., Lefur, Y., Devoulon, P., Benabid, A.L., & Decorps, M. (1991). Proton spectroscopic imaging: A tool for studying intracerebral tumor models in rat. Magn. Reson. Med. 21, 10–20.

Brand, A., Engelmann, J., & Leibfritz, D. (1992). A ^{13}C NMR study on fluxes into the TCA cycle of neuronal and glial tumour cell lines and primary cells. Biochemie 74, 941–948.

Brindle, K.M. (1988a). NMR methods for measuring enzyme kinetics *in vivo*. Prog. NMR Spectrosc. 20, 257–293.

Brindle, K.M. (1988b). ^{31}P NMR magnetization transfer measurements of flux between inorganic phosphate and ATP in yeast cells genetically modified to over-produce phosphoglycerate kinase. Biochemistry 27, 6187–6196.

Brindle, K.M., Blackledge, M.J., Challiss, R.A.J., & Radda, G.K. (1989). ^{31}P NMR magnetization transfer measurements of ATP turnover during steady state isometric contraction in the rat hindlimb *in vivo*. Biochemistry 28, 4887–4893.

Brindle, K., Braddock, P., & Fulton, S. (1990). ^{31}P NMR measurements of the ADP concentration in yeast cells genetically modified to express creatine kinase. Biochemistry 29, 3295–3302.

Brindle, K.M., Brown, F.F., Campbell, I.D., Gratwohl, C., & Kuchel, P.W. (1979). Application of spin-echo nuclear magnetic resonance to whole cell systems. Biochem. J. 180, 37–44.

Brindle, K.M. & Campbell, I.D. (1984). ^{1}Hydrogen nuclear magnetic resonance studies of cells and tissues in Biomedical Magnetic Resonance (James, T.L. & Margulis, A.R., ed.), Radiology Research and Education Foundation, San Francisco.

Brindle, K.M. & Campbell, I.D. (1987). NMR studies of kinetics in cells and tissues. Quart. Rev. Biophys. 19, 159–182.

Brindle, K.M., Fulton, A.M., & Williams, S.-P. (1994). A combined NMR and molecular genetic approach to studying enzymes *in vivo*. In: NMR in Physiology and Medicine (Gillies, R.J., ed.), Academic Press, Orlando, Florida.

Brown, F.F., Campbell, I.D., Kuchel, P.W., & Rabenstein, D.L. (1977). Human erythrocyte metabolism studies by ^{1}H spin echo NMR. FEBS Lett. 82, 12–16.

Cadoux-Hudson, T.A., Blackledge, M.J., & Radda, G.K. (1989). Imaging of human brain creatine kinase activity *in vivo*. FASEB J. 3, 2660–2666.

Callies, R., Altenburger, R., Abarzua, S., Mayer, A., Grimme, L.H., & Leibfritz, D. (1992). In situ nuclear magnetic resonance of ^{15}N pulse labels monitors different routes for nitrogen assimilation. Plant Physiol. 100, 1584–1586.

Callies, R., Sri-Pathmanathan, R.M., Ferguson, D.Y.P., & Brindle, K.M. (1993). The appearance of neutral lipid signals in the ^{1}H NMR spectra of a myeloma cell line correlates with the induced formation of cytoplasmic lipid droplets. Magn. Reson. Med. 29, 546–550.

Chance, B., Eleff, S., Bank, W., Leigh, J.S., & Warnell, R. (1982). ^{31}P NMR studies of control of mitochondrial function in phosphofructokinase-deficient human skeletal muscle. Proc. Natl. Acad. Sci. USA 79, 7714–7718.

Clore, G.M. & Gronenborn, A.M. (1991). Structures of larger proteins in solution: Three- and four-dimensional heteronuclear NMR spectroscopy. Science 252, 1390–1399.

Cohen, J.S. & Lyon, R.C. (1987). Multinuclear NMR study of the metabolism of drug-sensitive and drug-resistant human breast cancer cells. Ann. N.Y. Acad. Sci. 508, 216–228.

Cohen, J.S., Lyon, R.C., Chen, C., Faustino, P.J., Batist, G., Shoemaker, M., Rubalcaba, E., & Cowan, K.H. (1986). Differences in phosphate metabolite levels in drug-sensitive and -resisitant human breast cancer cell lines determined by ^{31}P magnetic resonance spectroscopy. Cancer Res. 46, 4087–4090.

Cohen, S.M., Ogawa, S., & Shulman, R.G. (1979). ^{13}C NMR studies of gluconeogenesis in rat liver cells: Utilization of labeled glycerol by cells from euthyroid and hyperthyroid rats. Proc. Natl. Acad. Sci. USA 76, 1603–1607.

Cohen, S.M., Rognstad, R., Shulman, R.G., & Katz, J. (1981). A comparison of ^{13}C nuclear magnetic resonance and ^{14}C tracer studies of hepatic metabolism. J. Biol. Chem. 256, 3428–3432.

Daly, P.F. & Cohen, J.S. (1989). Magnetic resonance spectroscopy of tumors and potential *in vivo* clinical applications: A review. Cancer Res. 49, 770–779.

Daly, P.F., Lyon, R.C., Faustino, P.J., & Cohen, J.S. (1987). Phospholipid metabolism in cancer cells monitored by ^{31}P NMR spectroscopy. J. Biol. Chem. 262, 14875–14878.

Daly, P.F., Lyon, R.C., Straka, E.J., & Cohen, J.S. (1988). ^{31}P-NMR spectroscopy of human cancer cells proliferating in a basement membrane gel. FASEB J. 2, 2596–2604.

Derome, A.E. (1987). In: Modern NMR techniques for Chemistry Research. Pergamon Press, New York.

Desmoulin, F., Galons, J.-P., Canioni, P., Marvaldi, J., & Cozzone, P.J. (1986). ^{31}P nuclear magnetic resonance study of a human colon adenocarcinoma cultured cell line. Cancer Res. 46, 3768–3774.

Drury, D.D., Dale, B.E., & Gillies, R.J. (1988). Oxygen transfer properties of a bioreactor for use within a nuclear magnetic resonance spectrometer. Biotechnol. Bioeng. 32, 966–974.

Dunn, J.F., Frostick, S., Adams, G.E., Stratford, I.J., Howells, N., Hogan, G., & Radda, G.K. (1989). Induction of tumour hypoxia by a vasoactive agent. A combined NMR and radiobiological study. FEBS Lett. 249, 343–347.

Eakin, R.T., Morgan, L.O., Gregg, C.T., & Matwiyoff, N.A. (1972). Carbon-13 nuclear magnetic resonance spectroscopy of living cells and their metabolism of a specifically labeled ^{13}C substrate. FEBS Lett. 28, 259–264.

Evanochko, W.T., Ng, T.C., & Glickson, J.D. (1984). Application of *in vivo* NMR spectroscopy to cancer. Magn. Reson Med. 1, 508–534.

Fantini, J., Galons, J.-P., Marvaldi, J., Cozzone, P.J., & Canioni, P. (1987). Growth of a human colonic adenocarcinoma cell line (HT 29) on microcarrier beads: Metabolic studies by ^{31}phosphorus nuclear magnetic resonance spectroscopy. Int. J. Cancer 39, 225–260.

Farrow, N.A., Kanamori, K., Ross, B.D., & Parivar, F. (1990). A ^{15}N-n.m.r. study of cerebral, hepatic and renal nitrogen metabolism in hyperammonaemic rats. Biochem. J. 270, 473–481.

Fernandez, E.J., Mancuso, A., & Clark, D.S. (1988). NMR spectroscopy studies of hybridoma metabolism in a simple membrane reactor. Biotechnol. Prog. 4, 173–183.

Foxall, D.L. & Cohen, J.S. (1983). NMR-studies of perfused cells. J. Magn. Reson. 52, 346–349.

Frahm, J., Michaelis, T., Merboldt, K.D., Bruhn, H., Gyngell, M.L., & Hänicke, W. (1990). Improvements in localized proton NMR spectroscopy of human brain. Water suppression, short echo times, and 1 ml resolution. J. Magn. Reson. 90, 464–473.

Freeman, R. (1990). In: A Handbook of Nuclear Magnetic Resonance. Wiley, New York.

Freyer, J.P., Schor, P.L., Jarrett, K.A., Neeman, M., & Sillerud, L.O. (1991). Cellular energetics measured by phosphorous nuclear magnetic resonance spectroscopy are not correlated with chronic nutrient deficiency in multicellular tumor spheroids. Cancer Res. 51, 3831–3837.

Gadian, D.G. (1982). In: Nuclear Magnetic Resonance and its Applications to Living Systems. Clarendon Press, Oxford.

Garlick, P.B., Soboll, S., & Bullock, G.R. (1992). Evidence that mitochondrial phosphate is visible in ^{31}P NMR spectra of isolated, perfused rat hearts. NMR Biomed. 5, 29–36.

Gerweck, L.E., Koutcher, J.A., Zaidi, S.T.H., & Seneviratne, T. (1992). Energy status in the murine FSaII and MCaIV tumors under aerobic and hypoxic conditions: An *in-vivo* and *in-vitro* analysis. Int. J. Radiation Oncology Biol. Phys. 23, 557–561.

Gillies, R.J., Chresand, T.J., Drury, D.D., & Dale, B.E. (1986). Design and application of bioreactors for analysis of mammalian cells by NMR. Rev. Magn. Reson. Med. 1, 155–179.

Gillies, R.J., Scherer, P.G., Raghunand, N., Okerlund, L.S., Martinez-Zaguilan, R., Hesterberg, L., & Dale, B.E. (1991). Iteration of hybridoma cell growth and productivity using ^{31}P NMR. Magn. Reson. Med. 18, 181–192.

Gründer, W., Krumbiegel, P., Buchali, K., & Blesin, H.J. (1989). Nitrogen-15 NMR studies of rat liver *in vitro* and *in vivo*. Phys. Med. Biol. 34, 457–463.

Gupta, R.K. & Gupta, P. (1984). NMR studies of intracellular metal ions in intact cells and tissues. Annu. Rev. Biophys. Bioeng. 13, 221–246.

Haase, A., Frahm, J., Haenicke, W., & Matthei, D. (1985). ^{1}H-NMR chemical-shift selective (CHESS) imaging. Phys. Med. Biol. 30, 341–344.

Häussinger, D. (1990). Nitrogen metabolism in liver: Structural and functional organization and physiological relevance. Biochem. J. 267, 281–290.

Hutson, S.M., Berkich, D., Williams, G.D., La Noue, K.F., & Briggs, R.W. (1989). ^{31}P NMR visibility and characterization of rat liver mitochondrial matrix adenine nucleotides. Biochemistry 28, 4325–4332.

Jans, A.W.H. & Leibfritz, D. (1988). A ^{13}C-NMR study on the influxes into the tricarboxylic acid cycle of a renal epithelial cell line, LLC-PK$_1$/Cl$_4$: The metabolism of [2-^{13}C]glycine, L-[3-^{13}C]alanine and L-[3-^{13}C]aspartic acid in renal epithelial cells. Biochim. Biophys. Acta 970, 241–250.

Jans, A.W.H. & Leibfritz, D. (1989). A ^{13}C NMR study on fluxes into the Krebs cycle of rabbit renal proximal tubular cells. NMR Biomed. 1, 171–176.

Jans, A.W.H. & Willem, R. (1988). ^{13}C-NMR study of glycerol metabolism in rabbit renal cells of proximal convoluted tubules. Eur. J. Biochem. 174, 67–73.

Jeffrey, F.M.H., Rajagopal, A., Malloy, C.R., & Sherry, A.D. (1991). ^{13}C NMR: A simple yet comprehensive method for analysis of intermediary metabolism. Trends Biochem. Sci. 16, 5–10.

Jelicks, L.A. & Gupta, R.K. (1989). Double-quantum NMR of sodium ions in cells and tissues. Paramagnetic quenching of extracellular coherence. J. Magn. Reson. 81, 586–592.

Jeener, J. (1971). Ampère International Summer School, Basko Polje, Yugoslavia, unpublished.

Kanamori, K., Legerton, T.L., Weiss, R.L., & Roberts, J.D. (1982). Effect of the nitrogen source on glutamine and alanine biosynthesis in Neurospora crassa. J. Biol. Chem. 257, 14168–14172.

Kanamori, K., Parivar, F., & Ross, B.D. (1993). A ^{15}N NMR study of in vivo cerebral glutamine synthesis in hyperammonemic rats. NMR Biomed. 6, 21–26.

Kaplan, O., Navon, G., Lyon, R.C., Faustino, P.J., Straka, E.J., & Cohen, J.S. (1990). Effects of 2-deoxyglucose on drug-sensitive and drug-resistant human breast cancer cells: Toxicity and magnetic resonance spectroscopy studies of metabolism. Cancer Res. 50, 544–551.

Kennedy, E.P. & Weiss, S.B. (1956). The function of cytidine coenzymes in the biosynthesis of phospholipids. J. Biol. Chem. 222, 193–214.

Kirschenlohr, H.L., Metcalfe, J.C., Morris, P.G., & Rodrigo, G.C. (1988). Ca^{2+} transient, Mg^{2+}, and pH measurements in the cardiac cycle by ^{19}F NMR. Proc. Natl. Acad. Sci. USA 85, 9017–9021.

Koretsky, A.P., Brosnan, M.J., Chen, L., Chen, J., & van Dyke, T. A. (1990). NMR detection of creatine kinase expressed in liver of transgenic mice: Determination of free ADP levels. Proc. Natl. Acad. Sci. USA 87, 3112–3116.

Koretsky, A.P. & Williams, D.S. (1992). Application of localized in vivo NMR to whole organ physiology in the animal. Ann. Rev. Physiol. 54, 799–826.

Kriat, M., Confort-Gouny, S., Vion-Dury, J., Viout, P., & Cozzone, P.J. (1992). Two-dimensional ^{1}H NMR spectroscopy of normal and pathological human plasma: Complete water suppression and further assignment of resonances. Biochimie, 74, 913–918.

Lawson, J.W.R. & Veech, R.L. (1979). Effects of pH and free Mg^{2+} on the Keq of the creatine kinase reaction and other phosphate hydrolyses and phosphate transfer reactions. J. Biol. Chem. 254, 6528–6537.

Leach, M.O. (1988). In: The Physics of Medical Imaging (Webb, S., ed.), pp. 389–487, Hilger, Philadelphia.

Le Fur, Y., Ziegler, A., Bourgeois, D., Decorps, M., & Remy, C. (1993). Phased spectroscopic images: Application to the characterization of the ^{1}H 1.3-ppm resonance in intracerebral tumors in the rat. Magn. Reson. Med. 29, 431–435.

Legerton, T.L., Kanamori, K., Weiss, R.L., & Roberts, J.D. (1981). ^{15}N NMR studies of nitrogen metabolism in intact mycelia of Neurospora crassa. Proc. Natl. Acad. Sci. USA 78, 1495–1498.

Lin, P.-S., Blumenstein, M., Mikkelsen, R.B., Schmidt-Ulrich, R., & Bachovchin, W.W. (1987). Perfusion of cell spheroids for study by NMR spectroscopy. J. Magn. Reson. 73, 399–404.

Live, D.H., Davis, D.G., Agosta, W.C., & Cowburn D. (1984). Observation of 1,000-fold enhancement of ^{15}N NMR via proton-detected multiquantum coherences: Studies of large peptides. J. Am. Chem. Soc. 106, 6104–6105.

Llinás, M., Wüthrich, K., Schwotzer, W., & von Philipsborn, W. (1975). [15]N nuclear magnetic resonance of living cells. Nature, 257, 817–818.

Lyon, R.C., Cohen, J.S., Faustino, P.J., Megnin, F., & Myers, C.E. (1988). Glucose metabolism in drug-sensitive and drug-resistant human breast cancer cells monitored by magnetic resonance spectroscopy. Cancer Res. 48, 870–877.

Macdonald, J.M., Kurkanewicz, J., Dahiya, R., Espanol, M.T., Chang, L.-H., Goldberg, B., James, T.L., & Narayan, P. (1993). Effect of glucose and confluency on phosphorus metabolites of perfused human prostatic adenocarcinoma cells as determined by ^{31}P MRS. Magn. Reson. Med. 29, 244–248.

Malet-Martino, M.C. & Martino, R. (1992). Magnetic resonance spectroscopy: A powerful tool for drug metabolism studies. Biochimie 74, 785–800.

Mancuso, A., Fernandez, E.J., Blanch, H.W., & Clark, D.S. (1990). A nuclear magnetic resonance technique for determining hybridoma cell concentration in hollow fibre bioreactors. Bio/Technology 8, 1282–1285.

Mansfield, P. & Morris, P.G. (1982). NMR imaging in biomedicine. Adv. Magn. Reson. S2, 1–343.

Matthews, P.M., Bland, J.L., Gadian, D.G., & Radda, G.K. (1982). A ^{31}P NMR saturation transfer study of the regulation of creatine kinase in the rat heart. Biochim. Biophys. Acta 721, 312–320.

May, G.L., Wright, L.C., Holmes, K.T., Williams, P.G., Smith, I.C.P., Wright, P.E., Fox, R.M., & Mountford, C.E. (1986). Assignment of methylene proton resonances in NMR spectra of embryonic and transformed cells to plasma membrane triglyceride. J. Biol. Chem. 261, 3048–3053.

McLean, J.S. (1993). Improved techniques for immortalizing animal cells. Trends Biotech. 11, 232–238.

Minichiello, M.M., Albert, D.M., Kolodny, N.H., Lee, M.-S., & Craft, J.L. (1989). A perfusion system developed for ^{31}P NMR study of melanoma cells at tissue-like density. Magn. Reson. Med. 10, 96–107.

Moon, R.B. & Richards, J.H. (1973). Determination of intracellular pH by ^{31}P magnetic resonance. J. Biol. Chem. 248, 7276–7278.

Moonen, C.T.W. & Van Zijl, P.C.M. (1990). Highly effective water suppression for *in vivo* proton NMR spectroscopy (DRYSTEAM). J. Magn. Reson. 88, 28–41.

Mountford, C.E., Grossman, G., Reid, G., & Fox, R.M. (1982). Characterization of transformed cells and tumors by proton nuclear magnetic resonance spectroscopy. Cancer Res. 42, 2270–2276.

Mountford, C.E. & Wright, L.C. (1988). Organization of lipids in the plasma membranes of malignant and stimulated cells: A new model. Trends Biochem. Sci. 13, 172–177.

Murphy, E.J., Brindle, K.M., Rorison, C.J., Dixon, R.M., Rajagopalan, B., & Radda, G.K. (1992). Changes in phosphatidylethanolamine metabolism in regenerating rat liver as measured by ^{31}P-NMR. Biochim. Biophys. Acta 1135, 27–34.

Murphy, E.J., Rajagopalan, B., Brindle, K.M., & Radda, G.K. (1989). Phospholipid bilayer contribution to ^{31}P NMR spectra *in vivo*. Magn. Reson. Med. 12, 282–289.

Neeman, M., Rushkin, E., Kaye, A.M., & Degani, H. (1987). ^{31}P-NMR studies of phosphate transfer rates in T47D human breast cancer cells. Biochim. Biophys. Acta 930, 179–192.

Negendank, W. (1992). Studies of human tumors by MRS: A review. NMR Biomed. 5, 303–324.

Newell, D.R., Maxwell, R.J., & Golding, B.T. (1992). *In vivo* and *ex vivo* magnetic resonance spectroscopy as applied to pharmacokinetic studies with anticancer agents: A review. NMR Biomed. 5, 273–278.

Nicholson, J.K. & Wilson, I.D. (1989). High resolution proton magnetic resonance spectroscopy of biological fluids. Progr. Nucl. Magn. Reson. Spectrosc. 21, 449–501.

Pianet, I., Merle, M., Labouesse, J., & Canioni, P. (1991). Phosphorus-31 nuclear magnetic resonance of C6 glioma cells and rat astrocytes. Eur. J. Biochem. 195, 87–95.

Post, J.F.M., Baum, E., & Ezell, E.L. (1992). ^{13}C NMR studies of glucose metabolism in human leukemic CEM-C7 and CEM-C1 cells. Magn. Reson. Med. 23, 356–366.

Post, J.F.M. & Varma, R.S. (1992). Growth inhibitory effects of bioflavonoids and related compounds on human leukemic CEM-C1 and CEM-C7 cells. Cancer Lett., 67, 207–213.

Potts, J.R., Hounslow, A.M., & Kuchel, P.W. (1990). Exchange of fluorinated glucose across the red cell membrane measured by ^{19}F n.m.r. magnetisation transfer. Biochem. J. 266, 925–928.

Preece, N.E. & Timbrell, J.A. (1990). Use of NMR spectroscopy in drug metabolism studies: Recent advances. Prog. Drug. Metab. 12, 147–203.

Purcell, E.M., Torrey, H.C., & Pound, R.V. (1946). Resonance absorption by nuclear magnetic moments in a solid. Phys. Rev. 69, 37–38.

Radda, G.K. (1986). The use of NMR spectroscopy for the understanding of disease. Science 233, 640–645.

Radda, G.K. (1992). Control, bioenergetics, and adaptation in health and disease: Noninvasive biochemistry from nuclear magnetic resonance. FASEB J. 6, 3032–3038.

Rao, B.D.N., Cohn, M., & Scopes, R.K. (1978). ^{31}P NMR study of bound reactants and products of yeast 3-phosphoglycerate kinase at equilibrium and the effect of sulfate ion. J. Biol. Chem. 253, 8056–8060.

Robitaille, P.M., Merkle, H., Sublett, E., Hendrich, K., Lew, B., Path, G., From, A.H.L., Bache, R.J., Garwood, M., & Ugurbil, K. (1989). Spectroscopic imaging and spatial localization using adiabatic pulses and applications to detect transmural metabolite distribution in the canine heart. Magn. Reson. Med. 10, 14–37.

Ronen, S.M., Rushkin, E., & Degani, H. (1992). Lipid metabolism in large T47D human breast cancer spheroids: ^{31}P- and ^{13}C-NMR studies of choline and ethanolamine uptake. Biochim. Biophys. Acta 1138, 203–212.

Ronen, S.M., Stier, A., & Degani, H. (1990). NMR studies of the lipid metabolism of T47D human breast cancer spheroids. FEBS Lett. 266, 147–149.

Rothman, D.L., Novotny, E.J., Shulman, G.I., Howseman, A.M., Petroff, O.A.C., Mason, G., Nixon, T., Hanstock, C.C., Prichard, J.W., & Shulman, R.G. (1992). ^{1}H-[^{13}C] NMR measurements of [4-^{13}C]glutamate turnover in human brain. Proc. Natl. Acad. Sci. USA 89, 9603–9606.

Ruiz-Cabello, J. & Cohen, J.S. (1992). Phospholipid metabolites as indicators of cancer cell function. NMR Biomed. 5, 226–233.

Sanders, J.K.M. & Hunter, B.K. (1987). In: Modern NMR Spectroscopy. Oxford University Press, New York.

Seguin, F., Le Pape, A., & Williams, S.R. (1992). In: Magnetic Resonance Spectroscopy in Biology and Medicine (Bovée, W.M.M.J., de Certaines, J.D., & Podo, F., eds.), pp. 479–491, Pergamon Press, Oxford.

Shulman, R.G., Brown, T.R., Ugurbil, K., Ogawa, S., Cohen, S.M., & den Hollander, J.A. (1979). Cellular applications of ^{31}P and ^{13}C nuclear magnetic resonance. Science 205, 160–166.

Sillerud, L.O., Han, C.H., Bitensky, M.W., & Francendese, A.A. (1986). Metabolism and structure of triacylglycerols in rat epididymal fat pad adipocytes determined by ^{13}C nuclear magnetic resonance. J. Biol. Chem. 261, 4380–4388.

Sonawat, H.M., Leibfritz, D., Engel, J., & Hilgard, P. (1990). Biotransformation of mafosfamide in P388 mice leukemia cells: Intracellular ^{31}P-NMR studies. Biochim. Biophys. Acta. 1052, 36–41.

Sri-Pathmanathan, R.M., Braddock, P., & Brindle, K.M. (1989). ^{31}P NMR studies of glucose and glutamine metabolism in cultured mammalian cells. Biochim. Biophys. Acta 1051, 131–137.

Street, J.C., Delort, A.-M., Braddock, P.S.H., & Brindle, K.M. (1993). A ^{1}H/^{15}N n.m.r. study of nitrogen metabolism in cultured mammalian cells. Biochem. J. 291, 485–492.

Styles, P., Blackledge, M.J., Moonen, C.T.W., & Radda, G.K. (1987). Spatially resolved ^{31}P NMR spectroscopy of organs in animal models and man. Ann. N. Y. Acad. Sci. 508, 349–359.

Styles, P., Gratwohl, C., & Brown, F.F. (1979). Simultaneous multinuclear NMR by alternate scan recording of ^{31}P and ^{13}C spectra. J. Magn. Reson. 35, 329–336.

Summers, M.F., Marzilli, L.G., & Bax, A. (1986). Complete ^1H and ^{13}C assignments of coenzyme B_{12} through the use of new two-dimensional NMR experiments. J. Am. Chem. Soc. 108, 4285–4294.

Sze, D.Y. & Jardetzky, O. (1990a). Determination of metabolite and nucleotide concentrations in proliferating lymphocytes by ^1H-NMR of acid extracts. Biochim. Biophys. Acta 1054, 181–197.

Sze, D.Y. & Jardetzky, O. (1990b). Characterization of lipid composition in stimulated human lymphocytes by ^1H-NMR. Biochim. Biophys. Acta 1054, 198–206.

Szwergold, B.S. (1992). NMR spectroscopy of cells. Ann. Rev. Physiol. 54, 775–798.

Tanaka, Y., Epstein, L.B., Brecher, G., & Stohlman, F., Jr. (1963). Transformation of lymphocytes in cultures of human peripheral blood. Blood 22, 614–629.

Taylor, J. & Deutsch, C. (1988). ^{19}F-nuclear magnetic resonance: Measurements of $[O_2]$ and pH in biological systems. Biophys. J. 53, 227–233.

Tsien, R.Y. & Poenie, M. (1986). Fluorescence ratio imaging: A new window into intracellular ionic signaling. Trends Biochem. Sci. 11, 450–455.

Ugurbil, K., Guernsey, D.L., Brown, T.R., Glynn, P., Tobkes, N., & Edelman, I.S. (1981). ^{31}P NMR studies of intact anchorage-dependent mouse embryo fibroblasts. Proc. Natl. Acad. Sci. USA, 78, 4843–4847.

Urenjak, J., Williams, S.R., Gadian, D.G., & Noble, M. (1993). Proton nuclear magnetic resonance spectroscopy unambiguously identifies different neural types. J. Neurosci. 13, 981–989.

van Zijl, P.C.M., Moonen, C.T.W., Faustino, P., Pekar, J., Kaplan, O., & Cohen, J.S. (1991). Complete separation of intracellular and extracellular information in NMR spectra of perfused cells by diffusion-weighted spectroscopy. Proc. Natl. Acad. Sci. USA 88, 3228–3232.

Vion-Dury, J., Favre, R., Sciaky, M., Kriat, M., Confort-Gouny, S., Harlé, J.R., Grazziani, N., Viout, P., Grisoli, F., & Cozzone, P.J. (1993). Graphic-aided study of metabolic modifications of plasma in cancer using proton magnetic resonance spectroscopy. NMR Biomed. 6, 58–65.

Vion-Dury, J., Nicoli, F., Torri, G., Torri, J., Kriat, M., Sciaky, M., Davin, A., Viout, P., Confort-Gouny, S., & Cozzone, P.J. (1992). High resolution NMR spectroscopy of physiological fluids: From metabolism to physiology. Biochimie 74, 801–807.

von Recklinghausen, I.R., Scott, D.M., & Jans, A.W.H. (1991). An NMR spectroscopic characterization of a new epithelial cell line, TALH-SVE, with properties of the renal medullary thick ascending limb of Henle's loop. Biochim. Biophys. Acta 1091, 179–187.

Walker, T.E., Han, C.H., Kollman, V.H., London, R.E., & Matwiyoff, N.A. (1982). ^{13}C nuclear magnetic resonance studies of the biosynthesis by *Microbacterium ammoniaphilum* of L-glutamate selectively enriched with carbon-13. J. Biol. Chem. 257, 1189–1195.

Williams, S.-P., Fulton, A.M., & Brindle, K.M. (1993). Estimation of the intracellular free ADP concentration by ^{19}F NMR studies of fluorine-labeled yeast phosphoglycerate kinase in vivo. Biochemistry 32, 4895–4902.

Wüthrich, K. (1986). In: NMR of Proteins and Nucleic Acids. Wiley, New York.

RECOMMENDED READINGS

Gadian, D.G. (1982). Nuclear Magnetic Resonance and its Applications to Living Systems. Clarendon Press, Oxford.

Bovée, W.M.M.J., de Certaines, J.D., & Podo, F. (eds.) (1992). Magnetic Resonance Spectroscopy in Biology and Medicine. Pergamon Press, Oxford.

Gillies, R.J. (ed.) (1994). NMR in Physiology and Medicine. Academic Press, Orlando, Florida. (In press).

Chapter 10

Calorimetric Techniques

INGEMAR WADSÖ

Principles of Medical Biology, Volume 4
Cell Chemistry and Physiology: Part III, pages 271–301.
Copyright © 1996 by JAI Press Inc.
All rights of reproduction in any form reserved.
ISBN: 1-55938-807-2

INTRODUCTION

Calorimeters are instruments used for the direct measurement of heat quantities including heat production rates and heat capacities. Different measurement principles are employed and a very large number of calorimetric designs have been described since the first calorimetric experiments were reported more than 200 years ago. The amount of heat evolved in a chemical reaction is proportional to the amount of material taking part in the reaction and the heat production rate; the "thermal power," is proportional to the rate of the reaction. Calorimeters can therefore be employed as quantitative analytical instruments and in kinetic investigations, in addition to their use as thermodynamic instruments. Important uses of calorimeters in the medical field are at present in research on the biochemical level and in studies of living cellular systems. Such investigations are often linked to clinical applications but, so far, calorimetric techniques have hardly reached a state where one may call them "clinical (analytical) instruments."

Thermodynamics, Thermochemistry, and Calorimetry

It is beyond the scope of this chapter to discuss thermodynamics and thermochemistry in any depth. However, thermodynamic investigations form one of the major application areas of calorimetry including calorimetric investigations in biochemistry and on living systems. Further, when calorimeters are used as kinetic instruments or in different analytical applications, it is no doubt important that the experimenters are aware that all calorimetric experiments result in "thermodynamic" measurements. The general and lasting value of calorimetric results will therefore increase if the experiments are conducted and recorded in a way that makes it meaningful to express results in terms of well-defined thermodynamic properties. This does not mean that analytical users of calorimeters need to be experts in thermodynamics. But they should be encouraged to be acquainted with the nature of heat and of other thermodynamic properties which can be determined by calorimetry. In chapter 11 of this book, Heat Dissipation and Metabolism in Isolated Mammalian Cells, Richard B. Kemp gives a brief but rather broad introduction to equilibrium and nonequilibrium thermodynamics of particular importance for biology. Below, a few basic thermodynamic properties and conventions of particular importance for calorimetry and thermochemistry are introduced.

The experimental study of heat produced or absorbed in chemical reactions is usually called *thermochemistry*. Such investigations are often best conducted by direct calorimetric measurements, but values for heat quantities and their time and temperature derivatives can also be obtained from other kinds of thermodynamic experiments. A heat quantity, Q, which is lost or gained in a process conducted under constant pressure, p (in chemistry and biology most frequently the atmospheric pressure), is defined as the *enthalpy* change, ΔH, accompanying the process.

If a process takes place at constant volume, the corresponding heat quantity is called the change in *internal energy*, ΔU. An example of a constant volume process is provided by a combustion calorimetric experiment where the sample is burned in a closed vessel (a calorimetric bomb) charged with pressurized oxygen. The relationship between ΔH and ΔU is given by

$$\Delta H = \Delta U + p\Delta V \tag{1}$$

where ΔV is the net increase in volume of the reaction components. For processes without gaseous reaction components or where the amount of gaseous components does not change, ΔV is small and changes in ΔH and ΔU can usually be taken as equal. The $p\Delta V$ term represents the work conducted by the reaction system against the atmosphere.

An *exothermic* process means that heat is evolved; in an *endothermic* process heat is absorbed by the reaction system. Thus, if an exothermic chemical process takes place in an insulated vessel, e.g., a Dewar vessel, the temperature of the system will increase. Correspondingly, if the process is endothermic the temperature will decrease. For the case where there is no heat exchange between the vessel and the surroundings, the system (the vessel and its content) is said to be *adiabatic*. In that case, the energy of the system has not changed as a result of the process (provided that no work is supplied to or taken out from the system except for the "volume work," $p\Delta V$), i.e., the enthalpy change is zero. When an energy value (energy change) is reported for a process it will always refer to the *isothermal* process conducted at a specified temperature. In order to keep an exothermic reaction system isothermal, it is necessary to allow the heat evolved to dissipate to the surroundings (usually a thermostatted bath or a hypothetical, infinitely large heat sink). Correspondingly, for an endothermic process heat is allowed to flow from the surroundings to the reaction system. *The exothermic process, conducted isothermally, will thus be accompanied by a loss of enthalpy, whereas the endothermic reaction system will gain enthalpy.* Enthalpy values for exothermic processes are therefore negative and the values for endothermic processes are positive. For example, the enthalpy of combustion of crystalline α-D-glucose to form liquid water and gaseous carbon dioxide is strongly exothermic:

$$\alpha\text{-D-glucose (cr)} + 6O_2 \text{ (g)} = 6\,H_2O \text{ (l)} + 6\,CO_2 \text{ (g)} \tag{2}$$

$$\Delta_c H° = -2816 \text{ kJ/mol (25 °C)}$$

and the enthalpy of dissolution of sodium chloride is weakly endothermic:

$$NaCl \text{ (s)} = NaCl \text{ (aq)} \tag{3}$$

$$\Delta_{sol} H^\infty \text{ (aq)} = 3.89 \text{ kJ/mol (25 °C)}$$

Thermochemical equations like (2) and (3) are used to express, accurately, the conditions to which reported calorimetric results refer. The symbol for the process (c for combustion and sol for dissolution) is in accordance with IUPAC[*] recommendations placed as an index to the Δ-operator. Likewise, the symbol $\Delta_{sol}H^{\infty}$(aq) is shorthand for enthalpy of dissolution in water to form an infinitely dilute solution. The $^{\circ}$-sign in $\Delta_c H^{\circ}$ indicates that the value refers to a *standard state*, in the present case defined by the thermochemical equation (2).

Accurate thermochemical data compiled in the literature are usually obtained from calorimetric experiments. But the values found in thermochemical tables are normally not simply mean values derived from results of a series of calorimetric measurements. For example, the $\Delta_c H^{\circ}$ value for glucose was derived from results of combustion in a calorimetric bomb at constant volume leading to a $\Delta_c U$-value to which several correction terms were added, accounting for minor deviations from the ideal process described by equation (2). The $p\Delta V$ term (equation (1)) is in this case small, as the amount of gas does not change as a result of the combustion process. For reactions taking place in solution it is common that the results must be corrected for dilution effects. As an example, we may consider a ligand binding experiment: the binding of an inhibitor to an enzyme in aqueous solution. Such measurements are usually carried out by use of a microtitration calorimeter, cf. p. 289. Typically a small volume (e.g., 5 μl) of the inhibitor solution is injected into a much larger volume of enzyme solution (e.g., 1 ml). But the injected inhibitor solution will normally be relatively concentrated, and in addition to the heat effect from the binding process, there can be a significant enthalpy of dilution of the inhibitor solution. The appropriate correction term is obtained from a separate blank dilution experiment. Other effects which sometimes must be corrected for include ionization, incomplete reaction, mechanical effects due to operation of an injection system, etc. Before accurate thermochemical data can be reported, it is thus necessary to analyze the gross process recorded in the calorimetric experiment. After that, corrections should be applied so that the reported value refers to the (simple) process described by the thermochemical equation. See also the paragraph on systematic errors, p. 299.

The thermodynamic equilibrium constant, K°, for a process

$$A + B + ... \rightleftharpoons C + D + ...$$ (4)

is given by

$$K^{\circ} = \frac{a_C \cdot a_D \cdots}{a_A \cdot a_D \cdots}$$ (5)

where a_C, etc. are the *activities* of the reaction components in the equilibrium mixture. In biothermodynamics, activities are usually not known and correspond-

---

[*]IUPAC is the International Union of Pure and Applied Chemistry. Their recommendations for symbols and units are given in Mills et al. (1993). Some IUPAC recommendations and guidelines with particular reference to biocalorimetry are given in Belaich et al. (1982).

ing values for *concentration* are used. The recommended symbol for concentration equilibrium constants is K_c but more commonly the plain symbol K is used.

The standard Gibbs energy change, $\Delta G°$ (still often called standard free energy, but this term should be avoided), is related to the equilibrium constant by equation (6).

$$\Delta G° = -R\,T\ln K \tag{6}$$

where R is the gas constant ($R = 8.3145\ \text{J·K}^{-1}\ \text{mol}^{-1}$) and T the thermodynamic temperature (unit: Kelvin, symbol: K; 25 °C = 298.15 K, 37 °C = 310.15 K). The standard state symbol (°) means in this case that $\Delta G°$ refers to the process where all reaction components have unit activity or the concentration 1 mol·l^{-1}. In biochemistry the symbol $\Delta G°'$ is often used to indicate that the standard Gibbs energy value refers to certain experimental conditions (which should be specified), usually pH 7.0 and ionic strength 0.1.

A Gibbs energy change is related to the corresponding change in enthalpy by

$$\Delta G° = \Delta H° - T\Delta S° \tag{7a}$$

or

$$\Delta G = \Delta H - T\Delta S \tag{7b}$$

for a non-standard state process.

ΔS is the *entropy* change (cf. chapter 11). As will be discussed on p. 290, the important thermodynamic properties in equation (5) – (7) can be determined by titration calorimetry for wide ranges of K_c provided that the value for the enthalpy change is not too small. The temperature derivative of the enthalpy change is the change in heat capacity, ΔC_p,

$$\Delta C_p = \frac{\mathrm{d}\Delta H}{\mathrm{d}T} \tag{8}$$

This property is useful for recalculation of a ΔH value from one temperature to another. It is also of great importance in biothermodynamics— processes characterized by changes in hydrophobic interaction or hydrophobic hydration have exceptionally large ΔC_p values expressing solute–water interactions. ΔC_p values are often determined for ligand binding processes involving biopolymers and for denaturation (unfolding) of biopolymers. The latter type of processes are mainly studied by differential scanning calorimetry (DSC), cf. p. 287.

CALORIMETRIC PRINCIPLES AND CONCEPTS

The mechanical and electronic design of calorimeters can be very different, partly because they can be used for so many different purposes. In some cases calorimeters are built for one specific type of experiment, for example combustion in a pressure vessel, measurement of low temperature heat capacities, or vaporization measurements. In other cases calorimeters are designed for a more flexible use, for example "reaction and dissolution calorimeters." In still other cases a modular concept is used making it possible to assemble different types of calorimeters from a limited number of mechanical and electronic units. From the point of view of heat measurement principles, calorimeters are essentially limited to three main groups: adiabatic type of calorimeters, heat conduction calorimeters (also called heat leakage or heat flux calorimeters), and power compensation calorimeters.

Adiabatic Calorimeters

In an ideal adiabatic calorimeter, there is no heat exchange between the calorimetric vessel and the surroundings. This implies that the temperature in the calorimetric vessel will increase (exothermic processes) or decrease (endothermic processes) during the measurement. The heat quantity, evolved or absorbed during an experiment, is in the ideal case equal to the product between the temperature change, ΔT, and the heat capacity of the calorimetric vessel (including its content), C:

$$Q = C \cdot \Delta T \tag{9}$$

In practice there will always be some heat conductance between the vessel and the surroundings and a "practical" heat capacity value, ε, is determined in a calibration experiment:

$$Q = \varepsilon \cdot \Delta T \tag{10}$$

The heat production rate or the thermal power, P, is thus

$$P = \varepsilon \cdot d\Delta T / dt \tag{11}$$

In particular, calorimeters used for studies of chemical reactions or dissolution processes of short duration are often of a semiadiabatic or quasiadiabatic type, meaning that the adiabatic conditions are far from being perfect. Figure 1B shows schematically the design of a simple instrument of this kind. These instruments are often called isoperibolic, a term which hardly has any merit. The reaction vessel, usually made from glass or a noncorrosive metal, is surrounded by an insulation (air or vacuum) outside of which there is a thermostatted body, in most cases a water bath. The vessel is equipped with a thermometer which normally is a resistance thermometer, e.g., a thermistor. In the figure is also indicated a stirrer and an electrical calibration heater. If the instrument is designed as a reaction and dissolution calorimeter, the vessel will also be equipped with a device for initiating the process, e.g., an injection tube or an ampoule-breaking mechanism. Semiadiabatic conditions are provided by the low heat conductivity space between the vessel and the surrounding thermostatted bath. In order to reach close to ideal adiabatic

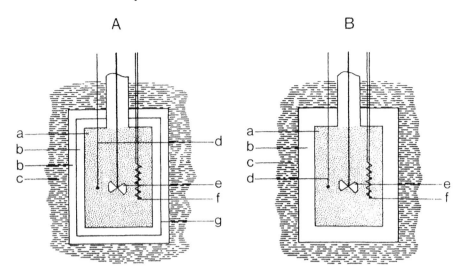

Figure 1. Schematic diagrams of sections through adiabatic-type calorimeters. **A:** Adiabatic shield calorimeter. **B:** Semiadiabatic calorimeter. a, calorimetric vessel; b, air or vacuum; c, thermostatted bath; d, thermometer; e, stirrer; f, calibration heater; g, adiabatic shield.

conditions during an exothermic process, an adiabatic shield can be inserted between the vessel and the thermostat, Figure 1A. Typically this shield consists of a thin-walled metal envelope fitted with a heater winding. When heat is evolved in the vessel and its temperature increases, the shield temperature will follow, i.e., there will be no driving force for a net heat flow from the vessel. The zero temperature difference between the shield and the vessel is usually indicated by thermocouples or by resistance thermometers placed on the shield and on the wall of the vessel.

A thermocouple is constructed from wires of different metals (or from semiconducting materials) which are connected at two junctions. If there is a temperature difference between the junctions, there will be an electrical potential between the two free ends, U_c. With good approximation the potential is proportional to the temperature difference between the junctions,

$$U_c = k \, \Delta T \tag{12}$$

where k is a material constant. If n thermocouples are connected in series to form a *thermopile* the potential, U, will be

$$U = k \, n \, \Delta T \tag{13}$$

In chemical and biological experiments processes are normally exothermic. In case an endothermic process is measured, adiabatic calorimeters can be operated

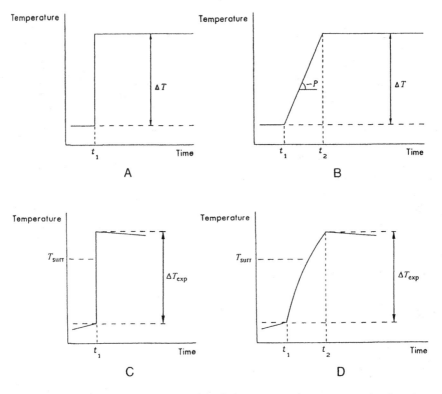

Figure 2. Temperature–time curves for adiabatic type calorimeters (with a low time constant). Curves **A** and **C** show curves following the release of a short heat pulse in an ideal adiabatic calorimeter and a semiadiabatic calorimeter, respectively. Curves **B** and **D** show the curves from experiments where a constant thermal power was released between t_1 and t_2 for an ideal adiabatic calorimeter and a semiadiabatic calorimeter, respectively. For the ideal adiabatic instrument the slope of the curve during the heating period is proportional to the thermal power, P.

as power compensation calorimeters (see below). Figure 2 shows temperature–time curves for an ideal adiabatic calorimeter and for a semiadiabatic calorimeter. For the ideal instrument there is no net heat exchange between the vessel and the surroundings. The temperature during the initial and the final parts of the curves will therefore not change (provided that the heat of stirring is insignificant). But for the semiadiabatic calorimeter a significant heat exchange between the vessel and the surroundings will cause the temperature of the vessel during the fore- and afterperiods to drift in the direction of the temperature of the surroundings, T_{surr}. For the ideal adiabatic instrument, the experimental ΔT-value will directly lead to the Q-value for the reaction (equation 10). But when semiadiabatic instruments are used in accurate work it is necessary to apply a correction to the experimental temperature change, ΔT_{exp}; see, e.g., Sturtevant (1971). For long reaction periods ($\geq$ 30 minutes) such corrections tend to be uncertain.

Heat Conduction Calorimeters

Most calorimeters currently used as monitors for biological processes are of the heat conduction type. In these instruments there is a controlled transfer of heat between the reaction vessel and a surrounding body, usually a metal block, serving as a heat sink. The heat transfer normally takes place through a thermopile wall placed between the vessel and the heat sink (Figure 3). Around the heat sink there is an insulation (usually air) and the assembly is placed in a thermostatted bath (not shown in the figure). The heat capacity of the heat sink should preferably be large so that its temperature stays essentially constant. In modern heat conduction calorimeters, the thermopile wall usually consists of commercially available semiconducting thermoelectric plates (Peltier effect plates, primarily designed for use in thermoelectric cooling devices; cf. the thermoelectric heat pump calorimeter below).

If a heat conduction calorimeter is left for some time and no process takes place in the reaction vessel, there will, ideally, be no temperature gradients in the system made up by vessel, thermopile, and heat sink. The thermopile potential, U, which is proportional to the temperature difference between vessel and heat sink will thus be zero. If a reaction takes place in the vessel and heat is produced (or absorbed), the temperature of the vessel will increase (decrease) leading to $U \neq 0$ (see Figure 4). The temperature gradient will cause the heat evolved in the vessel to flow through the thermopile to the heat sink or, in case of an endothermic process, in the

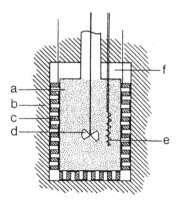

Figure 3. Schematic diagram of a section through a thermopile heat conduction calorimeter. a, calorimetric vessel; b, heat sink; c, thermopile; d, stirrer; e, calibration heater; f, air.

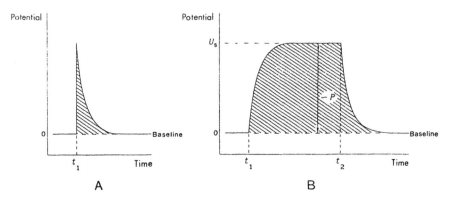

Figure 4. Potential–time curves from experiments with a thermopile heat conduction calorimeter. **A:** A short heat pulse released at time t_1. **B:** A constant thermal power released between time t_1 and t_2. The steady-state potential value, U_s, is proportional to the released thermal power.

reverse direction. If a short heat pulse is evolved, U will soon reach a maximum value and will then decrease exponentially to the baseline value where, ideally, the temperature difference between vessel and heat sink is zero. In case the heat production rate in the vessel is constant, the thermopile potential will asymptotically reach a steady-state value.

For the ideal case where no significant thermal gradients exist in the vessel, the heat production rate will be related to the heat flow through the thermopile, dQ/dt, and to the change of the vessel temperature, dT/dt:

$$P = dQ/dt + C\, dT/dt \tag{14}$$

where C is the heat capacity of the reaction vessel including its content and half of the thermopile wall. For the ideal case, where all heat flows through the thermopile wall, the following applies:

$$dQ/dt = G_c\, n\Delta T \tag{15}$$

where G_c is the thermal conductance associated with each thermocouple and n the number of thermocouples. The resulting thermopile potential is given by:

$$U = n\, e\, \Delta T \tag{16}$$

where e is a material constant for the thermocouples (the Seebeck coefficient). Combination of equations (14)–(16) will lead to

$$P = \varepsilon[U + \tau(dU/dt)] \tag{17}$$

where τ, the time constant for the instrument, is defined by equation (18).

$$\tau = C/(G_c n) \tag{18}$$

ε, which can be regarded as a calibration constant is (in the ideal case) given by equation (19).

$$\varepsilon = G_c/e \tag{19}$$

Equation (17) is usually called the Tian equation. In cases where significant temperature gradients are present within the reaction vessel, two or more time constants must be used. When the change in rate of a process is small, the value for $\tau(dU/dt)$ will often be insignificant compared to the value for U (equation (17)). With heat conduction calorimeters used in work on cellular systems, this is typically the case and the heat production rate is then, with a good approximation, given by the simple expression

$$P = \varepsilon U \tag{20}$$

i.e., the heat production rate is directly proportional to the thermopile signal. Equation (20) holds exactly for a steady-state process, cf. Figure 4B.

The heat released in the calorimetric vessel is obtained by integration of equations (17) or (20) which both will lead to the simple expression

$$Q = \varepsilon \int U dt \tag{21}$$

provided that the initial and the final potentials are the same (normally the baseline value). In such cases the heat quantity is thus proportional to the area under the potential–time curve.

The sensitivity (S) of a thermopile heat conduction calorimeter can be defined as

$$S = U/(dQ/dt) \tag{22}$$

or (cf. equations (15), (16), and (19))

$$S = e/k = \varepsilon^{-1} \tag{23}$$

Thus, in the ideal case and for a given type of thermopile, *the sensitivity of the calorimeter* is *independent of the number of thermocouples in the thermopile wall.* Furthermore, and in contrast to adiabatic instruments, the sensitivity of a thermopile heat conduction calorimeter is *independent of the heat capacity of the reaction vessel and its content.*

The thermal inertia of a heat conduction calorimeter is described by its time constant τ (equation (18)). In practice, τ is given by

$$\tau = C/G \tag{24}$$

where G is the total thermal conductance of the thermopile wall, plus that of leads, supports, surrounding air, etc. τ can be determined from the decay rate of the thermopile signal as the thermal power produced by an electrical calibration heater is turned off. When no heat is evolved in the vessel, the decay rate of the thermopile potential is (equation (17); P = 0)

$$dU/U = -\tau^{-1}\,dt \tag{25}$$

or

$$U_t = U_0 e^{-t/\tau} \tag{26}$$

where U_t and U_0 are the thermopile potential at time t and time zero, respectively. The half-time of the exponential decay, $t_{1/2,}$ is thus

$$t_{1/2} = \tau \ln 2 \tag{27}$$

Time constant for thermopile heat conduction calorimeters used in biochemistry–cell biology are usually in the range 1–10 minutes. The decay signal will approximate the baseline value by 99.99% in about 9 τ. If the differential term in equation (17) is significant, the thermopile signal (U) must be corrected for the thermal inertia of the calorimeter in order to be directly proportional to the actual thermal power (dynamic correction) (see, e.g., Randzio and Suurkuusk, 1980). Note, however, that the expression for the heat quantity evolved, equation (21), only contains the potential–time integral and a calibration factor, ε, which solely depends on the design of the instrument and the properties of the thermopile.

Power Compensation Calorimeters

When an adiabatic calorimeter is used to monitor an exothermic process, there will be a lasting temperature increase of the reaction vessel. In a heat conduction calorimeter the exothermic process will cause a temporary temperature increase. In both cases it is possible to compensate for the exothermic process by applying a cooling power to the reaction vessel. Most conveniently this can be done by use of the Peltier effect principle: when an electrical current is allowed to pass through a thermocouple (or thermopile plate) there will be a cooling effect in one of the junctions (on one side of the thermopile plate) and a heating effect in the other junction (on the other side of the plate). In a practical case, one or more thermopile plates are positioned between the calorimetric vessel and a surrounding heat sink which usually consists of a thermostatted water bath or metal block similar to the arrangement in a thermopile heat conduction calorimeter (Figure 5). The cooling power is proportional to the electrical current, I. But the Peltier cooling is superimposed on the joule heating effect caused by the electrical resistance of the thermopile circuit. Part of this heating power is released where the Peltier cooling takes place and will thus reduce the value for the cooling power, $-P_c$, of the thermoelectric heat pump:

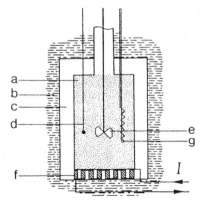

Figure 5. Schematic diagram of a section through a power compensation calorimeter. a, calorimetric vessel; b, heat sink (e.g., a thermostatted water bath); c, air or vacuum; d, thermometer; e, stirrer; f, thermopile; g, calibration heater; I is the current through the thermopile.

$$-P_c = \pi I - r I^2 \tag{28}$$

where π is a material constant and r is the effective resistance value for the thermopile circuit. The signal from a separate thermopile or from a thermometer is used to regulate the current in order to keep the temperature of the vessel constant and equal to that of the surrounding heat sink. Under such conditions the heat production rate in the vessel will be equal to P_c.

Heat can also be pumped out from a calorimeter by other principles, e.g., by use of a cooling liquid, but such procedures are now only of historical interest in connection with laboratory calorimeters. Mainly of historical interest is also compensation of exothermic processes by use of melting of a solid, e.g., ice, surrounding the calorimeter vessel. Lavoisier used such an "ice calorimeter" (later often called a Bunsen calorimeter) in his pioneering biocalorimetric work (see Kleiber, 1961). For endothermic processes, compensation is easily achieved by release of electrical energy in the vessel.

Further Definitions and Concepts

Most calorimeters used in biochemical work and in studies of living cells and tissue pieces are usually *microcalorimeters*. This term is not well defined but the micro- prefix is primarily used for calorimeters with a thermal power sensitivity of 1 μW or better. The volume of a microcalorimetric (batch) vessel is usually 1–25 ml. It is common, and frequently suitable, to use typical microcalorimeters at a reduced sensitivity, for example, in work on fast growing microbial suspensions or on small animals where the thermal power evolved may be several orders of magnitude larger than the sensitivity of the instrument.

The term "macrocalorimeter" is not used very often but will usually imply a vessel volume of 0.1–1 l and a power sensitivity significantly lower than what is typical for microcalorimeters. A "whole body calorimeter" is a term used for large, up to room-size, calorimeters employed for direct measurement of heat produced by humans and large animals, sometimes under conditions where the objects are conducting some sort of physical exercise.

The term "indirect calorimetry" is often used in biology when the sum of metabolic processes in an animal, e.g., a human being, is investigated by measurement of rates of consumption of oxygen and the production of carbon dioxide and

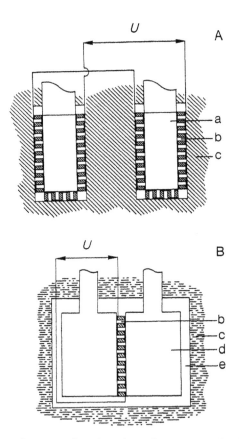

Figure 6. Schematic diagrams of sections through two types of twin calorimeters. **A:** Twin heat conduction (micro)calorimeter (vessels not shown). **B:** Semiadiabatic twin calorimeter. a, vessel holder; b, thermopile; c, heat sink; d, vessel (stirrer, thermometer, heater not shown); e, air or vacuum.

other metabolic products. From known heat of formation data for foodstuff and metabolic products, it is then possible to estimate the heat production rate.

Practically all microcalorimeters currently used in work of biological interest are arranged as twin instruments. Such calorimeters have two identical, or similar, reaction vessels. Two different types of twin-calorimetric arrangements are shown schematically in Figure 6. The instrument in Figure 6A is a typical twin thermopile heat conduction calorimeter. The two vessels are surrounded by thermopiles (as in Figure 3) and the two calorimetric units are positioned in a common heat sink. The thermopiles are connected in opposition and the differential signal is thus recorded. In Figure 6B one thermopile is placed between the twin vessels which is surrounded by a common thermostatted bath (heat sink). The thermopile will thus measure the temperature difference between the vessels. For the instrument type indicated in Figure 6B there is poor thermal contact between the vessels and the heat sink and the instrument arrangement may thus be called a semiadiabatic twin calorimeter.

When a twin calorimeter is used, one of the vessels in a twin instrument is charged with the reaction system that will be studied. The other will usually contain a nonreacting system (a reference system). One may expect that thermal and electrical disturbances from the surroundings will affect both vessels to nearly the same extent and their contributions to the differential signal will thus cancel out or at least be reduced significantly. This feature is of particular importance in experiments of long duration. Twin calorimeters can also be used to study two reaction systems simultaneously and thus record their difference in heat production. For example, if both vessels are charged with the same amount of a certain cell suspension, the differential signal is expected to be zero. But if a drug which interferes with the metabolic activity of the cells is added to one of the vessels, the differential signal will directly reflect the effect of the drug. It should be pointed out that more information is obtained if such experiments are conducted as two separate measurements (in microcalorimetry preferably conducted by use of two twin instruments).

Calorimeters of any type in twin arrangements can also be used as power compensation calorimeters: an exothermic process in one of the vessels can be simulated by evolution of electrical energy in the other (a "thermal balance").

In a batch microcalorimetric experiment, reaction components usually are brought together by means of an injection technique or by mixing the contents of two separate compartments through rotation or rocking of the calorimeter. In flow (perfusion) calorimeters employing continuous flow or stopped flow techniques, mixing may take place in a small mixing chamber or in a stirred vessel (Wadsö, 1987). Experiments with cell suspensions are sometimes conducted as "flow-through experiments," i.e., the cell suspension is pumped or sucked through a flow calorimeter (a perfusion calorimeter).

Isothermal Calorimetry and Temperature Scanning Calorimetry

Thermodynamic data such as enthalpy changes and equilibrium constants are to different degrees temperature-dependent and their temperature derivatives are often of great intrinsic interest. Heat production rates of chemical processes are also temperature-dependent; for living systems the metabolic rate typically doubles for a temperature change of 10 °C up to a maximum value. In all kinds of calorimetric experiments it is therefore important to know with sufficient accuracy the measurement temperature(s) and to which temperature a derived thermodynamic value will refer. As was pointed out earlier, the thermodynamic value for a process reported in the literature, e.g., the molar enthalpy change always will refer to the isothermal process conducted at a specified temperature. If a ΔH-value is determined by use of an isothermal calorimeter, such as a thermoelectric heat pump calorimeter (Figure 5), the determined enthalpy change simply will refer to the temperature of the thermostatted bath. Experiments with thermopile heat conduction microcalorimeters can, in practice, normally also be considered as isothermal: the initial and the final state of the system are very nearly at the same temperature if the heat sink is large. In experiments where a more or less continuous heat evolution is monitored, e.g., from living cell systems, the measurement temperature will be higher than that of the heat sink. However, semiconducting thermopile plates have a high heat conductance and the temperature change will normally not cause any significant effect on the enthalpy value or on the kinetics of the process.

For adiabatic type calorimeters, the initial and the final temperatures are by definition different. To which temperature will a derived ΔH-value then refer? In connection with microcalorimetric measurements, the problem may merely be academic as the temperature change may be very small. However, it is at this point instructive to analyze in some detail what we do when we calibrate an adiabatic calorimeter. Let the initial and the final state of the experimental process be represented by A and B and the corresponding temperatures by T_A and T_B, respectively. The experimental process is thus:

$$A(T_A) \rightarrow B(T_B) \tag{29}$$

As the system is adiabatic, no heat is lost and the enthalpy change of the system is zero. We can, hypothetically, conduct the same process by many other routes, e.g., those indicated in Figure 7. Route a consists of two steps: the reaction conducted isothermally at temperature T_A

$$A(T_A) \rightarrow B(T_A) \tag{30}$$

followed by a temperature change of the product system B from temperature T_A to temperature T_B

$$B(T_A) \rightarrow B(T_B) \tag{31}$$

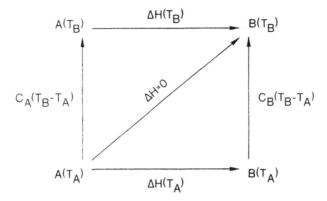

Figure 7. One experimental and two hypothetical routes from state A at temperature T_A to state B at temperature T_B, in an adiabatic calorimeter.

This temperature change, which in practice may be caused by release of electrical energy, is equivalent to a calibration process. The enthalpy change for the isothermal step is $\Delta H(T_A)$ and for the heating step the enthalpy change is C_B ($T_B - T_A$) where C_B is the average heat capacity of the product system in the temperature range T_A to T_B. We usually employ a practical heat capacity value which we call calibration constant, ε_B (equation (10)). From Figure 7 it is clear that

$$\Delta H(T_A) + \varepsilon_B (T_B - T_A) = 0 \qquad (32)$$

or

$$-\Delta H(T_A) = \varepsilon_B \Delta T_{exp} \qquad (33)$$

where ΔT_{exp} is the temperature increase in an ideal adiabatic experiment. Thus, if an adiabatic calorimeter is calibrated with the final reaction system, the derived enthalpy change will refer to the initial temperature of the experimental process. If the corresponding calibration process would have been conducted on the initial system (A), the enthalpy change would have referred to temperature T_B.

Temperature scanning calorimeters are used in experiments where the temperature changes continuously, not due to some chemical process but rather by electrical heating or cooling. Special instruments have been designed for such measurements. They are usually twin instruments and are therefore called differential (temperature) scanning calorimeters (DSCs) and are primarily used for studies of processes initiated by an increase of temperature, e.g., melting, decomposition, or denaturation. This group of calorimeters has become very important as general analytical tools (in particular in industrial laboratories) and as thermodynamic instruments. In biophysical chemistry the technique has become of first rate importance for studies of thermal transition (denaturation) processes involving proteins, nucleic

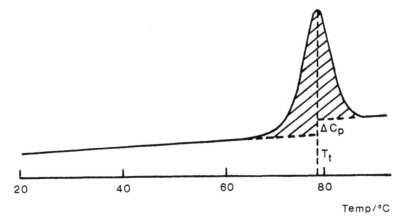

Figure 8. A DSC record (a thermogram) from a thermally induced transition of lysozyme in dilute aqueous solution. T_t = transition temperature; ΔC_p = change in heat capacity accompanying the unfolding process. The hatched area is proportional to the enthalpy of transition. Adapted from Privalov (1980).

acids, and lipids. So far their use in work on living systems has been marginal except for work on plant materials. Work on biochemical systems is usually conducted on dilute solutions or suspensions and high sensitivity is needed. Instruments used for such high sensitivity DSC work are often microcalorimeters.

In order to distinguish DSCs from nonscanning calorimeters, it has become common to use the term isothermal microcalorimeter for nonscanning microcalorimeters even if they, in a strict sense, are not isothermal, e.g., thermopile heat conduction microcalorimeters. The two calorimetric vessels in a DSC are heated at a uniform rate, e.g., in the order of 1 °C per minute. In an adiabatic-type DSC the temperature difference between the two vessels is monitored and kept at zero by use of power compensation of the vessel lagging behind. If the basic heating power for the two vessels is identical, this compensation power—which forms the calorimetric baseline signal—is proportional to the heat capacity difference between the two vessels, provided that no process takes place in the reaction vessel. During a process in the reaction vessel (Figure 8), the power compensation signal above that of the baseline corresponds to the thermal power of the process and the integrated area (dashed line in Figure 8) is proportional to the enthalpy change.

Also DSC instruments of the heat conduction type exist (see, for example, Ross and Goldberg, 1974). If the temperature of the heat sink is increased, heat will flow through the two thermopiles into the two vessels. If heat conductance and heat capacity of the two calorimetric units are the same, we can expect that the differential potential signal from the thermopiles will be zero. However, if the heat capacity differs, for example, between an aqueous solution in the sample vessel and pure water in the reference vessel, the differential potential will be proportional to the heat capacity difference. Obviously the power compensation principle can also be applied on this kind of instrument.

SOME APPLICATION AREAS

In this section several practical applications of interest for the medical field, in a broad sense, will be illustrated.

Thermodynamics of Purified Biochemical Systems

A significant part of medical and pharmacological research is conducted on the biochemical level. Calorimetric work on such systems usually has the character of thermodynamic measurements and can be considered as part of biophysical chemistry. Two types of experiments currently seem to be the most important: studies of binding processes using titration microcalorimetry and investigations of thermal transitions involving high sensitivity DSCs.

Enzyme-inhibitor binding (where the inhibitor might be a drug), and association reactions involving biopolymers are important examples of binding studies. Figure 9 shows results from a microcalorimetric titration experiment where aliquots of a solution of 2' cytidine monophosphate is injected stepwise into a vessel charged with ribonuclease A. The experiments were conducted by a twin calorimeter of the type indicated in Figure 6B which employs the power compensation principle (Wieseman et al., 1989). It is seen that the areas under the peaks (which are proportional to the heat evolution accompanying each injection), will become smaller as the equilibrium gradually is approached, i.e., decreasing fractions of the added inhibitor will bind to the protein. The pattern formed by the titration peaks depends both on the molar enthalpy change and on the equilibrium constant for the reaction which in this case is moderately high, about 10^5 l/mol. Values for $\Delta H°$ and

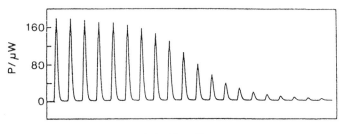

Injection Sequence

Figure 9. Calorimetric record from a microcalorimetric titration experiment. Four μl of 2'CMP were injected stepwise into the calorimetric vessel (volume 1.4 ml) which was charged with 0.65 mM RNase solution. Adapted from Wieseman et al. (1989).

K (and thus also for $\Delta G°$ and $\Delta S°$, equations (6), (7), can be derived using an iterative procedure provided that the K-value is not higher than about 10^7 (for a 1:1 binding process) and, of course, that the ΔH is not close to zero (see, e.g., Wieseman et al. (1989), Freire et al. (1990), and Bäckman et al. (1994)). If the binding constant is very high (>10^8 for a 1:1 binding process), the injected substance will be almost completely bound at each injection until the equivalence is reached. Experimental titration curves must frequently be corrected for effects of dilution and ionization/protonation. In order to calculate the thermodynamic quantities it is necessary to assume a binding model (1:1, 1:2, 2:1, etc.) and to test if the derived values agree with the assumed model.

The calorimetric technique used in the titration experiment illustrated in Figure 9 allows short time intervals between the injections due to a comparatively low time constant for the instrument in combination with the electrical compensation technique. Rather, slow heat conduction microcalorimeters can be used in fast titration experiments if a dynamic correction, based on the Tian equation (equation (17)), is employed (Bastos et al., 1991; Bäckman et al., 1994).

Figure 8 shows a very simple transition curve determined by a high sensitivity DSC. For many proteins much more complex thermograms are obtained. By different deconvolution techniques, the experimental curves can be resolved into several peaks which may be linked to the transition of different cooperative regions in the protein (see Privalov (1989)).

Quantitative Analyses of Biochemical Compounds

As has been pointed out earlier, calorimetry can be employed as a general, quantitative analytical principle. This was recognized during earlier phases in the development of modern microcalorimetry, in particular, systems involving enzymatic reactions where the enzyme specificity can compensate for lack of specificity of heat measurements. However, it soon turned out that "real" microcalorimeters had a small chance to become competitive in that field. But there exists an interesting group of instruments, sometimes called calorimetric biosensors, which has some potential as analytical tools in the clinical area and in biotechnology (Danielsson, 1990), especially in cases where changes during the enzymatic processes cannot be measured easily by some convenient spectrophotometric technique. In calorimetric biosensors, a flow of substrate is pumped through a miniaturized column charged with an immobilized enzyme. The temperature of the liquid flow is measured before and after the passage of the column. The temperature difference will be proportional to the substrate concentration provided that the enzyme will not become saturated with substrate. These instruments can usually be characterized as semiadiabatic flow calorimeters. Thermistors are used for the temperature measurements and the instruments are often named enzyme thermistors (which is a misleading term). Although calorimetric biosensors are much less sensitive than "real microcalorimeters" in terms of detectable heat production rate, it is likely that their power sensitivity can be significantly increased.

Growth and Metabolism of Microorganisms

Many calorimetric investigations have been reported on microbial systems in fundamental metabolic studies and in work directed towards applications in biotechnology, ecology, pharmacology, and the clinical areas. Two examples are given here.

When bacteria and mycoplasma grow on complex media, their heat production rates show very complex patterns reflecting the gradual change of the media. Such growth curves are typically very different for different strains growing on the same media and it has been suggested, and also demonstrated, that such thermal power time curves can be used for the identification of microorganisms. However, it has also been shown that several experimental parameters, in particular the history of the inoculum, must be strictly controlled (Newell, 1980; James, 1987b). This appears to significantly reduce the value of calorimetric finger-print identification, but not the use of calorimeters as monitors for growth and growth phases of microorganisms. An interesting application of that kind is provided by tests of antibiotic efficiencies which have been demonstrated in several investigations. One

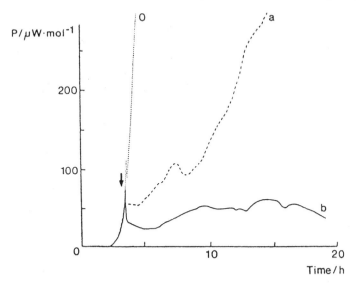

Figure 10. Microcalorimetric growth experiments with *Escherichia coli* in the presence of antibiotics (1.6 μg/ml = twice the minimum inhibitory concentration), added at the time indicated by the arrow. a, tetracycline; b, doxycycline; 0, control experiment. Obviously doxycycline suppressed the growth more efficiently than tetracycline. Adapted from Mårdh et al. (1976).

example is shown in Figure 10. The value of microcalorimetric techniques in work concerned with the development and the control of antibiotics and growth media is now well documented but has not been widely used in practice.

Characterization of Human and Animal Cellular Systems

Much methodological work has been conducted during the last two decades on microcalorimetric techniques for the investigation of cell preparations of human and animal origin. To a large extent that work has been motivated by the possible use of microcalorimetry in clinical and pharmacological research and development work (cf. chapter 11 by Kemp). In particular, reference should be made to the work on the major fractions of blood cells: erythrocytes, granulocytes, lymphocytes, and platelets (Monti, 1987). Many method-oriented studies have also been conducted on macrophages, adipocytes, hepatocytes, sperm cells, and on several types of cultured tissue cells (Wadsö 1988; Kemp 1991, 1993). In most cases microcalorimeters equipped with simple static ampoules were used. Such measurements are very simple to perform but part of the cells will be sedimented, or at least the cell concentration will be different in different parts of the calorimetric vessel. The pH and the concentration of metabolites and medium constituents, including oxygen, will also vary. Further, the technique does not allow addition of reagents during a measurement. Flow-through calorimetry has also been used on these cell systems but is in general less sensitive and requires more material than batch measurements. For some types of cells, such as granulocytes, adhesion/sedimentation in the flow lines and in the flow vessel may cause problems.

More recently, vessels with stirring allowing uniform and well-defined cell concentration are being used (Wadsö, 1987). Such vessels are equipped with devices for injection (of reagents or of cells) and they may also have miniaturized electrodes for the continuous recording of pH and O_2 concentration, with simultaneous calorimetric measurement (Bäckman and Wadsö, 1991) (Figure 11A). As an example, Figure 11B shows a record from an experiment with cultured T-lymphoma cells. The vessel was charged with growth medium (no gas phase present) and introduced to the measurement position in the calorimeter via a heat exchange tube. Following an equilibration period, 100 μl of cell suspension was injected (about 10^6 cells) at time zero. The calorimetric curve increased momentarily to about 30 μW after which a gradual increase due to aerobic cell growth was observed. The signal from the oxygen electrode showed a fast linear decrease during this phase, reaching zero after about 5 hours. The pH also decreased. The resulting changes in the metabolic pattern during the anaerobic state of the cells are reflected by changes of the heat production rate: first a rapid decrease, then a recovery followed by a slow decrease. Results from experiments of the type indicated in Figure 11B are used in the analysis of the energy metabolism of cellular systems. It is essential that such calorimetric measurements be accompanied by

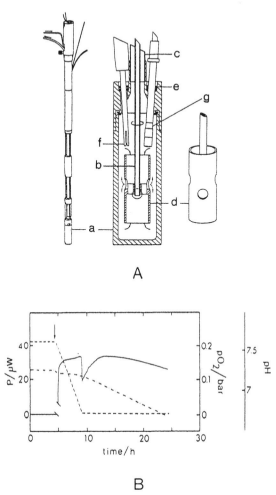

A

B

Figure 11. Electrodes in microcalorimetric vessels. **A:** Schematic diagram of a section through a titration–perfusion microcalorimetric vessel equipped with a polarographic oxygen electrode and a pH electrode. a, sample compartment, volume 3 ml; b, hollow stirrer shaft; c, steel tube; d, turbine stirrer; e, O-rings; f, combination pH electrode protected by a steel tube; g, polarographic oxygen sensor (Clark electrode). **B:** Record from a growth experiment with T-lymphoma cells. The vessel was completely filled with medium. Once the baseline had been established, the experiment was started (as indicated by the arrow) by the injection of 100 μl concentrated cell suspension. _____ heat production rate; ----- oxygen pressure; —·—·— pH. From Bäckman and Wadsö (1991).

accurate pH-measurements—heat production rates are very pH-dependent. In the present example, the pH decrease caused the aerobic growth to be essentially linear instead of exponential.

With the type of calorimetric vessel shown in Figure 11A, it is also possible to perfuse the medium through the vessel. This is particularly useful in studies of cells adherent to a solid support (for example a plastic film or "microcarriers") and in work with tissue pieces (see below). Under such conditions it is possible to conduct the calorimetric measurements over long periods of time since fresh medium, with oxygen, can be supplied continuously. The experiments may also be conducted with a gas phase present, but under such conditions the oxygen electrode cannot be used for a quantitative measurement of oxygen consumption. Its role is then mainly to warn the experimenter when the oxygen concentration gets too low.

Long-term experiments (> 3 hours) with human and animal cells are usually conducted in the presence of antibiotics in order to prevent bacterial growth. Control experiments should then be made in order to show that the antibiotic, at the concentration used, does not significantly interfere with the metabolism of the mammalial cells.

The same strain of cultured T-lymphoma cells as was used in the experiments discussed above has been tested for its sensitivity against antineoplastic drugs (Figure 12). Such experiments are believed to become important in drug developments and, possibly, in predictive clinical tests but have not yet been analyzed or evaluated in much detail.

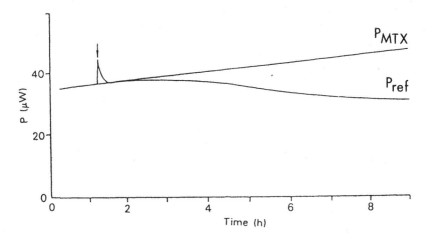

Figure 12. Calorimetric record from an experiment with growing T-lymphoma cells where the antineoplastic drug methotrexate (MTX) is injected at the time indicated by the arrow. The curve is superimposed on the record from a reference experiment, conducted simultaneously on an identical cell sample. From Schön and Wadsö (1988).

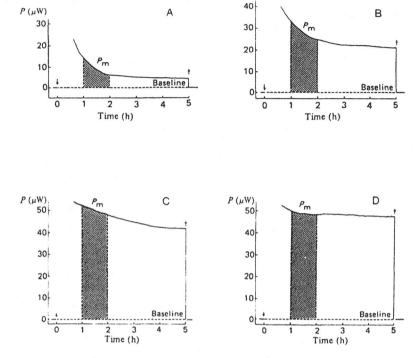

Figure 13. Calorimetric curves from experiments with 50 mg of human skeletal muscle using different calorimetric techniques. **A**, Static ampoule used as reaction vessel. **B**, Perfusion vessel without stirring. **C** and **D**, Perfusion vessels with stirring. In all cases the measurement temperature was 37 °C but in experiments **A**, **B**, and **C** sample preparation and storage (1 hour) was at 4 °C. In experiment **D** the sample was prepared and stored at 37 °C. Adapted from Fagher et al. (1986).

Pieces from muscle, nerve, and fat tissue have been investigated in many microcalorimetric experiments. Among some recent work of clinical and pharmacological interest are several studies of muscle fiber bundles of human and animal origin (see, e.g., Fagher et al., 1993). Figure 13 shows results from four microcalorimetric experiments carried out with almost identical 50 mg samples of human muscle fiber bundles. The calorimetric techniques employed were different and the results clearly illustrate the importance of arranging for appropriate experimental conditions when calorimetric experiments are performed on cellular systems. The results shown in Figure 13A were obtained by use of a static vessel, in this case the sample rested on the bottom of the vessel (a sealed steel ampoule). A low and decreasing calorimetric signal was seen. Most probably pH close to the tissue decreased significantly compared to the rest of the medium. Further, it is possible that oxygen locally became depleted. Figure 13B shows a record from the same type of experiment conducted with a perfusion vessel. A flow of medium (5 ml/hour) passed through the vessel. The content was not stirred. Also in this case the tissue rested at the bottom of the vessel but the continuous flow of fresh medium led to more stable and thus more favorable physiological conditions, resulting in an increased rate of heat production. Figure 13C shows the result of a corresponding experiment conducted in a different type of perfusion vessel. The medium flow was the same but in this case the fiber bundles were contained in a rotating cage, giving them better contact with the medium. Further, in this case, there should be no risk in the development of concentration gradients in the vessel. It is seen that the heat production rate is slightly higher and declining at a slower rate than in Figure 13B. In experiments A–C preparation and storage (1 hour) of the samples prior to the calorimetric experiments were conducted at about 5 °C. The results shown in Figure 13D were obtained exactly as in Figure 13C, except that the preparation and storage temperature was about 37 °C, resulting in a more constant heat production rate (also found with other mammalian cell systems).

Whole Body Calorimetry

Frequently typical microcalorimeters are used in work on small animals in basic physiology and in ecology applications. Of more direct medical interest are investigations carried out in large calorimeters primarily designed for measurements of human subjects, known as "whole body calorimeters." Figure 14 schematically shows a comparatively small human calorimeter. The instrument can be characterized as a heat conduction calorimeter; the measurement chamber is surrounded by a "gradient layer" consisting of 2.4 mm thick epoxy with a copper circuit printed on both sides. The copper circuits serve as resistance thermometers which will record the temperature gradient over the epoxy layer. This gradient is proportional

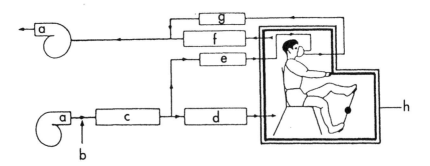

Figure 14. Schematic picture of a whole body calorimeter with its ventilatory system. a, Fan; b, steam; c, saturator; d, heater; e, respiratory heater; f, condenser; g, respiratory condenser; h, gradient layer. Adapted from Jéquier et al. (1975).

to the heat flow from the chamber. A number of corrections, such as evaporative heat loss from the subject, have to be applied.

The instrument reported by Dane et al. (1985) was designed for use with newborn babies and has a comparatively small measurement chamber. Very large human calorimeters have also been designed, e.g., the 24 m^3 room calorimeter described by Jacobsen et al. (1985). Currently only very few human calorimeters appear to be in use.

Use of Calorimetric Techniques in the Pharmaceutical Industry and in Medical Techniques

The development, manufacturing, and storage control of drugs has direct bearings on medicine, and some important uses of calorimetry in the pharmaceutical industry will therefore be pointed out. As a result of recent developments in microcalorimetry, techniques for thermodynamic characterization of binding reactions between drugs and biopolymers have become readily accessible. To an increasing extent, titration microcalorimetry is now used in the pharmaceutical industry.

Surprisingly, as yet very little calorimetric work on drugs seems to be conducted on the cellular or animal level. By far most calorimeters employed in the pharmaceutical industry are used for measurements of different physical and chemical properties of drugs and of materials used in the preparation of drugs. Currently, the most important instrument type is the moderately sensitive DSC instrument. But more recently, many industries have started to use "isothermal" microcalorimeters for the characterization of the physical and chemical stability of drug materials, including effects of their hygroscopic and crystallinic properties, and of the compatibility between drugs and excipients. Figure 15 shows results from a comparison at 50 °C between solid samples from different batches of a drug. Lots 1 and 2 are obviously inferior to lot 3.

The use of artificial materials in contact with blood and other tissues is essential in today's medical practice: Some are extracorporeal circulation devices such as artificial kidneys and oxygenators, cardiovascular prosthesis, blood vessel replacements, blood access devices, and pacemakers. Although great progress has been made in these areas during the last few decades, many fundamental problems remain, including noncompatibility between human tissues and the different materials employed, in particular organic polymers and metals. Problems of this nature are studied by microcalorimetric techniques both on the protein level (adsorbed layers of plasma proteins can initiate cascade reactions leading to the formation of blood clots) and on the level of metabolic studies of cells. A prime example of work in the latter area is provided by a study of human granulocytes in contact with a few organic polymers which are used as hemodialysis membranes: polyacrylonitrile, polyether polycarbonate, and regenerated cellulose. Fluorinated ethylene

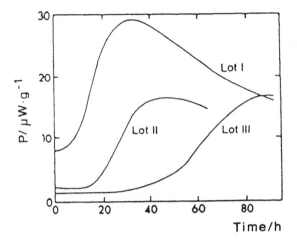

Figure 15. Thermal power–time curves for different lots of a derivative of the drug lovastatine. Experiments were conducted at 50 °C. Adapted from Hansen et al. (1989).

propylene was used as reference material. The experiments were conducted in static microcalorimetric vessels lined with the polymers. The cells were suspended in plasma. Basal values for the heat production rates were determined as well as peak values following stimulation of the cells with zymosan. The results obtained are summarized in Table 1.

It is seen that the basal values increase in the order fluoroethylene-polypropylene < polyacrylonitrile < polyetherpolycarbonate < regenerated cellulose which was interpreted as a sign of increasing cell–polymer interaction. (The fluorinated reference polymer is assumed to be compatible with the cells.) When the functional activity of the cells was tested by stimulation with zymosan, the values varied in the reverse order, i.e., the interactions are harmful as it will lead to a decreased ability of the cells to phagocytoze. (It should be noted that uremic patients on hemodialysis are particularly prone to infections.) Another interesting use of

Table 1. Human Granulocytes in Contact With Organic Polymers; Heat Production Rates (P) and Stimulation by Zymosan

Polymer	Basal Value, P/pW per Cell	Stimulation, % of Basal Value
FEP	1.5 ± 0.3	1240 ± 470
Polyacrylonitrile	3.1 ± 0.6	590 ± 160
Polyetherpolycarbonate	5.5 ± 2.0	300 ± 80
Regenerated cellulose	8.9 ± 6.0	130 ± 190

Note: Fluoroethylene-polypropylene (FEP) was used as reference material. pW = pico watt = 10^{-12} W. From Monti et al., 1993.

microcalorimetry connected with medical techniques is the testing of pacemaker batteries. In case a battery is impaired by internal corrosion, which will decrease its lifetime, heat will be produced and can be quantitatively assessed by calorimetry. Only calorimetry can measure effects from corrosion or any other process taking place inside a sealed container!

ARE CALORIMETRIC RESULTS ACCURATE?

All calorimeters must be calibrated, the instrument signal must be standardized by use of a well-defined heat quantity or thermal power. In most cases this is done by use of evolution of heat in an electrical resistor (an "electrical heater"). It is easy to measure electrical power and electrical energy more accurately than needed for any process of biological interest, but it can be difficult to arrange for a strict comparison between the heat flow pattern in a calibration experiment and that in a chemical or biological process. In fact, due to different practical considerations, the design of many microcalorimeters in current use is far from ideal with respect to the employed electrical calibration technique. As may be expected, many users look upon their instruments as black boxes operated by means of a few push buttons and a computer. Sometimes this approach can be perfectly acceptable but in other cases it may lead to significant errors or misinterpretations, even when the instruments are in good condition and used as recommended. Users of all kinds of calorimeters should be aware of the fact that practically all processes—physical, chemical or biological— give rise to heat effects. This property is most valuable as it makes calorimeters very powerful tools for the discovery and quantitative characterization of unknown or unexpected processes, as in biology. But this property can also lead to systematic errors and serious misinterpretations of calorimetric signals. In advertisements for commercial instruments it is rarely pointed out that precise calorimetric measurements can also be inaccurate measurements! Calorimetry is more vulnerable to systematic errors and misinterpretation than most other measurement techniques. This is particularly important in microcalorimetry where heat effects due to, for instance, mechanical disturbances or evaporation effects, can easily be of the same magnitude or larger than the process investigated. Users of calorimeters should therefore be acquainted with these problems, as well as with the calibration procedures and test processes suitable in different experimental situations. Problems with electrical calibrations and the use of chemical calibrations and chemical test processes, with particular reference to microcalorimetric measurements of biological significance, have recently been discussed (Briggner and Wadsö, 1991; Wadsö, 1993).

REFERENCES

Bäckman, P. & Wadsö, I. (1991). Cell growth experiments using a microcalorimetric vessel equipped with oxygen and pH-electrodes. J. Biochem. Biophys. Meth. 23, 283–293.

Bäckman, P., Bastos, M., Hallén, D., Lönnbro, P., & Wadsö, I. (1994). Heat conduction calorimeters: Time constants, sensitivity and fast titration experiments. J. Biochem. Biophys. Meth. 28, 85–100.

Bastos, M., Hägg, S., Lönnbro, P., & Wadsö, I. (1991). Fast titration experiments using heat conduction calorimeters. J. Biochem. Biophys. Meth. 23, 255–258.

Beezer, A.E. (ed.) (1980). Biological Microcalorimetry. Academic Press, London.

Belaich, J.P., Beezer, A.E., Prosen, E., & Wadsö, I. (1982). Calorimetric measurements on cellular systems: Recommendations for measurements and presentation of results. Pure Appl. Chem. 54, 671–679.

Briggner, L.-E. & Wadsö, I. (1991). Test and calibration processes for microcalorimeters, with special reference to heat conduction instruments used with aqueous systems. J. Biochem. Biophys. Meth. 22, 101–118.

Dane, H.J., Holland, W.P.J., Sauer, P.J.J., & Visser, H.K.A. (1985). A calorimetric system for metabolic studies of newborn babies. Clin. Phys. Physiol. Mess. 6, 37–46.

Danielsson, B. (1990). Calorimetric biosensors (a minireview). J. Biotech. 15, 187–200.

Fagher, B., Monti, M., & Wadsö, I. (1986). A microcalorimetric study of heat production in resting skeletal muscle from human subjects. Clinical Science 70, 63–72.

Fagher, B., Liedholm, H., Sjögren, A., & Monti, M. (1993). Effects of terbutaline on basal thermogenesis of human skeletal muscle and Na-K pump after 1 week of oral use—a placebo controlled comparison with propranolol. Br. J. Clin. Pharmac. 35, 629–635.

Freire, E., Mayorga, O.L., & Straume, M. (1990). Isothermal titration calorimetry. Anal. Chem. 62, 950A–959A.

Hansen, L.D., Lewis, E.A., Eastough, D.J., Bergstrom, R.G., & DeGraft-Johnson, D. (1989). Kinetics of drug decomposition by heat conduction calorimetry. Pharm. Res. 6, 20–27.

Hinz, H.-J. (1986). In: Thermodynamic Data for Biochemistry and Biotechnology. Springer-Verlag, Heidelberg.

Jacobsen, S., Johanson, O., & Garby, L. (1985). A 24-m^3 direct heat-sink calorimeter with on-line data acquisition, processing and control. Am. J. Physiol. 249, E416–E432.

James, A.M. (1987a). Growth and metabolism of bacteria. In: Thermal and Energetic Studies of Cellular Biological Systems (James, A.M., ed.), pp. 68–105, Wright, Bristol.

James, A.M. (ed.) (1987b). In: Thermal and Energetic Studies of Cellular Biological Systems. Wright, Bristol.

Jéquier, E. (1975). Direct calorimetry. A new clinical approach for measuring thermoregulatory responses in man. Bibl. Radiol. 6, 185–190.

Jones, M.N. (1988). In: Biochemical Thermodynamics. Elsevier, Amsterdam.

Kemp, R.B. (ed.) (1990). In: Biological Calorimetry. Thermochim. Acta 172 (special issue), 1–271.

Kemp, R.B. (1991). Calorimetric studies of heat flux in animal cells. Thermochim. Acta 193, 253–267.

Kemp, R.B. (1993). Developments in cellular microcalorimetry with particular emphasis on the valuable role of the energy (enthalpy) balance method. Thermochim. Acta 219, 17–41.

Kleiber, M. (1961). In: The Fire of Life. Wiley, New York.

Lamprecht, I., Hemminger, W., & Höhne, G.W.H. (eds.) (1991). In: Calorimetry in the Biological Sciences. Thermochim. Acta 93 (special issue), 1–452.

Mårdh, P.-A., Andersson, K.-E., Ripa, T., & Wadsö, I. (1976). Microcalorimetry as a tool for evaluation of antibacterial effects of doxycycline and tetracycline. Scand. J. Infect. Dis. Suppl. 9, 12–16.

Marsh, K.N. & O'Hare, P.A.G. (eds.) (1994). In: Experimental Chemical Thermodynamics. Vol. 2. Solution Calorimetry. Blackwell, Oxford.

Mills, I., Cvitas, T., Homan, K., Kallay, N., & Kuchitsu, K. (1993). In: Quantities, Units and Symbols in Physical Chemistry. Blackwell, Oxford.

Monti, M. (1987). In vitro thermal studies of blood cells. In: Thermal and Energetic Studies of Cellular Biological Systems (James, A.M., ed.), pp. 131–146, Wright, Bristol.

Monti, M., Ikomi-Kumm, J., Ljunggren, L., Lund, U., & Thysell, H. (1993). Medical application of microcalorimetry in human toxicology. A study of blood compatibility of haemodialysis membranes. Pure Appl. Chem. 65, 1979–1981.

Newell, R.D. (1980). The identification and characterization of microorganisms by microcalorimetry. In: Biological Microcalorimetry (Beezer, A.E., ed.), pp. 163–186, Academic Press, London.

Parrish, W.R. & Lewis, E.A. (eds.) (1996). Handbook of Calorimetry. Marcel Dekker, New York. In press.

Privalov, P.L. (1980). Heat capacity studies in biology. In: Biological Microcalorimetry (Beezer, A.E., ed.), pp. 413–451, Academic Press, London.

Privalov, P.L. (1989). Thermodynamic problems of protein structure. Ann. Rev. Biophys. Chem. 18, 47–69.

Randzio, S. & Suurkuusk, J. (1980). Interpretation of calorimetric thermograms and their dynamic corrections. In: Biological Microcalorimetry (Beezer, A.E., ed.), pp. 311–341, Academic Press, London.

Ross, P.D. & Goldberg, R.N. (1974). A scanning microcalorimeter for thermally induced transitions in solution. Thermochim. Acta 10, 143–151.

Schön, A. & Wadsö, I. (1988). The potential use of microcalorimetry in predictive tests of the action of antineoplastic drugs on mammalian cells. Cytobios 55, 33–39.

Sturtevant, J.M. (1971). Calorimetry. In: Physical Methods of Chemistry (Weissberger, A. & Rossiter, B.W., eds.), Vol. 1, Part V, pp. 347–425, Wiley, New York.

Wadsö, I. (1987). Calorimetric techniques. In: Thermal and Energetic Studies of Cellular Biological Systems (James, A.M., ed.), pp. 34–67, Wright, Bristol.

Wadsö, I. (1988). Thermochemistry of living cell systems. In: Biochemical Thermodynamics (Jones, M.N., ed.), pp. 241–309, Elsevier, Amsterdam.

Wadsö, I. (1993). On the accuracy of results from microcalorimetric measurements on cellular systems. Thermochim. Acta 219, 1–15.

Wieseman, T., Williston, S., Brandts, J.F., & Lin, L.-N. (1989). Rapid measurement of binding constants and heats of binding using a new titration calorimeter. Anal. Biochem. 179, 131–137.

RECOMMENDED READINGS

Some monographs have been published with chapters authored by leading specialists in areas of interest for the subjects treated in this chapter. Beezer (1980), Hinz (1986), James (1987a), Jones (1988), Marsh and O'Hare (1993), and Parrish and Lewis (1994). Two special issues of Thermochimica Acta edited by Kemp (1990) and by Lamprecht et al. (1991) contain many papers of interest for the medical field. These publications are listed under References.

Chapter 11

Heat Dissipation and Metabolism in Isolated Mammalian Cells

RICHARD B. KEMP

Principles of Medical Biology, Volume 4
Cell Chemistry and Physiology: Part III, pages 303–329.
Copyright © 1996 by JAI Press Inc.
All rights of reproduction in any form reserved.
ISBN: 1-55938-807-2

INTRODUCTION

"Respiration is therefore a combustion, a very slow one to be precise." With this (translated) conclusion written with LaPlace over two hundred years ago, Lavoisier "gave [the science of life] a decisive and lasting direction" (Bernard writing in French, 1878) towards explaining physiological phenomena on the basis of physicochemical laws. Using an ice calorimeter, they quantified the heat dissipated per unit CO_2 produced by a (shivering!) guinea pig combusting carbonaceous food and corrected it for endogenous respiration. This study pioneered a succession of calorimetric experiments on animals, tissues, and cells which paralleled developments in the concepts of thermodynamics and led to a greater understanding of bioenergetics. The purpose of this chapter, however, is to concentrate on the contribution of calorimetric measurements to our knowledge of the metabolism of mammalian cells and thereby provide a foundation to the developing field of thermodynamics in cellular bioenergetics.

THERMODYNAMIC BACKGROUND

Systems

A system is a bounded space containing whatever is going to be studied in thermodynamics. The three types of systems are physically separated from, but related to, the environment that surrounds them. In an *isolated* system, there is no exchange of either matter or energy with the surroundings (e.g., a closed, completely insulated vessel or space—the universe, for example). A *closed* system is materially self-contained, but there can be a free exchange of energy across its boundary (e.g., a hot water bottle). An *open* system may exchange both energy and matter with its environment (e.g., all living organisms).

Equilibrium and Nonequilibrium

Classical thermodynamics applies only to systems in equilibrium and thus cannot be applied precisely to open biological systems because there is a continual exchange of substrates and products with the environment, which precludes establishment of a true equilibrium. Open systems, however, cannot disobey the laws of classical thermodynamics; so it is possible to disqualify, rather than prove, a proposed mechanism. It is also feasible to calculate by classical thermodynamics the conditions for establishing the equilibrium of an energy transformation and thus define the displacement from equilibrium, which will give the capacity to perform work. Under nonequilibrium conditions in an open system, the equilibrium parameters can be treated by irreversible thermodynamics to give the rate of energy flow (Westerhoff and Van Dam, 1987). It should also be stated that equilibrium thermodynamicists can employ an intellectually and physically valid trick to enclose an open system so that only energy is exchanged with the environment (Battley, 1987), e.g., temporarily, a goldfish in a sealed bowl! The goldfish is an open system but an aquarium containing it can be considered a closed system and then the room in which it sits is part of the environment.

States

In chemistry, a transformation can be started in some way and will stop when all the possible changes in matter and energy have occurred within the system. The point before which any change in matter or energy has happened is called the *initial* state, the termination is the *final* state and the transition between the two is the reaction. The composition of the initial and final states is represented by the chemical reaction equation, which is an expression of the Law of the Conservation of Mass: "Matter can neither be created nor destroyed." As such the equation should balance electrically as well as chemically, unlike the shorthand biochemical equations seen in many textbooks.

First Law

It was Joule who first concluded that heat, measured by calorimetry (see Wadsö, chapter 10), is a form of energy. This means that energy is thermal (heat) or nonthermal (chemical, electrical, mechanical, or radiant) in nature. It is conserved in that the quantities of nonthermal and thermal energy at the end of a process is exactly equal to the quantity of energy available at the beginning. This is the First Law of Thermodynamics—The Law of the Conservation of Energy—which states that "Energy can neither be created nor destroyed." In an isolated system, the amount of energy is constant, though it may undergo a change in form or distribution. Energy is conserved in a closed system as the sum of that within the system and that exchanged with the environment; the closed system plus the surroundings is an isolated system in this respect.

Internal Energy

Energy has been defined as "the capacity to perform work." Traditionally, mechanical work has been defined as the product of a force through a distance. In certain situations, it is possible to convert the energy stored in a temperature gradient to a form of work: a heat engine. Indeed, many 19th century physiologists thought that biological systems such as muscle were engines of this type with (absurdly great) temperature gradients to give the necessary mechanical work. In 1912, Hill finally dismissed this idea because of the need for physiologically impossible temperatures. In general, it is important to distinguish between work (W) which is nonthermal energy, and heat (Q) which is thermal energy, the sum of which is the *internal energy* (U) of the system. This is a function of a system that depends only on its physical state at the time it is being considered: a snapshot or, more formally, a function of state. The actual amount of energy within it cannot be measured but it is frequently possible to quantify differences in the amount of energy at two states of the same closed system, ΔU^a. In changing from an initial state having an internal energy U_1, to a final state with an internal energy U_2, a closed system may perform various kinds of work (chemical, mechanical, etc.) on the environment and/or exchange heat with the surroundings. By convention, loss of energy from the system is denoted by a minus sign. The mathematical expression of the 1st Law for a closed system where work is performed, is

$$\Delta U = U_2 - U_1 = (Q + W) \tag{1}$$

It is important to realize that the change in energy of a system depends on the functions of state (U_1 and U_2) not on the path of the transformation. This is known as Hess's Law of Constant Heat Summation.

Work is most widely defined as the nonthermal energy that is exchanged between two masses because of a force which is exerted between them. In a closed system one mass is contained within the system and the other is within the environment. These must change in some way associated with the exchange of nonthermal energy and thus the quantity of work can be determined as a function of this difference. Suppose that a rubber water bottle has a thin partition to make two compartments, one containing a 0.1 M solution of hydrochloric acid and the other a 0.1 M solution of sodium bicarbonate (Figure 1). It is at the same temperature (isothermal) and pressure (isobaric) as the environment and is sealed to create a closed system. The boundary is arbitrarily defined as the thin rubber between the solutions and the environment. After setting up this initial state with an internal energy U_1, the solutions are allowed to mix and react, with the result that both heat and CO_2 gas are produced. The heat causes the temperature of the solution to rise in the closed system. Heat is defined as thermal energy that is exchanged between a hotter and a colder mass because of a difference in temperature between them, so, in this case, heat flows ($\Phi = dQ/dt$)[b] from the mass of solution in the closed system (water bottle) to the mass of the environment. At the same time, CO_2 production causes an increase in volume of the closed system to expand it against the environment by an amount V. The quantity of work performed on the environment would be $p \, \Delta V$ to compress the environment. When the reactants are exhausted, the internal energy U_2 at the final state of the system will change by

---

[a] the relevant operators are Δ = finite change and δ = infinitesimal change.

[b] dQ/dt is an inexact differential quotient. In such cases, the symbol of the operator should be strictly followed by a subscript to distinguish it from an exact differential. Thus heat flow should be $d_{th}Q/dt$ (th = thermal). Other suffixed subscripts for this purpose are e = external and i = internal.

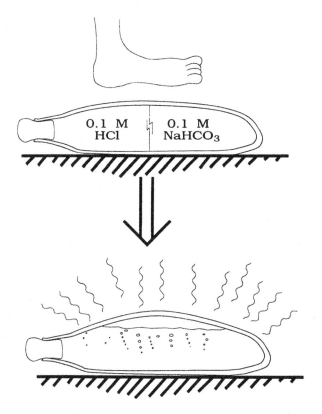

Figure 1. A representation of the change in internal energy, ΔU of a closed system, when a 0.1M solution of sodium bicarbonate is allowed to mix with a 0.1M solution of hydrochloric acid by breaking the infinitesimally thin partition in a "Battley-Kemp" rubber water bottle using virtual "foot" power. As detailed in the text, carbon dioxide is evolved in the reaction to expand the bottle and thus work, W, is performed on the environment. At the same time heat, Q, is dissipated to the environment. When the reaction is complete, the internal energy at the final state of the system U_2, will change from the initial state, U_1, by exactly the sum of the pressure-volume work, $p\,\Delta V$, and the heat released by the system (see equation (1)).

exactly the sum of the pressure-volume work ($p\,\Delta V$) performed and the heat (Q) released by the system, ΔU (see equation (1)).

If no work is performed by a closed, isothermal system other than that associated with the change in volume, then

$$\Delta U = \Delta_e Q_p - p\,\Delta V \qquad (2)$$

where $\Delta_e Q_p$ (see footnote b) represents the thermal energy exchanged under conditions such that the system remains at constant pressure.

Enthalpy

In an isobaric closed system which only performs work on the environment associated with a change in volume, it is evident from equation (2) that

$$\Delta_e Q_p = \Delta U + p\,\Delta V \tag{3}$$

It is seen that the heat exchanged by such a system is greater than the change in internal energy by an amount determined by the pressure and volume. As a function of state, this is called the enthalpy (H) of the system. As with internal energy, only changes in enthalpy (H) can be measured and expressed as quantities. It is defined as the quantity of heat exchanged with the environment by an isothermal, isobaric, closed system undergoing a change in state during which the only work performed on the environment is that associated with expansion. Thus, the quantity ΔH is equivalent to the quantity $\Delta_e Q_p$ and may be substituted for it in equation (3) to give:

$$\Delta H = \Delta U + p\,\Delta V \tag{4}$$

As previously stated for internal energy, the change in enthalpy is solely a reflection of the initial and final states of the system. It does not matter whether the change is the result of one reaction or many intermediate reactions. For instance, the "test tube" combustion of glucose to carbon dioxide and water has the same enthalpy of reaction in an isothermic, isobaric, closed system as the algebraic sum of the enthalpies for all the reactions in the respiratory degradation of the sugar by cells in a similar system. This is an example of Hess's Law.

Entropy

Entropy (S) is a function of state which relates the total quantity of heat exchanged by a closed system to the absolute temperature (T) of that system,

$$S = \frac{Q}{T} \tag{5}$$

As with the other functions of state, only the change in entropy from state X to Y can be measured experimentally; in the case of a finite, but irreversible change by,

$$\Delta S = S_Y - S_X = \frac{\Delta Q}{T} \tag{6}$$

in which ΔS is the sum of internal entropy production and the external entropy exchanged with the surroundings (see footnote b),

$$\Delta S = \Delta_i S + \Delta_e S \tag{7}$$

Entropy is a measure of the degree of disorder of a state because heat energy causes a change in the random motion of molecules. Heat dissipated solely within a closed system increases the entropy of that system; but, heat exported from a closed system increases the entropy of the surroundings by an amount precisely equal to the increase in heat of the surroundings. The Second Law of Thermodynamics states that a system changes spontaneously from a state of greater order (lower probability) to one of lesser order (higher probability); that is, when there is an increase in the entropy of the system. So, a process only occurs spontaneously if the net entropy of the system (sys)[c] and its surroundings (sur)[c] increases.

$$\Delta S_{sys} + \Delta S_{sur} > O \tag{8}$$

In order to know whether or not a process in a closed system will occur spontaneously, it would be necessary to measure the entropy change in the surroundings (the rest of the universe) and obviously this is not possible. A device is needed to allow deduction of the entropy change of the universe from the process in the closed system.

Gibbs Energy

In a closed, isothermal system at constant pressure, the maximum amount of energy that is potentially free to do something chemical, mechanical, electrical, or radiant as the system proceeds from a given initial to a given final state is defined as Gibbs energy. It is a function of state which can only be measured and expressed as a quantity by an irreversible change in it (ΔG). Its relationship to changes in enthalpy and entropy under isothermal conditions is given mathematical form by combining the 1st and 2nd Laws of Thermodynamics in the following way.

In an isolated system there can be no net change in enthalpy. A net increase in entropy, so that a system can change spontaneously, is related to a decrease in Gibbs energy,

$$\Delta G = -T\Delta S \tag{9}$$

In a closed system the change in entropy of the universe including the closed system (Figure 2) is

$$\Delta S = \Delta S_{sys} - \Delta S_{sur} \tag{10}$$

From equation (9) and (10), the Gibbs energy of a closed system is

$$\Delta G = T\Delta S_{sur} - T\Delta S_{sys} \tag{11}$$

In an isothermal closed system at constant pressure:

$$\Delta H = -Q \tag{12}$$

---

[c] The entropy change of the system, defined in equation (7), and the entropy change of the surroundings are equal, of course, to the entropy change of the universe.

Figure 2. Relationship between Gibbs energy and entropy changes in a closed system. The goldfish is an open system but it is enclosed in a sealed bowl which, like a calorimeter, constitutes a closed system in thermodynamic terms. The room in which the bowl sits is part of the environment, which is the universe: an isolated system. The Gibbs energy change, ΔG, in the isothermal and isobaric bowl is determined using only those parameters appertaining to the closed system: its entropy change, ΔS_{sys}, and the enthalpy change, ΔH. A change in Gibbs energy in the closed system is a direct measure of the change in entropy of the universe, ΔS_{sur}. A decrease in ΔG in the system means an increase in ΔS_{sys} and a given reaction will occur spontaneously.

Therefore, from equation (6) the entropy exchanged with the environment is

$$\Delta S_{sur} = \frac{\Delta H}{T} \tag{13}$$

Substituting equation (13) into equation (11) gives

$$\Delta G = \Delta H - T\Delta S_{sys} \tag{14}$$

The change in Gibbs energy in the closed system is a direct measure of the change of entropy of the universe when no work is done. To demonstrate this from equation (12), equation (14) can be written as

$$\Delta G = -Q - T\Delta S_{sys} \tag{15}$$

and, in terms of absolute temperature, as

$$\frac{\Delta G}{T} = \frac{-Q}{T} - \Delta S_{sys} \tag{16}$$

But Q/T is equal to the entropy change of the environment, ΔS_{sur} (equation 6). Therefore,

$$\frac{\Delta G}{T} = \Delta S_{sur} - \Delta S_{sys} \tag{17}$$

The terms on the right-hand side of equation (17) equal the entropy change of the universe (equation 10).

By convention, thermodynamic functions of state refer to the system and not the environment, so $-\Delta G$ (exergonic) represents the Gibbs energy potentially available for expenditure and potentially dissipated to the environment. Under suitable conditions, this energy could be made to perform work. An endergonic reaction $(+\Delta G)$ cannot proceed spontaneously and requires an input of Gibbs energy to proceed from its initial to its final state.

Gibbs energy is often called free energy but that term includes Helmholtz energy (A), an irreversible change which represents the change in free energy of a closed system, resulting from an isothermal reaction at constant volume. A is the maximum potential for performing all kinds of work (total work) whereas ΔG is the potential for performing only that work (useful work) which is not associated with expansion. Most biochemical reactions, including those in living systems, take place in dilute aqueous solutions at constant temperature, pressure and volume, so $\Delta A = \Delta G$. However, in experiments with living material, it is usual only explicitly to control temperature and pressure, so in these cases Gibbs energy is the correct term. In some studies, changes in enthalpy measured as heat by calorimetry are related to changes in free energy determined from actual concentrations of reactants. ΔH is explicitly Q_p (equation 4) and so, once again, the free energy should be called Gibbs energy. Indeed, some biochemical reactions have very small entropy changes, so that the changes in Gibbs energy and enthalpy are very similar (Prigogine, 1967). This means that the former can be estimated fairly accurately from measuring the latter by calorimetry. A good example is the combustion of glucose by cells (respiration).

The values for enthalpy and Gibbs energy changes quoted in textbooks for biochemical reactions are standard values, $\Delta H^{o\prime}$ and $\Delta G^{o\prime}$, when the reactions take place in solution under standard conditions. All components are in their standard states at standard concentration of 1 mol.dm^{-3} and the H^+ ion activity is taken to be 10^{-7}. Actual ΔG and ΔH values in cells and tissues may be markedly different; in the case of ΔG, even to the extent of a different sign depending on the concentrations of the reactants. It must be emphasized that ΔG for a particular reaction provides no information about the rate of that reaction.

Units

Scientists must talk to one another in common quantities, units, and symbols. The International Union of Pure and Applied Chemistry (IUPAC) has drawn up tables of "chemspeak" (Mills et. al., 1993). The Système International (SI) unit of energy is the joule (J) and that of power is the watt (W). In terms of SI base units the former is N m and the latter, J s^{-1}. Older textbooks and scientific papers use the calorie (= 4.184 J) but today's scientists must use the multiplier to be correct when quoting earlier work using this antiquated unit.

ENTHALPY BALANCES

According to the 1st Law of Thermodynamics and Hess's Law, measuring by calorimetry the net change of heat in living material gives the net sum of all the enthalpy changes in that material. This means that heat flow (Φ) is a reflection of metabolic activity of the cells, tissues, organs, or organisms in the closed system, the "aquarium" (see p. 310), of a calorimeter. Of itself, this may be sufficient reason for calorimetric study but the technique is far more powerful than simply acting as an analytical monitor.

Each of the reactions (r) contributing to the overall metabolic activity has a characteristic *molar enthalpy*, $\Delta_r H_B$ (J mol^{-1}), where subscript B indicates that any given *reaction stoichiometry* must be divided by v_B (v = stoichiometric number) to give a stoichiometric form of unity ($|v_B| = 1$). Each is calculated from the balanced reaction stoichiometry and enthalpy of formation. If this value is multiplied by the measured chemical reaction flux, J_B (mol s^{-1} m^{-1}), then the reaction enthalpy flux, J_H(J mol^{-3}), is obtained,

$$J_H = J_B \, {}_r H_B \tag{18}$$

This forms the basis of constructing an enthalpy budget in which the total enthalpy flux is compared with the scalar heat flux, J_Q(W m^{-3}), obtained from dividing heat flow by size (volume or mass) of the living matter. If account is made of all the reactions and side reactions in metabolism, the ratio of heat flux to enthalpy flux, the so-called *energy recovery* ($Y_{Q/H} = J_Q/J_H$) will equal 1. If it is more than 1, then the chemical analysis has failed fully to account for heat flux and if it is less than 1, then there are undetected endothermic reactions. "Account for all reactions" may seem a formidable task, but it should be borne in mind that anabolic processes dissipate insignificant amounts of heat compared with those of catabolism and that ATP production and utilization are balanced in cells at steady-state. Catabolism is generally limited to a relatively few well-known pathways with established overall molar enthalpies. So, as will be seen later, the task is by no means "mission impossible."

If the interest is solely in the contribution of anaerobic processes to overall catabolism, then it is possible simply to use the ratio between heat flux and oxygen reaction flux, the *calorimetric/respirometric ratio* (CR), $Y_Q/_{O_2} = J_Q/J_{O_2}$ (J mol^{-1} O_2). Gnaiger and Kemp (1990) have shown that the CR ratio of fully aerobic cells must be close to the theoretical *oxycaloric equivalent*, $\Delta_k H_{O_2}$, where the subscript k signifies a catabolic reaction or series of reactions constituting a pathway, most usually respiration. The equivalent is the calculated enthalpy change of respiratory oxygen uptake, expressed per unit amount of oxygen, and its value for a given catabolic substrate is obtained from known enthalpies of formation, $\Delta_f H$. The exact figure depends on the precise physical conditions within the cells and the environment, but Gnaiger and Kemp (1990) calculated that the oxycaloric equivalent for glucose in buffered aqueous solution with O_2 and CO_2 exchanged with the gaseous phase is -469 kJ mol^{-1} O_2. The equivalents for other sugars, fatty acids and amino acids used in respiration are similar to this figure, because the enthalpy of combustion of these organic compounds is essentially constant when expressed as kJ per equivalent of available electrons. This regularity is called Thornton's Rule after the scientist who discovered it in 1917. Thus, the oxycaloric equivalents for all catabolites is -450 kJ mol^{-1} $O_2 \pm 5\%$.

In calculating this average figure, it is assumed that no work is performed, the process being irreversible and the efficiency zero. In practice, however, many types of cells *in vitro* give CR ratios more exothermic than -500 kJ mol^{-1} O_2 (Kemp, 1991). This has been ascribed to lower efficiency caused by uncoupling or decoupling respiration from ATP generation. This cannot be; efficiency is already zero. Gnaiger and Kemp (1990) have shown that highly negative CR ratios are due to simultaneous aerobic and anaerobic catabolism. Once the glycolytic endproduct (i), usually lactate, is quantified, the molar amount of it produced per unit amount of oxygen consumed (i/O_2 ratio comparable to the molar gas exchange ratio or respiratory quotient, CO_2/O_2 ratio) indicates the relative extent of aerobic glycolysis. In this type of enthalpy balance study, the generalized form of the equation is

$$\Delta_k H_{O_2(ox + anox)} = \Delta_k H_{O2} + \sum_i i/O_2 \times \Delta_k H_i, \tag{19}$$

where $\Delta_k H_i$ is the dissipative catabolic enthalpy change of endproduct(s) *i*. It is important to use the correct value for $\Delta_k H_i$ because it is highly dependent on conditions. As a cautionary example, the catabolic enthalpy change for lactate is -80 kJ mol^{-1} Lac when the acid is buffered intracellularly, but varies substantially with the buffer when it is excreted: from -59 kJ mol^{-1} (phosphate buffer), through -63 kJ mol^{-1} (bicarbonate), to -77 kJ mol^{-1} (2 mM HEPES).

METABOLIC ACTIVITY OF CELLS

As is seen in Table 1, the heat flow of the cells so far subjected to measurement varied from 10 fW per cell for human erythrocytes to 329 pW per cell for rat

Table 1.   Heat Dissipation ($\phi = dQ/dt$) of Human and Other Mammalian Cells

Cell Type	pW per Cell[a]	Researchers[b]
Human erythrocyte	0.010	Monti and Wadsö, 1976[b]
Human platelet	0.061	Monti and Wadsö, 1977[b]
Bovine sperm	1.3	Inskeep and Hammerstedt, 1985[b]
Human lymphocyte	2.4	Monti et al., 1981[b]
Human neutrophil	2.5	Eftimiadi and Rialdi, 1982
Human non-Hodgkin's lymphoma lymphocytes	3.7	Monti et al., 1981[b]
Molt 4 human leukemic	21	Nittinger et al., 1990[b]
2C11-12 mouse macrophage hybridoma	32	Kemp, 1990
Mouse lymphocyte hybridoma	40	Nässberger et al., 1988[b]
Human foreskin fibroblast	40	Schaarschmidt and Reichert, 1981[b]
Human white adipocyte	49	Monti et al., 1980[b]
H1477 human melanoma	80	Görman-Nordmark et al., 1984[b]
Human keratinocyte	83	Reichert and Schaarschmidt, 1986[b]
Human brown adipocyte	110	Nedergaard et al., 1977
SV-K14 transformed human keratinocyte	134	Reichert and Schaarschmidt, 1986[b]
Rat hepatocyte	329	Nässberger et al., 1986[b]

Notes: [a]Standard errors or deviations omitted.
[b]Papers not quoted in the reference list can be found by looking at the reviews quoted in Kemp and Gnaiger, 1989; Kemp, 1991, 1993; Monti, 1987, 1991.

hepatocytes (reviewed in Kemp and Gnaiger, 1989; Kemp, 1991). This would mean that hepatocytes were about 30,000 times more metabolically active than erythrocytes only if both types of cell were the same size; that is, had the same scalar heat flux. When corrected for size, hepatocytes were 2,000 times metabolically more active, which is not surprising considering the respective biochemical physiologies of the two cell types. Erythrocytes have limited requirements for ATP generation whereas hepatocytes have considerable needs because the liver is the chemical factory of an organism.

This example gives a clue as to the precise meaning of *metabolic activity* in the context of calorimetric measurements. For a cell at steady-state (Gnaiger, 1983, 1990), the catabolic half-cycle (k) is coupled to ATP generation (p) which itself is coupled to ATP utilization (–p), making the ATP cycle (Figure 3). Thus, the catabolic half-cycle supplies the energy which is transformed into ATP. This, in turn, is demanded by the anabolic half-cycle (a) which is coupled to ATP utilization. The net enthalpic change of biosynthetic processes is negligible (Battley, 1987). At steady-state, the endothermic enthalpic change of ATP phosphorylation is balanced by the exothermic enthalpic change of ATP dephosphorylation. So, if there is no net work, then the heat dissipated by the cell is a concomitant of the rate of catabolic half-cycle (input), that rate being determined by the demand for ATP in anabolism, the direct output. If the supply of catabolites for energy provision is infinite, then the demand governs the rate of the ATP cycle in a fully coupled conservative

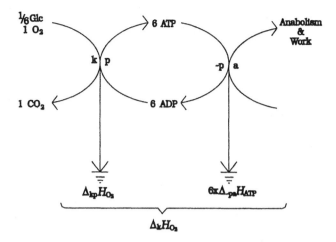

Figure 3. Energy transformation half-cycles in fully coupled conservative metabolism. The catabolic half-cycle (k) is coupled to the ATP cycle (p, –p) which is itself coupled to anabolism (a) and work. These processes demand ATP, the energy for the transformation of which is supplied by the oxidation of glucose in the catabolic half-cycle. The system is irreversible so the net work is zero.

system. Metabolic activity is the rate of that cycle and the dissipated heat is a reflection of the overall rate of the catabolic reactions supplying energy for transformation in the phosphorylation of ADP.

In order to ascertain whether one cell type is metabolically more active than another, it is necessary to standardize against size (volume or mass), i.e., scalar heat flux (W m^{-3}). This has rarely been done, so one can only make guesses from the data in Table 1. High metabolic rate is reasonably associated with cells that are actively growing and dividing *in vivo* or *in vitro*. The fully differentiated leukocytes (lymphocytes, neutrophils, etc.) in the blood do not grow and divide; they dissipate relatively little heat. Their cancerous equivalents are not differentiated and, in many cases, continue to grow in the blood stream. This is true, for instance, of lymphocytes from patients with non-Hodgkin's lymphoma which produce more heat than their normal counterparts (Table 1). Contrary examples have been discovered in which whole blood from patients with pathological conditions, for example, chronic lymphocytic leukemia (Monti, 1991), produced more heat than that from healthy patients, but the isolated lymphocytes dissipated less heat. In this case, the neoplastic transformation is at the stage of the hemopoietic stem cells where all the daughter cells continue to divide rather than half of them differentiating into mature lymphocytes. Growth in numbers does not actually continue in the blood stream; there are simply more of them.

Normal leukocytes are relatively short-lived in the blood and, because they are fully differentiated, cannot be grown *in vitro* in order to study their properties. This limitation has been overcome by hybridizing them with transformed (tumor) cells which immortalizes them while maintaining at least some of their distinctive properties. Retention of growth characteristics by the hybridoma cell was associated with greater heat dissipation in, for instance, lymphocyte and macrophage hybridoma cells (Table 1; cf. normal leukocytes).

Tumor cells in general have a high metabolic rate because they are undergoing rapid growth. There have been no studies on the thermogenesis of cells freshly obtained from tumor tissue, but cell lines derived from tumors were found to dissipate considerable amounts of heat, for example human melanoma (H1477) and leukemia (Molt 4) cells (Table 1). Many of the characteristics of tumor cells can be imitated by transformation with oncoviruses. It was shown that human keratinocytes transformed with the Sendai virus (SV) produced 60% more heat than their normal counterparts (Table 1), but the true significance of this finding cannot be ascertained because the measurements were of heat flow, not flux.

The most dramatic illustration of a mass-specific illusion is the comparative heat dissipation of the human erythrocyte and platelet. In mammals, both of these cell types are anucleate and discoid in shape, but the longest dimension of the former is four times that of the latter. Yet heat production of a human erythrocyte was shown to be 10 fW, a sixth that of a human platelet (61 fW; see Table 1). The relatively high metabolic activity of platelets is probably due to the need to maintain a considerable phosphagen (phosphocreatine) pool for actomyosin contraction at stimulation and clot retraction. Phosphocreatine is synthesized from creatine using ATP and acts as a demand on the ATP cycle to drive the coupled catabolic half-cycle. On the other hand, ATP requirements of the erythrocyte are relatively small, being mostly confined to active transport of ions at the plasma membrane.

Metabolic Inhibitors

Many inhibitors of catabolic pathways cause a decrease in cellular heat dissipation. They are therefore valuable tools to indicate the sources of the dissipation and give clues to the relative importance of each pathway in overall metabolic activity (see reviews by Kemp, 1987, 1993; Monti, 1987, 1991). To give a few examples from these reviews, sodium fluoride is a classical inhibitor of glycolysis and it has been shown to substantially reduce heat dissipation by human erythrocytes, lymphocytes, neutrophils, and murine macrophages, indicating the contribution of this pathway to metabolic activity. Cyanide inhibits oxidative phosphorylation by mitochondria at the cytochrome c oxidase complex (site 3) and studies revealed that it decreased heat production in a mouse LS-L929 fibroblast cell line but had no effect on human erythrocytes and neutrophils and murine macrophages, all of which lack mitochondria. Sodium azide inhibits at the same site and so it should come as no surprise that it had no effect on human neutrophils and lymphocytes, but it did reduce heat production by lymphocyte hybridoma cells, which contain mitochondria from the transformed partner in the hybrid. Rotenone is a specific inhibitor of electron transfer within the mitochondrial NADH-Q reductase complex (site 1) and thus inhibited heat production in hybridoma cells. It does not interfere with the oxidation of succinate, however, because its electrons enter the electron-transport chain after coenzyme Q. For a more complete picture of the importance of electron flow to metabolic activity, rotenone is combined with antimycin a which blocks flow between cytochromes b and c_1, (site 2). Hammerstedt's group (see Kemp, 1987) used this cocktail to evaluate the contribution of oxidative endogenous metabolism to the heat production of bovine sperm.

If a substrate is prevented from entering a cell, then its part in metabolic activity can be measured from the difference in heat flow. Loike et al. (1981) performed an elegant experiment in which murine macrophages were exposed to cytochalasins, either B or D. Both disrupt cytoskeletal microfilaments but, additionally, B inhibits glucose transport through the plasma membrane. Cells subjected to 0.01 mmol dm^{-3} B for 50 minutes in the presence of 5.5 mmol dm^{-3} glucose, generated 62% less heat than those incubated without cytochalasin or with D. Terminal glycolysis to lactate was the predominant pathway in these cells under the prevailing *in vitro* conditions and cytochalasin B inhibited lactate production by 88%. Interference with the cytoskeleton had no effect on cellular heat dissipation.

Little work has been done on the effect of anabolic inhibitors on cellular heat dissipation probably because there is empirical evidence that anabolic processes do not contribute significantly to it (see p. 312). Loike et al. (1981) found that 0.07 mmol dm^{-3} cycloheximide, an antibiotic inhibitor of protein synthesis, had no effect in 30 minutes on the heat production of murine macrophages. On the other hand, Krakauer and Krakauer (1976) showed that long-term exposure of lymphocytes from immunized horses to 1 mg dm^{-3} cycloheximide considerably reduced heat production. This was likely to be due to a secondary effect of the antibiotic arresting catabolism by inhibiting the turnover of short half-life enzymes.

Since growing cells should show a greater heat flux than those simply utilizing energy for maintenance, it would be reasonable to expect that anchorage-dependent cells at confluency would have a lower heat flux than those in log phase, because contact-inhibited cells become side-lined into a nongrowing phase, G_0 (Baserga, 1989). Technical problems associated with growing cells in monolayer for long periods in a calorimeter (but see chapter 10) seem to have precluded a systematic study, but evidence from human skin fibroblasts and keratinocytes (see Kemp, 1987), mouse lymphocyte hybridoma cells, and human leukemia cells (see Kemp, 1993) appear to support this hypothesis. Analysis of the results is complicated because some of the reduction may be due to nutrient depletion and/or accumulation of waste products. In addition, cells may overlap or heap together (the "Uriah" effect) exacerbating the problems of depletion and accumulation and, in particular, restricting the availability to the cells of dissolved oxygen in the bulk phase. Anchorage-dependent cells are usually grown in flasks with an air phase many times greater in volume than that of the medium, so oxygen is present in abundance. It is not available because the Uriah effect probably causes a restriction in local diffusion.

Subculture of cells in monolayer releases them from G_0 arrest imposed by intercellular contact. It was shown that freshly dissociated 2C11-12 mouse macrophage hybridoma cells had a higher rate of specific heat dissipation than confluent cells, presumably because they resumed growth (the G_1 phase of the cell cycle (Kemp, 1990)). Many fully differentiated and thus nongrowing types of cells still can be stimulated to divide by subjecting them to mitogens. For instance, Krakauer and Krakauer (1976) showed that horse lymphocytes stimulated by either the specific antigen, dinitrophenylated bovine serum albumin, or the relatively nonspecific lectin, concanavalin A, produced more heat than the controls.

Clinical Studies

Much of the research into the potential of calorimetric measurements both to diagnose and to scientifically investigate clinical conditions has been pioneered by Monti and his group (short reviews, 1987, 1991; earlier work, Levin, 1980). If you think you need the stimulation of caffeine to retain/restore attention to this text, then spare a thought for your platelets because Monti has shown that 1–2 cups of coffee will increase their heat production by 12% within an hour. This was considered to be due to an increase in circulating catecholamines which stimulate glycogen breakdown. If you eat several cookies while you are waiting for the caffeine stimulation, spare a thought for your white adipocytes (fat cells). This type of cell dissociated with collagenase from intradermal gluteal biopsies of young humans (23–31 years) of normal weight produced heat at 49 ± 15 pW per cell. Cells from long-term obese (132 ± 27 kg) individuals of similar age gave a significantly lower heat production of 26 ± 12 pW per cell. The difference was even more dramatic in terms of heat flux because white adipocytes from the obese were 68% larger than those from normal individuals; the figures were 129 pW μg^{-1} lipid for the former compared with 41 pW μg^{-1} lipid for the latter. Monti and his colleagues thought that the decreased heat flux was due to the restricted activity of "futile" substrate cycles in "obese" adipocytes (see Kemp, 1987). Certainly, the obese had considerable peripheral insulin resistance, exhibiting hyperinsulinemia, hyperglucosemia, and hyperglucagonemia.

Obese individuals were persuaded not to eat cookies (or much else), lose weight by controlled diet, and in five weeks lost an average of 17 kg (12%). Adipocyte heat flux increased from 41 ± 12 pW μg^{-1} lipid to 64 ± 17 pW μg^{-1} lipid and the "Alice" effect was a 10% reduction in cell size. Insulin sensitivity increased but there were no changes in variables reflecting thyroid status. Is there any hope in the long term? We do not know, because Monti was deflected to recruit surgeons who performed horizontal gastroplasty according to Gomez on another group of 15 obese individuals (127 ± 13 kg). The mean body weight was reduced by 30% to 97 ± 14 kg within 6–8 months after the operation and the Alice effect on the adipocytes was 33%. Heat flux was 112 pW μg^{-1} lipid, quite close to that of adipocytes from lean subjects. Blood insulin levels post-operation fell by 50% and fasting blood glucose concentration by 30%. Olsson et al. (1986) tentatively concluded that, although there appeared to be an inverse correlation between cellular heat flux and body weight, lowered heat production (metabolic activity) was probably a consequence of, rather than a cause of, obesity.

Thyroid conditions are a natural for calorimetric study because the hormones influence the basal energy requirements of the body, i.e., the metabolic activity (Dauncey, 1990). Studies in Lund, Sweden (see Monti, 1987, 1991) showed that heat production was greater than in the erythrocytes, platelets, and leukocytes from patients with clinical symptoms of hyperthyroidism and less than normal in the same cell types from hypothyroid humans. The lymphocytes of patients with acromegaly, a disease associated with excessive production of growth hormone, produced more heat than did those from normal individuals (euthyroid). The disease is often caused by tumors of the pituitary gland and is usually treated surgically or by radiotherapy. Recurrence of the tumor happens in 20% of cases, however, and this is not easy to detect at a sufficiently early stage for successful treatment. Monti's group was able to show that heat production in lymphocytes was significantly correlated with the degree of disease activity and suggested that calorimetry may be a way to secure early detection. It is of interest that cells from other body tissues also have a changed heat production in response to altered levels of thyroid hormones. For instance, Clark et al. (1982), using rats as a model for humans, showed that hepatocytes from animals treated with triiodothyronine dissipated 62% more heat than those from euthyroid individuals. Studies have indicated that thyroid hormones also influence thermogenesis in kidney and skeletal muscle cells (Monti et al., 1987).

The exact cellular mechanism by which thyroid hormones affect thermogenesis is uncertain, but a leading candidate for the site of action has been the plasma membrane Na^+-K^+-dependent adenosine triphosphatase (ATPase). However, it is difficult to be confident of this possibility because of the variations in estimates of energy expenditure by this membrane pump of between 5% and 60% in liver tissue from euthyroid animals and of between 5% and 90% in that of hyperthyroid animals (see Kemp, 1987). The likelihood of ATPase being involved was, nevertheless, explored using erythrocytes from patients with diffuse toxic goiter which produced 22% more heat than those from normal euthyroid individuals (Monti et al., 1987). Pump activity can be stopped by the cardiac glycoside, ouabain. At 3.75×10^{-4} M, it inhibited heat dissipation by 8% in both hyperthyroid and euthyroid cells. Therapy with radioiodine or carbimazole and thyroxine caused a decrease in heat production to a level similar to that of euthyroid erythrocytes and had no effect on ouabain sensitivity. Thus, activity of the pump had no bearing on thyroid-induced thermogenesis by these cells. Clark et al. (1982) examined the effect of ouabain on the heat production of hepatocytes from euthyroid rats and those treated with tri-iodothyronine; they found no difference in heat production between the two sets and concluded that Na^+-K^+-ATPase plays no major role in the increased thermogenesis of hyperthyroid rats.

It used to be thought that humans have a heat dissipation mechanism, termed luxuskonsumption, which allowed them to maintain a reasonably constant body weight despite changes in energy uptake (see Dauncey, 1991). It was said that the excess energy was dissipated as heat by brown adipose tissue, which is known to have an important thermoregulatory role in mammalian adaptation to cold, arousal from hibernation, and warming of newborns. The cells contain a natural protonophore or uncoupling protein (UCP - 32,000 M_r) which disengages mitochondrial oxidative phosphorylation by increasing the proton conductance of the inner membrane. Electron transport from NADH to O_2 proceeds normally but, because there is no protonmotive force, there is no mitochondrial ATP synthesis. The demand for ATP is still present, however, and this may be partly satisfied by substrate-level phosphorylation from the catabolic half-cycle (Figure 3). In addition, continuation of unregulated electron transport demands catabolic processes, as in the case of brown adipocytes from fatty acids. In general, the catecholamines stimulate fatty acid oxidation but, more specifically, norepinephrine (noradrenaline) is the physiological effector in brown adipose tissue, binding to β_3-adrenoceptors. Nedergaard et al. (1977) showed that this hormone (1 μM) stimulated heat dissipation in hamster brown adipocytes by 700% within five minutes of its addition. β-adrenergic blockers should prevent or reverse this effect and, indeed, propranolol depressed norepinephrine-stimulated heat dissipation in rat hepatocytes (Clark et al., 1986). The hypothesis was that the luxuskonsumption mechanism, now known as diet-induced thermogenesis, regulated body weight in humans by stimulating fatty acid oxidation in brown adipose tissue. Careful experimentation on overfeeding by human volunteers now has shown that the majority of the extra energy is stored in the body (see Dauncey, 1991)—QED obesity.

There remains the possibility that overfeeding in rodents causes an adaptive diet-induced thermogenesis. The interscapular brown adipose tissue of rats fed *ad libitum* on a diet of "cafeteria" food (hamburgers and candy) was found both to hypertrophy and have a 100-fold increase in the uncoupling protein (see Dauncey, 1991). Isolated multilocular brown adipocytes from cafeteria-fed rats had a volume 2.5 times greater than that of control rats (Clark et al., 1986) but, allowing for this difference, there appeared to be little distinction in heat production between cells from the two sets of animals. Despite "mettwurst and cocopops," the rats showed no significant weight gain over the controls. This evidence of diet-induced thermogenesis could not be traced to the isolated brown adipocytes.

Toxicological Aspects

Several researchers have realized that calorimetry is potentially a valuable nondestructive method to show the general effect of xenobiotics on metabolism (see reviews by Kemp, 1987, 1991, 1993). In this way, the antipsoriasis drug, anthralin, and its analogues and derivatives were shown to decrease heat production by human keratinocytes to varying degrees and thus calorimetry provided a screening test which was more sensitive than viability assays. Similar procedures using cultured cells have been evaluated successfully for the efficacy of phytohemagglutinin isolectins, antineoplastic drugs such as methotrexate, ARA-c, and cisplatin, heavy metals, and sedatives. Environmental health is an increasingly important field and Thorén (1992) used rabbit alveolar macrophages to assay airborne particles because these cells have an important role in protecting pulmonary tissue from invading microorganisms and inhaled dust. He showed that small particles of manganese dioxide strongly decreased heat production by the macrophages whereas those of titanium dioxide had no effect. Quartz particles increased metabolic activity, reflecting an altered cell function which, it was suggested, may be a prelude to pulmonary fibrosis. He concluded that "calorimetric methods with their high sensitivity but inherently low specificity may be advantageous for studying the total toxic effects on cell populations, whereas unexpected toxic mechanisms might not be observed using specific test methods."

SIMULTANEOUS AEROBIC AND ANAEROBIC PROCESSES

It is, of course, undeniable that the net metabolism of most animals is aerobic but this may not be true for all the tissues in an animal. The facultative glycolysis in striated muscle under stress, followed by the gluconeogenic conversion of lactate in the liver (the Cori cycle), could disguise the occurrence of anaerobic glycolysis in cells of other tissues and organs. Nevertheless, it came as something of a surprise to find that cells from normal animal tissues in many cases had a highly exothermic CR ratio when cultured *in vitro* (see p. 313; and Kemp and Gnaiger, 1989; Gnaiger and Kemp, 1990; Kemp, 1991). In some experiments, this could be ascribed to poor oxygenation of the bulk phase medium or the Uriah effect in the microenvironment. This explanation was broadened to encompass all reports of a high CR ratio and seemed to be supported by the fact that hamster brown adipocytes stimulated with norepinephrine had a CR ratio which showed them to be fully aerobic (Nedergaard et al., 1977) in medium in which they floated because of oil droplet inclusions. More careful attention was then given to making sure that dissolved oxygen was available to cells *in vitro* but many cell types still gave highly exothermic ratios (see Kemp, 1993).

It is accepted that the cells in tumors *in vivo* undergo high rates of aerobic glycolysis to produce lactate. A major substrate appears to be glutamine derived from the degradation of muscle proteins (Newsholme et al., 1985; McKeehan,

1986). This leads to cachexia in tumor-bearing individuals, the wasting often being the actual cause of death. Glutamine is required in normal tissues for protein synthesis and as an amino group donor for purine, pyrimidine, amino sugar, and asparagine biosynthesis but, additionally, glutaminolysis (oxidation of glutamine) itself appears to be necessary in growing cells, even in the presence of the glycolytic conversion of both glucose and glutamine to lactate.

It could be that unregulated growth with consequent high rates of proliferation in tumor cells *in vivo* would demand more production of ATP than is available by oxidative phosphorylation, necessitating resort to substrate-level phosphorylation by glycolysis. This would occur in tumor cells transferred to *in vitro* conditions (e.g., T-lymphoma cells) and could explain the highly exothermic CR ratios found for these cells and those utilized in forming experimental hybridomas (e.g., lymphocyte and macrophage hybridoma cells; see review by Kemp, 1991). The logarithmic growth of normal cells *in vitro* may make greater demands for ATP than can be satisfied by oxidative phosphorylation alone and the need for additional ATP from substrate-level phosphorylation may explain the highly negative CR ratios of many types of normal cell lines *in vitro*. It is certainly the case that wherever lactate assays have been performed on cells exhibiting highly negative CR ratios, the difference from the fully aerobic value ($-450 \pm 5\%$ kJ.mol^{-1} O$_2$) was explained satisfactorily by incorporating the molar enthalpy change for the amount of lactate produced into equation (19) (Kemp and Gnaiger, 1989; Kemp, 1991).

The liver as the chemical powerhouse of the body is the source of considerable heat and, before brown adipose tissue was recognized as the primary source of physiologically-regulated heat production, was itself a candidate for functional (rather than coincidental) thermogenesis. There was a search for the mechanism of facultative heat generation (see Clark et al., 1982) and substrate cycling was investigated as a possibility, specifically, cycling between glucose/glucose 6-phosphate, fructose 6-phosphate/fructose 1,6-bisphosphate, and pyruvate/phosphoenolpyruvate (see Kemp, 1987). It was shown by Clark's group (1986) that such cycling was associated with highly exothermic CR ratios in isolated rat hepatocytes. These ratios were ascribed to the metabolic inefficiency of ATP production (Figure 4). This cannot be the case because oxycaloric equivalents are calculated on the assumption of zero net efficiency, irrespective of intermittent efficiencies of ATP production in the phosphorylation reaction (Gnaiger, 1990; Gnaiger and Kemp, 1990). What, in fact, must have occurred in rat hepatocytes was that so much of the required ATP was being utilized in the futile cycles that total demand could not be met from oxidative phosphorylation. The highly exothermic CR ratios indicated that aerobic glycolysis had been activated in order to enhance ATP production by substrate-level phosphorylation.

Uncoupling in hamster brown adipocytes by a natural protonophore does not appear to cause simultaneous aerobic and anaerobic processes because Nedergaard et al. (1977) reported that these cells stimulated by norepinephrine (see p. 320) had

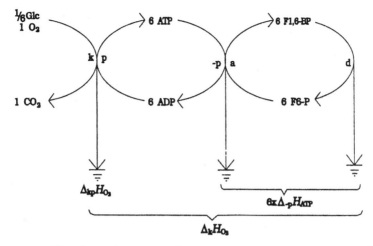

Figure 4. Half-cycles in dissipative (d) maintenance metabolism with steady-state ATP turnover, decoupled by futile cycling with, in this example, fructose 6-phosphate/fructose 1,6-bisphosphate. The net enthalpy change is calculated exclusively due to the catabolic half-cycle because both the ATP and the futile cycles contain equal but opposite exothermic and endothermic components (after Gnaiger, 1990).

a fully aerobic CR ratio of -490 kJ mol^{-1} O_2. So, how is the demand for ATP satisfied? This is not known and one is left to speculate that either some mitochondria in brown adipocytes have no UCP or that the UCP in the inner membrane of some mitochondria is inactive owing to an unknown regulatory mechanism. Exogenous uncouplers, such as 2,4-dinitrophenol, not only increase heat production but cause a more negative CR ratio (see Kemp, 1987), because of a need for substrate-level phosphorylation in glycolysis to satisfy ATP demand.

METABOLIC PATHWAYS AND ENTHALPY RECOVERY

As indicated earlier, the enthalpy balance method is a valuable tool for making an account of the pathways operating during cellular metabolism (reviewed by Kemp, 1993). In this way, it was shown by Eftimiadi and Rialdi (1982) that 36% of the heat dissipation by human neutrophils was due to the pentose phosphate pathway and the remainder (64%) to glycolysis, resulting in lactate production. The enthalpy recovery of 1.0 indicated that there was no measurable respiration, which is consistent with the almost complete absence of mitochondria from these cells.

When cells such as neutrophils undertake phagocytosis, the process is associated with a metabolic burst in which NADPH oxidase catalyzes the divalent reduction of oxygen to hydrogen peroxide, with NADPH as the hydrogen donor. The source of the donor is the pentose phosphate pathway. Activation of phagocytic cells can be caused artificially by chemicals such as phorbol-12-myristate-13-acetate

(PMA). Neutrophils stimulated in this way produced a 40-minute heat burst of 23.5 nW per cell, approximately fourfold more than resting cells over a similar period. The rate of glycolysis, as measured by lactate evolution, was not altered substantially but the flux through the pentose phosphate pathway was increased by nearly 500% to give the necessary NADPH. All of the heat dissipated was accounted for by measurements of oxygen consumption, lactate production and CO_2 evolution in the glucose carbon 1/carbon 6 ratio. Enthalpy recovery was, therefore, 1.0 which indicated that no other pathway was induced by neutrophil activation—an enhanced flux in the pentose phosphate pathway to produce NADPH was responsible solely for the observed metabolic burst.

The thermochemical characterization of the human cancer cell line T-lymphoma, CCRF-CEM, is underway in Wadsö's laboratory (Bäckman et al., 1992; reviewed in Kemp, 1993). Bäckman showed that these cells dissipated 12.2 pW per cell in a culture medium containing glucose and glutamine, and supplemented with 10% fetal calf serum. Radioisotope studies indicated that glucose was preferentially catabolized by the glycolytic pathway to lactate, whereas glutamine was selectively oxidized to CO_2 in glutaminolysis. Applying known molar enthalpy changes and stoichiometries to data for substrate consumption and endproduct formation gave calculated heat flows which apportioned 37% of the heat dissipation by these cells to lactate formation from glucose and only 10% to respiration of this substrate. Glutaminolysis accounted for 17% of heat dissipation. The heat flux/enthalpy flux ratio therefore gave an enthalpy recovery of 1.56, meaning that account had not been made of all pathways contributing to metabolism. In particular, it was recognized that there were oxygen requirements additional to those for the respiration of glucose and glutamine. If Thornton's rule for the combustion of carbonaceous substrates ($-450 \pm 5\%$ kJ mol^{-1} O_2; see Gnaiger and Kemp, 1990) was applied to this surplus, then oxidation would account for 34% of the total heat flux, and give an enthalpy recovery of 1.02. Fetal calf serum was present in the culture medium which also contained all the other essential amino acids (as well as glutamine). So, there were many potential substrates, including fatty acids which are said to be preferred substrates for mammalian cells (Baserga, 1989).

These examples of applying the enthalpy balance method to cells illustrate three general points. First, although oxidative processes account for the majority of the heat dissipated by the cells, 95% of the catabolic carbon flux was due to glycolysis. This is because the enthalpy change for the conversion of glycolytic substrates to lactate is 20 times lesser than that for the oxidation of these same substrates. Second, the enthalpy recoveries showed that anabolic processes made a negligible contribution to overall cellular heat dissipation (see also p. 312). The third point arises from the fact that the entropy change for the oxidation of typical cellular substrates is small (Prigogine, 1967; see p. 311). This results in similar values for Gibbs energy and enthalpy changes. For instance, glucose has a $\Delta_k G_{O_2}$ of -484 kJ.mol^{-1} O_2 and a $\Delta_k H_{O_2}$ of -469 kJ.mol^{-1} O_2. This means that, for aerobic processes only, calorimetry gives a good approximation of Gibbs energy changes as well as an accurate measurement of enthalpy changes. Furthermore, since the stoichiometrics of ATP production from the various oxidative pathways are known, then it is possible to calculate the turnover time of ATP in cells from the determination of heat flux. For LS-L929 mouse fibroblasts, this has been estimated to be nine seconds (Kemp, 1993).

MUSCLE

A myofiber in striated muscle is a syncytium with many nuclei, and is not a cell; it is formed by fusion of myoblasts and deserves attention in this chapter as a highly energetic system. Myofibers contain myofibrils of thick (myosin) and thin (mainly actin) filaments which interact to give contraction in the presence of Ca^{2+} ions. The hydrolysis of ATP by myosin drives the sliding mechanism of the two filaments, the force being generated in the actin-myosin crossbridges. Even so, there is relatively little change in the level of ATP during contraction because normally a supply is available from a phosphagen reservoir. The tightly coupled conversion of the phosphagen, phosphocreatine, to ATP is maintained close to equilibrium by highly active creatine kinase. Continued contraction and thus ATP demand when the reservoir has been depleted results in adenylate kinase (myokinase)-catalyzed conversion of ADP to ATP and also, when oxygen tension falls, substrate-level phosphorylation to form ATP, with production of lactate. It was this event that led early muscle physiologists to conclude that the free energy for muscle contraction was obtained in the liberation of lactic acid from an unknown precursor, with the evolution of heat (see Hill, 1912).

It is now known that much of the heat in contraction is derived from the splitting of phosphocreatine and the reactions which resynthesize it (Woledge, 1971). In this type of study on an open system, the enthalpy balance method must take into account the mechanical work involved in contraction as well as the heat dissipation. Calculations revealed that a significant fraction of the heat evolved could not be explained by phosphocreatine splitting; the energy recovery was > 1.0. For instance, in the case of *Rana temporaria* sartorius muscle in isometric tetanic contraction at 0 °C, 21% of the expended energy could not be explained by this reaction (Woledge, 1971). Research then showed that some of this energy was attributable to the macromolecular activity of muscle proteins in the myofibers, while the remainder was ascribed to the movement of Ca^{2+} ions at the beginning of contraction from binding sites in the sarcoplasmic reticulum to binding sites on troponin C and parvalbumin (Woledge et al., 1985). These two sources of measured heat production illustrate the point that heat dissipation in living systems does not originate only from chemical reactions. Heat can be dissipated in the conformational changes that occur in polymeric protein complexes, as well as the interactions between macromolecules. In addition, mass movement of ions can be accompanied by changes in enthalpy, resulting from changes in Gibbs energy and entropy (see p.

310). This statement requires explanation in terms of the electrical double layer at the surface of all cells, including at the sarcolemma of myofibers and the plasma membrane of neurons. The inner part of this double layer (Stern layer) can be regarded as a condenser with its complement of ions largely giving it a certain numerical value for permittivity (dielectric constant). This is charged when the membranes of muscle and nerve are at rest (resting potential).

When the voltage gates open in response to membrane depolarization, ions flow into the cytoplasm (sarcoplasm in myofiber; axoplasm in neuron), and the condenser is discharged, giving a change in permittivity. The Gibbs energy stored in the condenser would be released as joule heating of the cytoplasm by the flow of the ionic current. For a membrane capacity (C) charged to a membrane potential (E) stores $\frac{1}{2}CE^2$ of Gibbs energy. According to Ritchie (1973), the membrane capacity of nerve fibers is insufficient to explain heat dissipation solely on the grounds of changes in Gibbs energy. The same objection may well be true for myofibers but it is possible to explain part of the heat changes in the excitation of muscle and all of that in nerve by entropy changes, additional to Gibbs energy changes. Ritchie (1973) states that change in enthalpy on discharging a condenser of capacity (C) at an absolute temperature (T) differs from the Gibbs energy change by an amount $T\Delta S$, which can be shown to equal $\Delta G(C/T)(dC/dT)$. If dC/dT is positive, the entropy change will lead to heat dissipation on discharge, which will add to the heat derived from the Gibbs energy change. The temperature coefficient for permittivity of giant squid axons has been measured and is indeed positive (see Ritchie, 1973). The value for $T\Delta S/\Delta G$ was found to be between +2 and +4. If the membrane capacity of a myofiber has a similar temperature dependence, it would be predicted that the total heat dissipation on discharge could be several times that derived from the energy stored in the condenser. Thus, it is possible to explain the heat changes associated with the movement of Ca^{2+} ions into the sarcoplasm of striated muscle.

A muscle relaxes when the sarcoplasmic voltage gates close and Ca^{2+} ions are pumped back into the sarcoplasmic reticulum. As with restoration of the resting potential in neurons, this is an active process requiring exothermic ATP hydrolysis to pump the ions (Woledge et al., 1985). It is a comparatively slow process subsequent to contraction, but part of the energetic cost. The accompanying heat dissipation is less than expected in value because hydrolysis, of course, is coupled to the transport of two Ca^{2+} ions for each ATP split in the sarcoplasmic reticulum. From the foregoing discussion on condensers, it is a reasonable expectation that this would be an endothermic process.

A considerable amount of fundamental physiological research on striated muscle has been done with the aid of calorimetry in the last 80 years but, only recently has the heat dissipation in human tissue been used as an indicator of clinical conditions. Monti's group has successfully developed a microcalorimetric technique for studying the metabolic activity of biopsies from the *vastus lateralis* muscle. It was found that the nonselective β-adrenoceptor blocker, propranolol, decreased heat production of muscle tissue by 25%, whereas the β_1-selective atenolol and β_2-agonist pindolol had no effect. Propranolol also affected temperature regulation in humans exposed to the cold, meaning that it is not a good idea to take this drug and sleep under the stars in Alaska! Among other findings, the group showed that the metabolic activity of muscle was reduced in hemodialysis patients, probably because of subnormal levels of the thyroid function variables, triiodothyronine and thyroxine. Heat production was also low in biopsies from patients with anorexia nervosa but not bulimia, possibly because of the reduced calorific intake in anorexia sufferers (Fagher et al., 1989). Clearly, the technique holds some promise in this field.

Little calorimetric research has been done on smooth muscle. Recently, however, Lonnbrö and Hellstrand (1991) showed that chemically skinned muscle from guinea pig *taenia coli* produced threefold more heat on activation (pCa 4.8) than at rest (pCa 9) (pCa being $-\log Ca^{2+}$). With stepwise increments in $[Ca^{2+}]$ from pCa 9 to 4.8, the energetic cost of force maintenance tended to rise at higher $[Ca^{2+}]$. Even after Ca^{2+} activation, force still increased beyond the point at which heat dissipation reached its maximum.

CONCLUSIONS

Calorimetric measurements of heat dissipation by animal tissue and cells have a firm foundation in the rigorous Laws of Thermodynamics. In essence, the metabolic activity of the material is being measured and detection of alterations to this activity in physiological, clinical, and pathological conditions could become a valuable addition to our medical armory. But calorimetry can also be a deeper probe to investigate human diseases, by combining it with conventional biochemical assays to ensure that account is made of all pathways in the metabolism of normal cells and those in a diseased state with altered metabolism—the enthalpy balance method. Above all, confidence in understanding the Laws of Thermodynamics enables us, as perennial students, to more clearly appreciate the Laws of Nature which all obey.

ACKNOWLEDGMENTS

The theoretical parts of this chapter were written while on sabbatical leave in the Department of Zoology, University of Innsbruck, Austria, and supported by a joint grant from The British Council and the Austrian Ministry for Science. This facilitated fruitful discussions with Dr. Erich Gnaiger and resulted, I hope, in a more accurate and intelligible account. I am also grateful to Dr. Ted Battley (SUNY at Stony Brook, New York) for many thoughtful, indeed thought-provoking, comments, some of which led synergistically to the evolution of the Battley-Kemp water bottle. My local colleagues Drs. Mervyn Jones and Mustak Kaderbhai and several seniors studied and helpfully criticized the section on thermodynamics.

REFERENCES

Bäckman, P., Kimura, T., Schön, A., & Wadsö, I. (1992). Effects of pH-variations on the kinetics of growth and energy metabolism in cultured T-lymphoma cells. J. Cell Physiol. 150, 99–103.

Baserga, R. (1989). Cell Growth and Division, p. 91, IRL Press, Oxford.

Battley, E.H. (1987). Energetics of Microbial Growth, Wiley, New York.

Clark, D.G., Brinkman, M., Filsell, O.H., Lewis, S.J., & Berry, M.N. (1982). No major thermogenic role for ($Na^+ + K^+$)-dependent adenosine triphosphate apparent in hepatocytes from hyperthyroid rats. Biochem. J. 202, 661–665.

Clark, D.G., Brinkman, M., & Neville, S.D. (1986). Microcalorimetric measurements of heat production in brown adipocytes from control and cafeteria-fed rats. Biochem. J. 235, 337–342.

Dauncey, M.J. (1990). Thyroid hormones and thermogenesis. Proc. Nutr. Soc. 49, 203–215.

Dauncey, M.J. (1991). Whole-body calorimetry on man and animals. Thermochim. Acta 193, 1–40.

Eftimiadi, C. & Rialdi, G. (1982). Increased heat production proportional to oxygen consumption in human neutrophils activated with phorbol-12-myristate-13-acetate. Cell Biophys. 4, 231–244.

Fagher, B., Monti, M., & Theander, S. (1989). Microcalorimetric study of muscle and platelet thermogenesis in anorexia nervosa and bulimia. Amer. J. Clin. Nutr. 51, 121–140.

Gnaiger, E. (1983). Heat dissipation and energetic efficiency in animal anoxibiosis: Economy contra power. J. Exp. Zoo, 228, 471–490.

Gnaiger, E. (1990). Concepts on efficiency in biological calorimetry and metabolic flux control. Thermochim. Acta 172, 31–52.

Gnaiger, E. & Kemp, R.B. (1990). Anaerobic metabolism in aerobic mammalian cells: Information from the ratio of calorimetric heat flux and respirometric oxygen flux. Biochim. Biophys. Acta 106, 328–332.

Hill, A.V. (1912). The heat-production of surviving amphibian muscle during rest, activity, and rigor. J. Physiol. 44, 466–513.

Kemp, R.B. (1987). Heat dissipation and metabolism in isolated animal cells and whole tissues/organs. In: Thermal and Energetic Studies of Cellular Biological Systems (James, A.M., ed.), pp. 147–166, Wright, Bristol.

Kemp, R.B. (1990). Importance of the calorimetric-respirometric ratio in studying intermediary metabolism of cultured animal cells. Thermochim. Acta 172, 61–73.

Kemp, R.B. (1991). Calorimetric studies of heat flux in animal cells. Thermochim. Acta 193, 253–267.

Kemp, R.B. (1993). Developments in cellular microcalorimetry with particular emphasis on the valuable role of the energy (enthalpy) balance method. Thermochim. Acta 219, 17–41.

Kemp, R.B. & Gnaiger, E. (1989). Aerobic and anaerobic energy flux in cultured animal cells. In: Energy Transformations in Cells and Organisms (Wieser, W. & Gnaiger E., eds.), pp 91–97. Georg Thieme Verlag, Stuttgart.

Krakauer, T. & Krakauer, H. (1976). Antigenic stimulation of lymphocytes. I. Calorimetric exploration of metabolic responses. Cell. Immunol. 26, 242–253.

Levin, K. (1980). In: Biological Microcalorimetry (Beezer, A.E., ed.), pp. 131–143, Academic Press, London.

Loike, J.D., Silverstein, S.L., & Sturtevant, J.M. (1981). Application of differential scanning calorimetry to the study of cellular processes: Heat production and glucose oxidation of murine macrophages. Proc. Natl. Acad. Sci. USA 78, 5958–5962.

Lönnbro, P. & Hellstrand, P. (1991). Heat production in chemically skinned smooth muscle of guinea-pig Taenia coli. J. Physiol. 440, 385–401.

McKeehan, W.L. (1986). Glutaminolysis in animal cells. In: Carbohydrate Metabolism in Cultured Cells (Morgan, M.J., ed.), pp. 111–150, Plenum Press, New York.

Mills, I., Cvitas, T., Homann K., Kallay, N., & Kuchitsu, K. (1988). IUPAC: Quantities, Units and Symbols in Physical Chemistry, Blackwell, Oxford.

Monti, M. (1987). *In vitro* thermal studies of blood cells. Thermal and Energetic Studies of Cellular Biological Systems (James, A.M., ed.), pp. 131–146, Wright, Bristol.

Monti, M. (1991). Calorimetric studies of lymphocytes and hybridoma cells. Thermochim. Acta 193, 281–285.

Monti, M., Hedner, P., Ikomi-Kumm, J., & Valdemarsson, S. (1987). Erythrocyte thermogenesis in hyperthyroid patients: Microcalorimetric investigation of sodium/potassium pump and cell metabolism. Metabolism 36, 155–159.

Nedergaard, .J., Cannon, B., & Lindberg, O. (1977). Microcalorimetry of isolated mammalian cells. Nature 267, 518–520.

Newsholme, E.A., Crabtree, B., & Ardawi, S.M. (1985). Glutamine metabolism in lymphocytes: Its biochemical, physiological and clinical importance. Quart. J. Exp. Physiol. 70, 473–489.

Olsson, S.-Å., Monti, M., Sörbris, R., & Nilsson-Ehle, P. (1986). Adipocyte heat production before and after weight reduction by gastroplasty. Int. J. Obesity 10, 99–105.

Prigogine, I. (1967). Introduction to Thermodynamics of Irreversible Processes, p. 147, 3rd Ed., Wiley, New York.

Ritchie, J.M. (1973). Energetic aspects of nerve conduction. Prog. Biophys. Mol. Biol. 26, 147–187.

Thorén, S.A. (1992). Calorimetry: A new quantitative *in vitro* method in cell toxicology. A dose/effect study of alveolar macrophages exposed to particles. J. Toxicol. Environ. Health 36, 307–318.

Westerhoff, H.V. & van Dam, K. (1987). Thermodynamics and Control of Biological Free-Energy Transduction. Elsevier, Amsterdam.

Woledge, R.C. (1971). Heat production and chemical change in muscle. Prog. Biophys. Mol. Biol. 22, 37–74.

Woledge, R.C., Curtin, N.A., & Homsher, E. (1985). Energetic Aspects of Muscle Contraction. Academic Press, London.

NOTES ADDED IN PROOF

Several of the papers in Kemp and Schaarschmidt (1995) are relevant to this chapter. It has long been known that the metabolic rate of mammals is not proportional to body mass (Kleiber's Rule) but Kemp and Gnaiger (1989) showed from measurements of cellular heat dissipation and protein biomass, that the mass exponent of isolated mammalian cells was also less than 1.0 (see p. 314). Singer et al. (1995) extended this type of study to include mammalian blood and human renal carcinoma cells and found that the exponent of the allometric power function (–0.25) could not be explained by a decrease in the surface-to-volume ratio (–0.33). They mathematically simulated the reduction in cellular metabolism as measured by heat flow (W per cell) with increasing density of cell number, explaining it in terms of greater intensity of anaerobiosis. It should be checked that the average size of the cells was not smaller (scalar heat flux, W m^{-3}—see p. 313) for it is possible that the Uriah cells were diverted into G_0 and consequently smaller in size (see p. 317). The Singer group (see Kemp and Schaarschmidt, 1995) also found that heat production by the blood of human neonates followed the allometric relationship by being correspondingly higher than that of adults; as well as demonstrating the value of microcalorimetry in monitoring the malignancy and cytostatic treatment of urological tissue samples and the recovery of rat liver tissue after ischemia.

Kemp et al. (1995) made an enthalpy balance for activated 2C11-12 mouse macrophage hybridoma cells, showing that glucose respiration and glycolysis plus glutaminolysis were the major catabolic pathways with a minor contribution from fatty acid oxidation. The respiration of glucose accounted for 87% of the enthalpy change associated with the metabolic burst of these cells induced by phorbol-12-myristate-13-acetate. The partial oxidation glutamine to lactate and carbon dioxide (see p. 322) causes an overestimate of the contribution of anaerobic lactate formation in mixed aerobic and anaerobic metabolism (see equation 19).

An important assumption in constructing an enthalpy balance is that the ΔH for anabolism is close to zero (see p. 312 and p. 314). Recently, Battley (1995) again investigated this assumption using *Saccharomyces cerevisiae* as a model system and found that it may not always hold true. While the anabolic changes accompanying the growth of the yeast on ethanol and acetic acid were small (less than 1% of the enthalpy of combustion), those associated with aerobic or anaerobic growth on glucose were considerable at 6%. These findings are important when applying enthalpy and free energy balances to the study of cellular growth (von Stokar et al., 1993).

Battley, E.H. (1995). A reevaluation of the thermodynamics of growth of *Saccharomyces cerevisiae* on glucose, ethanol and acetic acid. Can. J. Microbiol. 41, 388–398.

Kemp, R.B., Belicic-Kolsek, A., Hoare, S., Schmalfeldt, M., Townsend, C., & Evans, P.M. (1995). A thermochemical study of metabolic pathways in activated and triggered 2C11-12 mouse macrophage hybridoma cells. Thermochim. Acta 250, 259–276.

Kemp, R.B., & Schaarschmidt, B. (Eds.) (1995). Calorimetric and Thermodynamic Studies in Biology: Special Issue. Thermochim. Acta 251, 1–402.

Singer D., Schunck, O., Bach, F., & Kuhn, H.-J. (1995). Size effects on metabolic rate in cell tissue and body calorimetry. Thermochim. Acta 251, 227–240.

von Stockar, U., Gustafsson, L., Larsson, C., Marison, I., Tissot, P., & Gnaiger, E. (1993) Thermodynamic considerations in constructing energy balances for cellular growth. Biochim. Biophys. Acta 1183, 221–240.

INDEX

Printed and bound by CPI Group (UK) Ltd, Croydon, CR0 4YY

03/10/2024

01040431-0014